EXAM PREPARATION
Fire Pump Operations

EXAM PREPARATION
Fire Pump Operations

Dr. Thomas B. Sturtevant

THOMSON

DELMAR LEARNING

Australia Canada Mexico Singapore Spain United Kingdom United States

THOMSON

DELMAR LEARNING

Exam Preparation for Fire Pump Operations

Dr. Thomas B. Sturtevant

Vice President, Technology and Trades SBU:
Dave Garza

Director of Learning Solutions:
Sandy Clark

Managing Editor:
Larry Main

Acquisitions Editor:
Alison Pase

Product Manager:
Jennifer A. Starr

Marketing Manager:
Erin Coffin

Marketing Coordinator:
Patti Garrison

Production Manager:
Stacy Masucci

Content Project Manager:
Jennifer Hanley

Technology Project Manager:
Kevin Smith

Editorial Assistant:
Maria Conto

ISBN: 1-4180-2088-5

NOTICE TO THE READER

Publisher does not warrant or guarantee any of the products described herein or perform any independent analysis in connection with any of the product information contained herein. Publisher does not assume, and expressly disclaims, any obligation to obtain and include information other than that provided to it by the manufacturer.

The reader is expressly warned to consider and adopt all safety precautions that might be indicated by the activities herein and to avoid all potential hazards. By following the instructions contained herein, the reader willingly assumes all risks in connection with such instructions.

The publisher makes no representation or warranties of any kind, including but not limited to, the warranties of fitness for particular purpose or merchantability, nor are any such representations implied with respect to the material set forth herein, and the publisher takes no responsibility with respect to such material. The publisher shall not be liable for any special, consequential, or exemplary damages resulting, in whole or part, from the readers' use of, or reliance upon, this material.

Contents

Preface

Congratulations to you as you embark on the next step in your career in the fire service! Countless hours of work and training have led you to this point. The questions within these pages, as well as in the CD in the back of this book, will help prepare you for the driver/operator certification exam.

How To Use This Book

Based on the 2003 Edition of NFPA 1002 for driver/operator training and *Introduction to Fire Pump Operations, Second Edition*, each exam was created to ensure that all competencies for driver/operator were covered. As such, each exam contains questions from the chapters in the book, as well as references to applicable NFPA standard requirements. In addition, because the questions within this book are also designed based on the NFPA standard, the book will prepare you for any driver/operator exam.

The book is organized in a logical sequence based on Bloom's Taxonomy of Learning, that progresses from test-taking strategies to questions about memorizing important terms and on to concepts and application of those concepts in a given scenario.

The best place to start is at the introductory chapter, **Acing Your Certification Exam**. Here you will find valuable test-taking strategies, as well as advice on how to set up an effective study schedule prior to the certification exam date.

- *Phase One: Knowledge and Comprehension* contains three practice exams of 100 questions each to gauge your knowledge and ability to retain important facts and concepts.

- *Phase Two: Application and Analysis* contains three practice exams of 100 questions each that evaluate your ability to solve problems and recognize how important concepts fit together.

- *Phase Three: Evaluation and Synthesis* is the most complex stage, also containing three practice exams of 100 questions each, and judges how accurately you can make decisions based on given scenarios and your knowledge of the subject matter.

- *Phase Four: Final Exam* is a 200-question test combining questions from all three phases, simulating the certification exam.

- *Back of Book CD:* contains *all* exams in a self-grading ExamView format, allowing you the option of self-study. Also included is a bonus final exam of 200 questions for additional practice in computer-based testing.

- *Answers to Questions, NFPA and Textbook Rationale & References* are provided for each question in all of the exams, allowing you to track your progress as well as the applicable sources for additional study as needed.

Successful candidates will take all of these exams in the order provided, using the answers, rationale, and references to grade their own work. The CD provides an excellent opportunity to retake tests as needed, as well as a bonus final exam. A copy of *Introduction to Fire Pump Operations, Second Edition* the reference material for the practice tests is also a useful tool for further study.
(Order #: 0-7668-5452-3)

Features of this Book

This book provides many features to enhance your learning experience and help lead you closer to your goal of successful completion of the certification exam:

- *Test Taking Strategies* and *Study Guidelines* are outlined in the introductory chapter, allowing you to effectively practice and fully prepare for the exam.

- *Questions of increasing complexity* are organized in a logical sequence of the book to ensure understanding of important concepts related to the officer role.

- *Answers to the Questions* allow you to track your progress.

- *NFPA References* correlate the questions to the applicable NFPA 1002 standard requirements, illustrating the competency which the question covers.

- *Textbook References* correlate the questions to the appropriate pages in the *Introduction to Fire Pump Operations, Second Edition* book, to allow for further study if needed.

- *Rationale* also accompanies each question, providing you with the necessary explanation to support the correct answer.

- *Questions in ExamView format* on the CD in the back of the book allow you to use this book as a self-study tool and to practice computer-based testing. A bonus final exam is also provided.

About the Author

Dr. Thomas B. Sturtevant is a Program Manager for the Emergency Services Training Institute (ESTI) at Texas A&M University System. He is currently responsible for ESTI's accreditation/certification and curriculum development program, the Department of Defense Emergency Services Training and Education program, as well as the Emergency Management Administration online Bachelor Degree program with West Texas A&M University. He is a certified Fire Protection Specialist and has an Education Doctorate in Leadership for Teaching and Learning, as well as a Masters in Public Administration from the University of Tennessee.

About the Series Advisor

Mike Finney is the head of the Department of Public Safety Services at Great Oaks Institute of Technology and Career Development. He has served on the NFPA 1041 committee in past years and is a current board member for the International Association of Fire Service Instructors. His area of expertise is teaching and curriculum development.

Acing the Certification Exam: An Introduction to Test-Taking Strategies

Introduction

Test time. Whether you are preparing for a certification test or a hiring test, the thought of an examination strikes fear in many people's hearts. The fear is so common that psychologists even have a diagnosis called test anxiety. However, testing does not have to be that way. Evaluations are simply an instrument to determine if you were effectively taught the information intended, or if you have the knowledge base necessary to do the job. That's all! If the purpose of testing is so simple, why do so many people become so anxious when test time comes. Several factors play into test anxiety and why so many people have such fears of testing. However, these can be overcome. With the assistance of this guide, you too can be better prepared and calmer on examination day.

Differences between hiring and certification tests

A hiring test is intended to determine if an applicant has the necessary knowledge base to perform the job. Different departments have different requirements for the job that is being filled, and those doing the hiring want to know that the applicant can meet the challenge. The outcome is very simple—get the most qualified applicant for the position. This can be the opportunity for you to "show your stuff." Hiring tests are designed to give each applicant an objective avenue to present their skill base. Careful preparation will give you the opportunity to excel. Every applicant has the same test, same questions, and same opportunity to prepare. As an applicant, take this opportunity to show the skills and knowledge base you have. Success on the hiring test is not necessarily measured on a set score or a "passing" score. Some departments set a cut-off score for applicants to proceed to the next phase, but many will take the top scores for the next phase. In this situation, it is imperative to get the highest score when compared to your competitors.

A certification test takes on a different dimension. Certification tests are designed to demonstrate mastery and are measured to national consensus standards. Since the other students do not measure the success, there is typically a set score that you must reach to be successful. The minimum score is set rather than being driven by all people taking the test.

How are test questions developed?

How questions are developed is critical to the understanding of how to take a test. Whether the examination is a hiring test, a promotional test, or a certification examination, the questions are developed around a clear set of objectives.

The learning objectives are a very important piece of the educational process. The objectives tell you several things:

- Under what conditions the student should be able to apply the knowledge. (condition)

- What type of knowledge the student should gain. (behavior)

- What depth or level of understanding the student should have. (standard)

While the order may vary based on the preference of the developer, this condition, behavior, and standard approach should be addressed in all learning objectives. When reviewing the learning objectives, take a moment to break it down into the individual pieces. This will provide insight into how the class should be taught. One will gain a wealth of information as to the intent of the designer by studying the learning objectives.

Domains of learning

Based on Bloom's Taxonomy, there are three primary areas or domains around which testing is designed. Knowing the domains will give an indication of the evaluation approach. The three domains of learning are:

• Cognitive domain—primarily deals with intellectual or knowledge skills, those that are intellectual processing or mental learning.

• Psychomotor domain—primarily deals with physical skills, those that require physically doing a task.

• Affective domain—primarily deals with a mindset, those that require a change in attitude or behavior.

Within the three domains were levels from the most basic understanding to advanced. Since we are primarily focusing on written tests, we will only deal with the cognitive domain. Table I-1 gives you the breakdown of the levels of understanding for the cognitive domain. The key to remember is that the higher the level you reach, the greater the understanding of the material must exist.

Table I-1

Knowledge: Recall of data	Examples: Recite a policy. Quote prices from memory to a customer. Know the safety rules. Keywords: defines, describes, identifies, knows, labels, lists, matches, names, outlines, recalls, recognizes, reproduces, selects, states
Comprehension: Understand the meaning, translation, interpolation, and interpretation of instructions and problems. State a problem in one's own words.	Examples: Rewrite the principles of test writing. Explain in one's own words the steps for performing a complex task. Translate an equation into a computer spreadsheet. Keywords: comprehends, converts, defends, distinguishes, estimates, explains, extends, generalizes, gives examples, infers, interprets, paraphrases, predicts, rewrites, summarizes, translates
Application: Use a concept in a new situation or unprompted use of an abstraction. Apply what was learned in the classroom into novel situations in the workplace.	Examples: Use a manual to calculate an employee's vacation time. Apply laws of statistics to evaluate the reliability of a written test. Keywords: applies, changes, computes, constructs, demonstrates, discovers, manipulates, modifies, operates, predicts, prepares, produces, relates, shows, solves, uses
Analysis: Separate material or concepts into component parts so that its organizational structure may be understood. Distinguishe between facts and inferences.	Examples: Troubleshoot a piece of equipment by using logical deduction. Recognize logical fallacies in reasoning. Gather information from a department and select the required tasks for training. Keywords: analyzes, breaks down, compares, contrasts, diagrams, deconstructs, differentiates, discriminates, distinguishes, identifies, illustrates, infers, outlines, relates, selects, separates

Synthesis: Build a structure or pattern from diverse elements. Put parts together to form a whole, with emphasis on creating a new meaning or structure.	Examples: Write a company operations or process manual. Design a machine to perform a specific task. Integrate training from several sources to solve a problem. Revise and processe to improve the outcome. Keywords: categorizes, combines, compiles, composes, creates, devises, designs, explains, generates, modifies, organizes, plans, rearranges, reconstructs, relates, reorganizes, revises, rewrites, summarizes, tells, writes
Evaluation: Make judgments about the value of ideas or materials.	Examples: Select the most effective solution. Hire the most qualified candidate. Explain and justify a new budget. Keywords: appraises, compares, concludes, contrasts, criticizes, critiques, defends, describes, discriminates, evaluates, explains, interprets, justifies, relates, summarizes, supports

Cognitive Domain (courtesy of http://www.nwlink.com/~donclark/hrd/bloom.html)

NFPA standards

The movement toward national standards for the fire service began in 1971 with the Joint Council of National Fire Service Organizations. The intent was to develop national performance standards. Today, the end result, is 67 levels for 16 standards dealing with the professional qualifications for the fire service outlined by standards developed by the NFPA. Committees made up of training leaders, educators, private industry, and technical specialists guide standards development and revision. The intent is to give a better standard that meets a greater need. The standards are reviewed and revised as needed every five years. (This is why it is important when referencing an NFPA standard to also reference the year.) As a result, performance standards are kept up-to-date and accurate to what is needed in the field. While NFPA standards are not federal mandates, they do provide a widely accepted standard to follow. They provide the most current expectations for the position referenced.

The NFPA standards also provide an excellent reference for developing job descriptions. The 1000-series are referred to as the Professional Qualifications standards. Each provides a solid outline of performance requirements for almost any given position with the fire service. They are also easily adaptable to most any department. While there are other standards that can be followed, those set forth by the NFPA are the most widely recognized and are considered foundational for any department.

How does this apply to me?

NFPA standards are important to someone taking a fire-related examination, because often they are the basis for the learning objectives for test development. For certification examinations, NFPA standards are the foundation for many certifying bodies. For hiring and promotional tests, they are a solid basis for creating objective tests. A good resource to begin preparing for certification and hiring tests is the appropriate NFPA standard. Review the specific level and the requirements for such a level. This will give you some indication of what the test developer will be considering. As well, look at the requisite knowledge and skills. This will give you an indication of the extent of understanding of the information. An excellent way of reviewing the standards is to look at the verbs of the job performance requirement (JPR) in the context of Bloom's Taxonomy. Much like the requisite information, the JPR, as it relates to Bloom's Taxonomy, will tell you the level of understanding you must have with the information.

Test obstacles

Test obstacles are issues that complicate test taking. If we view test taking as simply an avenue to determine the individual's comprehension of the material, then test obstacles are barriers to the process. There are many issues that may create test obstacles. We will discuss a few.

Mental

Mental test obstacles can sometimes be the greatest hurdles to overcome. Mental preparation for a test can be as important as intellectual preparation. So often, many people have failed an exam before they even begin. Issues that arise out of mental obstacles are:

- feeling unprepared.
- feeling incompetent.
- fear of taking tests.
- fear of failure.

Overcoming these obstacles can be your greatest asset when testing. Not allowing yourself to be beaten before entering the testing area can make the difference in success and failure on the exam.

Physical

Improper rest, poor eating habits, and lack of exercise can be some of the physical obstacles to overcome. When preparing for tests, always ensure that you get plenty of rest the night before, have a well-balanced meal before the test, and ensure you have a regiment of proper exercise. Physical obstacles are typically the easiest to overcome, however, the most overlooked.

Emotional

The emotional obstacles are often the most vague with which to deal. Much like mental obstacles, emotional obstacles can cause a person to do poorly on an exam well before they enter the room. Stress-related issues that can interfere with test taking are:

- family concerns.
- work-related concerns.
- financial concerns.

Emotional issues can cause a person to lose focus, cloud decision-making skills, and become distracted. Overcoming these obstacles requires a conscious effort to ensure that emotions do not interfere with the test.

Preparing to Take a Test

BEFORE the Test

1. Start preparing for the examination. For certification exams, start the first day of class. You can do this by reading your syllabus carefully to find out when your exams will be, how many there will be, and how much they are weighed into your grade. For hiring exams, it is recommended to begin studying at least eight weeks before the test.

2. For certification classes, plan reviews as part of your regular weekly study schedule; a significant amount of time should be used to review the entire material for the class.

3. Reviews are much more than reading and reviewing class assignments. You need to read over your class notes and ask yourself questions on the material you don't know well. (If your notes are relatively complete and well organized, you may find that very little rereading of the textbook for detail is needed.) You may want to create a study group for these reviews to reinforce your learning.

4. Review for several short periods rather than one long period. You will find that you are able retain information better and get less fatigued.

5. Turn the main points of each topic or heading into questions and check to see if the answers come to you quickly and correctly. Do not try to guess the types of questions; instead, concentrate on understanding the material.

DURING the Test

1. Preview the test before you answer anything. This gets you thinking about the material. Make sure to note the point value of each question. This will give you some ideas on how best to allocate your time.

2. Quickly calculate how much time you should allow for each question. A general rule of thumb is that you should be able to answer 50 questions per hour. This averages out to one question every 1.2 seconds. However, make sure you clearly understand the amount of time you have to complete the test.

3. Read the directions CAREFULLY. (Can more than one answer be correct? Are you penalized for guessing? etc.) Never assume that you know what the directions say.

4. Answer the easy questions first. This will give you confidence and a feel for the flow of the test. Only answer the ones for which you are sure of the correct answer.

5. Go back to the difficult questions. The questions you have answered so far may provide some indication of the answers.

6. Answer all questions (unless you are penalized for wrong answers).

7. Generally, once the test begins, the proctor can ONLY reread the question. He/she cannot provide any further information.

8. Circle key words in difficult questions. This will force you to focus on the central point.

9. Narrow your options on the question to two answers. Many times, a question will be worded with two answers that are obviously inaccurate, and two answers that are close. However, only one is correct. If you can narrow your options to two, guessing may be easier. For example, if you have four options on a question, then you have a 25% chance of getting the question correct when guessing. If you can narrow the options to two answers, then you increase to a 50% chance of selecting the correct choice.

10. Use all of the time allotted for the test. If you have extra time, review your answers for accuracy. However, be careful of making changes on questions of which you are not sure. Oftentimes, people change the answers to questions of which they were not sure, when their first guess was correct.

AFTER the Test

Relax. The test has been turned in. You can spend hours second-guessing what you "could" have done, but the test is complete. For certification tests, follow up to see if you can find out what objectives you did well and what areas you could improve. Review your test if you can; otherwise, try to remap the areas of question and refocus your studying.

Preparation Plan

Once you have acquired the reference texts for the examination, begin by reviewing the introduction, the table of contents, and review how the book is organized. The introduction will tell you how the book has been set up and how it is intended to assist the individual with the learning process. The table of contents will provide a snapshot of how the book is organized. Scanning the text will give you an overview of the book's design. Once you have done this, break the chapters into four review sections. Using Table I-2 on the next page, fill in week 8 with the first section, week 7 with the second section, week 6 with the third section, and week 5 with the fourth section. Focus your energy into 50-minute increments with 10-minute breaks. Set aside one hour for each chapter. However, on chapters in which you are competent, less time can be spent than with chapters that are less familiar.

Week 4 will be spent taking section one of the Exam Prep guide. Allow an hour for 50 questions. (100 questions should take 2 hours.) Do not check the answers until you have completed the entire test. Any questions missed should be reviewed to ensure you have an understanding of the answer.

Week 3 will be spent taking section two of the Exam Prep guide. Again, allow an hour for 50 questions. At the end of the test, check answers and correct wrong answers.

Week 2 should be spent on section three of the Exam Prep guide. Maintain time frames and check answers at the end of the test.

Week 1 should be spent with section four of the Exam Prep guide.

Five days before the test, go through the section one tests again.

Four days before the test, go through the section two tests again.

Three days before the test, go through the section three tests again.

Two days before the test, go through the section four tests again.

One day before the test do a light review of the text, focusing on areas you missed on the practice tests. However, take the evening and relax. Do something you enjoy, but make sure it is not a late night. Go to bed early and make sure you get a good night's sleep.

The day of the test make sure you have a well-balanced breakfast and arrive at the test site early.

Table I-2

PREPARATION GUIDE *Plan based on a two-month schedule**	STUDY NOTES
Week 8 – Reference Text, Section I	
Week 7 – Reference Text, Section II	
Week 6 – Reference Text, Section III	
Week 5 – Reference Text, Section IV	
Week 4 – Exam Prep, Section I • Exam One • Exam Two • Exam Three	
Week 3 – Exam Prep, Section II • Exam One • Exam Two • Exam Three	
Week 2 – Exam Prep, Section III • Exam One • Exam Two • Exam Three	
Week 1 – Exam Prep, Section IV • Final Exam • Bonus Final Exam (on CD)	
5 days before – Review Section I Tests	
4 days before – Review Section II Tests	
3 days before – Review Section III Tests	
2 days before – Review Section IV Tests	
1 day before – Light review • Relax • Go to bed early	
Day of Test • Good breakfast • Arrive early	

Summary

Test taking does not have to be overwhelming. The obstacles to testing can be overcome and conquered through solid strategies and preparation. Initiating an effective plan, following it, and mentally preparing for a test can be your greatest tools to test success. As you work through the sections of this book, use the time well to work through some of the obstacles you face. When taking the tests in each section, try to simulate the environment of the actual test as much as possible. Successful testing is not an art; it is a learned skill. Through planning and practice anyone can acquire this skill.

KNOWLEDGE & COMPREHENSION

Section one is designed to evaluate your basic understanding of the material. In this section, we are testing you understanding of definitions, recalling information, and identifying terms. Referring to Table I-1 (Bloom's Taxonomy, Cognitive Domain), we are covering the following levels:

- knowledge

- comprehension

Having mastered section one, you should be able to have a basic comprehension of the material.

Phase One, Exam One

1. Pump operators should be licensed to drive all vehicles they are expected to operate.
   a. True
   b. False

2. Prior to certification, pump operators must meet the requirements of NFPA 1001, Firefighter 1.
   a. True
   b. False

3. A visual inspection is all that is needed to ensure proper tire inflation.
   a. True
   b. False

4. Inspecting, servicing, and testing apparatus are the three primary activities conducted during preventive maintenance.
   a. True
   b. False

5. Doubling the speed of a vehicle increases the stopping distance an estimated four times.
   a. True
   b. False

6. When an apparatus stops abruptly, the on-board water supply attempts to stay in motion; this is referred to as centrifugal force.
   a. True
   b. False

7. Federal laws and regulations dictate that when backing apparatus, a minimum of two spotters or guides is required.
   a. True
   b. False

8. Posted speed limits are based on ideal driving conditions. Speed should be reduced when driving in adverse weather conditions.
   a. True
   b. False

9. Water is virtually noncompressible under normal conditions.
   a. True
   b. False

10. According to NFPA 1901, only 2 ½-inch or smaller discharge outlets can be located on pump panels.
    a. True
    b. False

11. The around-the-pump foam proportioning system uses an eductor to mix foam with water.
    a. True
    b. False

12. Cotton, rather than synthetic fiber, is still the most popular material used in the construction of woven-jacket hose.
    a. True
    b. False

13. According to NFPA 291, a hydrant with an orange bonnet and nozzle caps can flow 750 gpm.
    a. True
    b. False

14. A 1,250-gpm pump must achieve a prime within 30 seconds.
    a. True
    b. False

15. One method used to verify flow meter accuracy is to flow water through a hose and smooth-bore nozzle and then measure flow via a pitot tube reading.
    a. True
    b. False

16. Pumps typically are powered by either a separate engine or from the engine that drives the apparatus.
    a. True
    b. False

17. The pre-action sprinkler system requires an automatic or manual detection system to operate, which allows water to enter the system and then immediately discharge through all fused sprinkler heads on the system.
    a. True
    b. False

18. Standpipe systems are directly connected and solely supported by a fire department connection.
    a. True
    b. False

19. Water prepared for domestic purposes is a colorless, orderless, and tasteless liquid that readily dissolves many substances.
    a. True
    b. False

20. The pressure at any point in a liquid at rest is equal in every direction.
    a. True
    b. False

21. Friction loss varies inversely with hose length if all other variables are held constant.
    a. True
    b. False

22. Hydraulics is a branch of science dealing with the principles of fluid at rest or in motion.
    a. True
    b. False

23. The drop-ten method is a fireground method to estimate friction loss in 2 ½-inch hose. It can also be used for other sizes of hose, although is less accurate.
    a. True
    b. False

24. In common appliances used in pump operations, friction loss ranges from 5 psi to 15 psi.
    a. True
    b. False

25. NFPA _____ identifies the minimum job performance requirements for individuals responsible for driving and operating fire department vehicles.
    a. 1001
    b. 1002
    c. 1041
    d. 1901

26. Pump operators are subject to periodic medical evaluations as required by NFPA _____.
    a. 1001
    b. 1041
    c. 1200
    d. 1500

27. Which of the following titles has been used to describe the position responsible for driving and operating pumping apparatus?
    a. engineer
    b. driver pump operator
    c. chauffeur
    d. all are correct

28. _____ are rules that are legally binding and enforceable.
    a. Standards
    b. Laws
    c. Regulations
    d. Codes

29. Requirements to conduct preventive maintenance on emergency apparatus can be found in manufacturer's requirements and in NFPA _____.
    a. 1001
    b. 1500
    c. 1915
    d. both b and c are correct

30. Engine oils are classified using a rating system developed by API and SAE. API stands for
    a. American Petroleum Institute.
    b. American Petrochemical Industry.
    c. Association of Petroleum Industries.
    d. Association of Products for Industry.

31. When charging emergency vehicle batteries, the area should be ventilated because
    a. excessive heat can build up.
    b. phosgene can be produced.
    c. hydrogen gas can be produced.
    d. acetylene gas can be produced.

32. Inspecting the steering system on emergency vehicles includes which of the following:
    a. Turn the steering wheel to both the right and left to ensure front tires turn at least 45 degrees.
    b. Turn the steering wheel until just before the wheels turn if excessive play exists.
    c. Turn the steering wheel while the vehicle is in motion.
    d. Turn the steering wheel in one direction fully and back while ensuring the front tires track the steering wheel rotation.

33. NFPA 1500 suggests that all of the following should be inspected on a routine basis *except*
    a. tires.
    b. brakes.
    c. transmission oil.
    d. windshield wipers.

34. During preventive maintenance inspections, air tanks should
    a. be drained slightly to remove moisture.
    b. be drained fully every day.
    c. never be drained.
    d. be drained according to manufacturers recommendations.

35. Which of the following is correct concerning the testing of emergency warning devices on in-service apparatus?
    a. Audible warning devices should be tested on a weekly basis.
    b. Audible warning devices should be tested on a monthly basis.
    c. Audible warning devices should be tested as infrequently as possible because the sirens may disturb those living next to the station.
    d. Audible warning devices should be checked only after ensuring that no one is in close proximity to the audible warning device.

36. Each of the following is an important reason for keeping emergency vehicles clean at all times *except*
    a. it is required by NFPA 1901.
    b. it helps ensure the pump, systems, and equipment operate as intended.
    c. it helps ensure vehicles can be inspected properly.
    d. it helps maintain a good public image.

37. From 1990 to 2000, emergency vehicles were involved in more than _____ accidents while responding to or returning from an incident.
    a. 21
    b. 500
    c. 5,000
    d. 11,000

38. Each of the following is a common factor that appears in most emergency-vehicle accidents *except*
    a. not following laws and standards.
    b. not fully aware of driver and/or apparatus limitations.
    c. lacking appreciation of environment, such as weather and traffic.
    d. lacking driver licensure and certification.

39. All of the following address safe emergency-vehicle operations and requirements that drivers should be familiar with *except*
    a. NFPA 1500.
    b. local ordinances.
    c. state statutes.
    d. Federal Fire Act.

40. _____ are rules that drivers must obey.
    a. Laws
    b. Standards
    c. Regulations
    d. Guidelines

41. Total stopping distance is measured from the time a hazard is detected until
    a. the vehicle comes to a complete stop.
    b. the vehicle slows significantly.
    c. the driver begins to depress the brake.
    d. the driver engages the parking brake.

42. Friction between the tire and the road is called
    a. torsion.
    b. traction.
    c. grip.
    d. slip.

43. When parking an apparatus next to a curb, the rotation of the front tire should be
    a. toward the curb.
    b. away from the curb.
    c. straight.
    d. all of the answers are correct.

44. Hydraulic principles that deal with liquids at rest and the pressures they exert or transmit are known as
    a. hydrostatics.
    b. hydrodynamics.
    c. hydrodisplacement.
    d. hydrocentrifugal.

45. The following are examples of positive-displacement rotary pumps *except*
    a. gear.
    b. vane.
    c. piston.
    d. lobe.

46. Which of the following methods used to power pumps provides stationary pumping only?
    a. PTO
    b. directly from engine crankshaft
    c. split-shaft transmission
    d. no correct answer provided

47. Which of the following is *not* correct concerning pump priming?
    a. Positive displacement pumps are self-priming.
    b. Centrifugal pumps must be primed.
    c. Priming is a suction process that forces water into the pump.
    d. Priming can be referred to as replacing air in a pump with water.

48. For a two-stage centrifugal pump operating in volume mode, water
    a. enters both impellers at the same time.
    b. enters one impeller and then enters the second impeller.
    c. pressure from one impeller is added to the pressure generated in the second impeller.
    d. volume from one impeller is provided to the second impeller.

49. According to NFPA 1901, a pump must be able to deliver 100% of its rated capacity at _____ psi.
    a. 100
    b. 150
    c. 200
    d. 250

50. The _____ side of the pump is the point at which water enters the pump.
    a. intake
    b. discharge
    c. draft
    d. all are correct

51. The rate and quantity of water delivered by the pump is called _____.
    a. pressure
    b. slippage
    c. speed
    d. flow

52. This is a cut-away of a _____ pump.
    a. positive displacement
    b. rotary impeller
    c. two-stage centrifugal
    d. rotary vane

53. The two largest gauges located on a pump panel, usually together, are called
    a. pump intake gauge and pump discharge gauge.
    b. main pump gauges.
    c. Bourdon tube gauges.
    d. all are correct.

54. All of the following are correct concerning flow meters, *except* that flow meters
    a. measure both pressure and flow.
    b. measure both quantity and rate.
    c. require less hydraulic calculations during pump operations.
    d. use paddle wheels.

55. All of the following are considered manual-controlled *except*
    a. push-pull "T-handles."
    b. quarter turn control.
    c. pneumatic control.
    d. crank control.

56. A pressure _____ regulating device protects against excessive pressure buildup by controlling the speed of the pump engine.
    a. relief valve
    b. governor
    c. auxiliary system
    d. automatic

57. The type of cooling system that pumps water from the discharge side of the pump into the engine cooling system is known as a
    a. auxiliary cooling system.
    b. radiator-fill system.
    c. pump cooling system.
    d. primary cooling system.

58. Foam systems on pumping apparatus include all *except*
    a. around-the-pump proportioning.
    b. balanced pressure.
    c. in-line eductor.
    d. high expansion.

59. During the priming process, oil is used to
    a. help lubricate the priming pump.
    b. provide a tighter seal between the moving parts.
    c. increase the viscosity of initial water being pumped.
    d. both a and b are correct.

60. Examples of common instrumentation found on pump panels include all of the following *except*
    a. pressure gauges.
    b. flow meters.
    c. indicators.
    d. foam systems.

61. The _____ control valve allows water to flow from the on-board water supply to the intake side of the pump.
    a. tank-to-pump
    b. pump-to-tank
    c. transfer valve
    d. discharge

62. NFPA _____ establishes requirements for the design, construction, inspection, and testing of new fire hose.
    a. 1500
    b. 1961
    c. 1962
    d. 1965

63. According to NFPA 1961 and 1962, maximum operating pressures for supply and attack lines are
    a. 185 and 200 psi.
    b. 185 and 250 psi.
    c. 185 and 275 psi.
    d. 200 and 275 psi.

64. The most common length of one section of hose used in the fire service today is
    a. 50 feet.
    b. 100 feet.
    c. 150 feet.
    d. 200 feet.

65. The thread used in nearly all fire hose threaded coupling construction is referred to as _____.
    a. NH
    b. HN
    c. NTS
    d. Storz

66. The appliance in this picture is a _____.
    a. wye
    b. siamese
    c. can be either a wye or siamese
    d. no correct answer provided

67. Which of the following is an example of a double female adapter?

a.

b.

c.

d.

68. The recommended operating pressure for a smooth-bore nozzle on a hand-line is
    a. 50 psi.
    b. 80 psi.
    c. 100 psi.
    d. 125 psi.

69. A nozzle that is designed to maintain a constant nozzle pressure over a wide range of flows is called a(n)
    a. fixed-flow or constant-flow nozzle.
    b. smooth-bore nozzle.
    c. selectable-flow nozzle.
    d. automatic nozzle.

70. Common hose tools that a pump operator may use include which of following?
    a. spanner wrenches
    b. hose clamps
    c. hose bridges
    d. all are common hose tools

71. The basic parts of a hose include each of the following *except*
    a. outer protective shell.
    b. reinforced inner liner.
    c. couplings.
    d. safety-relief ring.

72. The Venturi principle is used by what device?
    a. priming device
    b. eductor
    c. pressure regulator
    d. pressure governor

73. The amount of water flowing from a nozzle best describes
    a. nozzle pressure.
    b. nozzle reaction.
    c. nozzle flow.
    d. nozzle reach.

74. According to NFPA 1901, an apparatus with a 200-gallon tank is classified as
    a. an initial attack apparatus.
    b. a pumper fire apparatus.
    c. a mobile water apparatus.
    d. a tanker apparatus.

75. A hydrant supplied from only one direction is called a
    a. one-way hydrant.
    b. dead-end hydrant.
    c. gridded hydrant.
    d. zoned hydrant.

76. The hydrant in this picture is a
    a. dry-barrel hydrant.
    b. wet-barrel hydrant.
    c. dead-end hydrant.
    d. static hydrant.

77. If the pump is located at the hydrant, which of the following has *not* occurred?
    a. reverse lay
    b. forward lay
    c. first pump in a relay
    d. boosting pressure using a four-way hydrant valve

78. During a drafting evolution, a priming pump that achieved a perfect vacuum would reduce the pressure to 0 psi in a pump and would force water to its theoretical lift of approximately _____ feet.
    a. 2.35
    b. 14.7
    c. 22.5
    d. 33.81

79. During relay operations, the intake pressure for each pump in the relay should not fall below
    a. 10 psi.
    b. 14.7 psi.
    c. 20 psi.
    d. 22.5 psi.

80. Chocking or blocking the wheels of an apparatus should be completed
    a. as soon as initial pump operations are complete.
    b. after leaving the driver's seat.
    c. as required by NFPA 1901.
    d. all are correct

81. The test that ensures the interior of the pump can maintain a vacuum is called the
    a. priming device test.
    b. draft test.
    c. vacuum test.
    d. pressure control test.

82. Water hammer can be caused by each of the following reasons *except*
    a. abruptly shutting off a nozzle.
    b. closing discharge control valves too quickly.
    c. by pressure generated by cavitation.
    d. opening a hydrant too quickly.

83. _____ is the process that explains the formation and collapse of vapor pockets when certain conditions exists during pumping operations.
    a. Pressure fluctuation
    b. Vapor pressure
    c. Cavitation
    d. Water hammer

84. In addition to an engine's cooling system and auxiliary cooling system, another method that may help keep the temperatures within normal operating parameters is to
    a. to use a fine water spray directly on the engine block.
    b. open the engine compartment to increase ventilation.
    c. place ice packs on the radiator.
    d. recirculate water between the pump and the tank.

85. The _____ sprinkler system maintains compressed air or inert gas within the pipes. When a sprinkler head is fused, the air/gas escapes, lowering the pressure. This allows water to enter the system and discharge through fused heads.
    a. dry-pipe
    b. pre-action
    c. deluge
    d. wet pipe

86. A water supply control valve, often located on a sprinkler or standpipe system, that allows a quick visual indication of the valve position (open or closed) is called a
    a. POV.
    b. OS&Y.
    c. ball valve.
    d. butterfly valve.

87. When supporting a sprinkler or standpipe system, the pump operator connects to the
    a. fire department connection, which is usually two or more 2 ½-inch female couplings.
    b. fire department connection, which is usually two or more 2 ½-inch male couplings.
    c. fire department connection, which is usually located near the control panel.
    d. OS&Y valve.

88. According to NFPA 14, a Class 1 standpipe:
    a. provides 2 ½-inch hose connections and is intended for use by firefighters or fire brigade members.
    b. provides 1 ½-inch and 2 ½-inch hose connections for use by firefighters or fire brigade members.
    c. provides 1 ½-inch hose station and is intended primarily for trained personnel during initial attack efforts.
    d. provides 1 ½-inch and 2 ½-inch hose stations and is intended primarily for trained personnel during initial attack efforts.

89. Water exists in a solid state at temperatures below _____ degrees Fahrenheit.
    a. −32
    b. 0
    c. 32
    d. none of the answers is correct

90. The physical change of state from a liquid to a vapor best describes
    a. vapor point.
    b. vapor pressure.
    c. evaporation.
    d. latent heat of vaporization.

91. The downward force exerted on an object by the Earth's gravity best describes which of the following terms?
    a. density
    b. weight
    c. volume
    d. pressure

92. The correct way to express pressure units is
    a. psi.
    b. lb/in^2.
    c. lb/ft^3.
    d. both a and b are correct.

93. Gauge pressure is typically expressed as
    a. in. Hg.
    b. psia.
    c. psi.
    d. psig.

94. The measurement of pressure that does not include atmospheric pressure is called
    a. vacuum.
    b. gauge pressure.
    c. absolute pressure.
    d. head pressure.

95. The pressure in a system when no water is flowing is called
    a. static pressure.
    b. residual pressure.
    c. pressure drop.
    d. normal pressure.

96. When using the hand method to calculate friction loss in 100-foot sections of 2 ½-inch hose, the ring finger has a value of _____ at the time and _____ at the base.
    a. 1 (100 gpm) and 2 (200 gpm)
    b. 2 (200 gpm) and 4 (400 gpm)
    c. 4 (400 gpm) and 8 (800 gpm)
    d. 5 (500 gpm) and 10 (1,000 gpm)

97. In the formula $P = \dfrac{F}{A}$, the unit F is
    a. force.
    b. friction.
    c. weight.
    d. both a and c are correct.

98. The weight of 1 gallon of water is _____, while the density of water is _____.
    a. 8.34 lb/ft^3, 62.4 lb/gal
    b. 8.34 lb/gal, 62.4 lb/ft^3
    c. 1 lb, 8.34 lb/gal
    d. 8.34 lb/gal, 1 lb

99. For the formula cq^2L, all of the following are correct *except*
    a. c = constant or coefficient for a specific hose diameter.
    b. it can also be expressed as $FL = c \times \left(\dfrac{GPM}{100}\right)^2 \times \left(\dfrac{length}{100}\right)$.
    c. L = length of hose in hundreds of feet.
    d. it should only be used for 2 ½-inch and 3-inch hose diameters.

100. Keeping emergency vehicles clean at all times is important because it helps
    a. maintain a good public image.
    b. ensure the pump, systems, and equipment operate as intended.
    c. ensure vehicles can be inspected properly.
    d. all the answers are correct.

Phase I, Exam I: Answers to Questions

1. T	26. D	51. D	76. A
2. T	27. D	52. C	77. B
3. F	28. B	53. D	78. D
4. T	29. D	54. A	79. C
5. T	30. A	55. C	80. B
6. F	31. C	56. B	81. C
7. F	32. B	57. B	82. C
8. T	33. C	58. D	83. C
9. T	34. D	59. D	84. B
10. T	35. D	60. D	85. A
11. T	36. A	61. A	86. B
12. F	37. D	62. B	87. A
13. T	38. D	63. C	88. A
14. T	39. D	64. A	89. C
15. T	40. A	65. A	90. C
16. T	41. A	66. A	91. B
17. T	42. B	67. C	92. A
18. F	43. A	68. A	93. D
19. F	44. A	69. D	94. B
20. T	45. C	70. D	95. A
21. F	46. C	71. D	96. C
22. T	47. C	72. B	97. A
23. T	48. A	73. C	98. B
24. T	49. B	74. A	99. D
25. B	50. A	75. B	100. D

Phase I, Exam I:
Rationale & References for Questions

Question #1
According to NFPA 1002, 1.4.1, all those who operate fire department vehicles must be licensed to drive those vehicles they are expected to operate. NFPA 1002 1.4.1. *IFPO, 2E:* Chapter 1, page 12.

Question #2
NFPA 1002, 5.1, requires pump operators to meet the objectives in NFPA1001, Firefighter 1 prior to being certified as driver pump operator. NFPA 1002 5.1. *IFPO, 2E:* Chapter 1, page 15.

Question #3
A visual inspection of a tire is not sufficient to determine tire pressure. A pressure reading should be taken to ensure adequate tire pressure. NFPA 1002 4.2.1. *IFPO, 2E:* Chapter 2, page 30.

Question #4
Preventive maintenance can be defined as proactive activities taken to ensure the apparatus, pump, and related components remain in a ready state and peak operating performance. These activities can be grouped into inspecting, servicing, and testing. NFPA 1002 4.2.1, 5.1.1. *IFPO, 2E:* Chapter 2, page 23.

Question #5
The faster the vehicle travels, the greater the distance it takes to stop. NFPA 1002 4.3.1, 4.3.6. *IFPO, 2E:* Chapter 3, page 55.

Question #6
The weight shift described is correct in that the water attempts to stay in motion, but it is not referred to as centrifugal force. Centrifugal force is the tendency to move outward from the center and occurs when navigating a curve. NFPA 1002 4.3.1, 4.3.2, 4.3.3, 4.3.4, 4.3.5, 4.3.6. *IFPO, 2E:* Chapter 3, page 57.

Question #7
Federal laws and regulations do not specify requirements for backing of emergency apparatus. In addition, some departments only require one spotter as opposed to two.
At a minimum, one spotter should be used when backing apparatus. NFPA 1002 4.3. *IFPO, 2E:* Chapter 3, page 58.

Question #8
Laws governing the rules of the road are established for ideal conditions such as dry roads and good visibility. Appropriate adjustments must be made when adverse environmental conditions are encountered. NFPA 1002 4.3. *IFPO, 2E:* Chapter 3, pages 50 – 60.

Question #9
This is a fundamental principle for the operation of positive-displacement pumps. Because this is true, forces applied to water will tend to push or move water rather than compress it. NPFA 1002 5.2.1, 5.2.2. *IFPO, 2E:* Chapter 4, page 77.

Question #10
NFPA 1901 specifies many safety requirements for pumping apparatus. One such requirement is the limitation of discharge outlets on the pump panel to 2 ½-inch or smaller. NFPA 1002 5.2.1, 5.2.2. *IFPO, 2E:* Chapter 5, pages 101 – 102.

Question #11
The eductor is located between the discharge and intake sides of the pump, giving it its name of "around-the-pump." NFPA 1002 5.2.3. *IFPO, 2E:* Chapter 5, pages 126 – 130.

Question #12
Actually, synthetic fiber is the most popular material used in the construction of woven-jacket hose. NFPA 1002 5.2.1, 5.2.2, 5.2.4. *IFPO, 2E:* Chapter 6, pages 140 – 141.

Question #13
NFPA 291 suggests the following classification and color coding of hydrants:

Class	Rated Capacity	Color of Bonnets and Caps
AA	1,500 gpm or greater	Light blue
A	1,000 to 1,499 gpm	Green
B	500 to 999 gpm	Orange
C	Less than 500 gpm	Red

NFPA 1002 5.2.1, 5.2.2. *IFPO, 2E:* Chapter 7, page 178.

Question #14
NFPA 1901 requires that pumps rated less than 1,500 achieve a prime within 30 seconds. NFPA 1002 5.1.1. *IFPO, 2E:* Chapter 8, page 214.

Question #15
Flow meter accuracy can be checked using pitot tube readings taken from a smooth-bore nozzle. NFPA 1901 suggest specific flows for various pipe diameters, such as 300 gpm for 2 ½-inch pipes and 700 gpm for 3-inch pipe. NFPA 1002 5.1.1. *IFPO, 2E:* Chapter 8, page 231.

Question #16
The two basic methods for providing power to a pump are either a separate engine or from the drive engine. NFPA 1002 5.2.1, 5.2.2, 5.2.4. *IFPO, 2E:* Chapter 8, page 210.

Question #17
The *deluge system* maintains all sprinkler heads in an open position. When a detection system is activated, water enters the system and is discharged through all the open heads.
Pre-action systems are similar to dry-pipe systems in that air or compressed gas is maintained in the system. However, an automatic detection system (smoke, heat, flame, etc.) or manual system (pull box) must operate to allow water to enter the system. At this point, the pre-action system is similar to a wet-pipe system – when a sprinkler head fuses, water is discharged. NFPA 1002 5.2.4. *IFPO, 2E:* Chapter 9, pages 248 – 253.

Question #18
A standpipe can be supplied in a similar manner to an automatic sprinkler system and through a direct connection to a fire department connection. NFPA 1002 5.2.4. *IFPO, 2E:* Chapter 9, page 253.

Question #19
Pure water is a colorless, orderless, and tasteless liquid that readily dissolves many substances. Water prepared for domestic purposes may have chemicals and minerals naturally occurring or added that can cause the water to taste, smell, and even look different from pure water. NFPA 1002 5.2.1, 5.2.2. *IFPO, 2E:* Chapter 10, pages 262 – 263.

Question #20
This concept is one of five pressure principles related to the manner in which liquids behave while under pressure.
Another way to explain this principle is that pressure in water is exerted in every direction: downward, outward, and upward. NFPA 1002 5.2.1, 5.2.2. *IFPO, 2E:* Chapter 10, page 271.

Question #21
Friction actually varies directly with hose length if all other variables are held constant. When hose length doubles, friction loss doubles as well.
This is one of four fundamental friction-loss principles. NFPA 1002 5.2.1, 5.2.2. *IFPO, 2E:* Chapter 10, pages 280 – 282.

Question #22
This is the correct definition for hydraulics.
Hydraulics is a branch of science dealing with the principles of fluid at rest or in motion.
Hydrodynamics is that branch of hydraulics that deals with the principles and laws of fluids in motion.
Hydrostatics is that branch of hydraulics that deals with the principles and laws of fluids at rest and the pressures they exert or transmit. NFPA 2001 5.2. *IFPO, 2E:* Chapter 10, page 261.

Question #23
To use this method, simply subtract 10 from the first two numbers of the gpm flow. For example, the friction loss in 100 feet of 2 ½-inch hose flowing at 250 gpm is 15 psi (25 – 10 = 15). NFPA 1002 5.2.1, 5.2.2. *IFPO, 2E:* Chapter 11, pages 303 – 304.

Question #24
Appliance friction loss is the reduction in pressure resulting from increase turbulence caused by the appliance. NFPA 1002, 5.2.1, 5.2.2, 5.2.4. *IFPO, 2E:* Chapter 11, pages 309 – 310.

Question #25
- *NFPA 1002* Fire Apparatus Driver/Operator identifies the minimum requirements for pump operators. *NFPA 1001* focuses on firefighter professional qualifications.
- *NFPA 1041* is the standard for fire instructor professional qualifications.
- *NFPA 1901* deals with requirements for new fire apparatus.

NFPA 1002 1.1. *IFPO, 2E:* Chapter 1, page 15.

Question #26
NFPA 1002 requires that pump operators be subject to medical evaluations as required by NFPA 1500. NFPA 1002 1.4.2. *IFPO, 2E:* Chapter 1, page 15.

Question #27
Over the years, several names or titles have been used for the position responsible for driving and operating pumping apparatus, including:

- Wagon driver or tender
- Chauffeur
- Apparatus operator
- Motorized pump operator
- Driver/pump operator
- Apparatus engineer
- Engineer
- Pump operator
- Driver

NFPA 1002 1.4. *IFPO, 2E:* Chapter 1, pages 5 – 6.

Question #28
Laws are rules that are legally binding and enforceable. *Standards* are guidelines that are not legally binding or enforceable unless adopted as such. NFPA 1002 4.2, 4.3, 5.1. *IFPO, 2E:* Chapter 1, page 15.

Question #29
- *NFPA 1500*, chapter 4, section 4.4, suggests requirements for the inspection, maintenance, and repair of emergency vehicles.
- *NFPA 1915* is the standard for fire apparatus preventive maintenance programs.
- *NFPA 1001*, Professional Qualifications for Fire Fighters, does not provide or suggest requirements for conducting preventive maintenance on emergency vehicles.

NFPA 1002 4.2.1, 5.1.1. *IFPO, 2E:* Chapter 2, pages 23 – 25.

Question #30

SAE stands for the Society of Automotive Engineers. SAE rates oil using a two-letter system. NFPA 1002 4.2.1. *IFPO, 2E:* Chapter 2, pages 32 – 33.

Question #31

While charging vehicle batteries, hydrogen gas can be generated. Ventilation is important to reduce the accumulation of this combustible gas. NFPA 1002 4.2.1. *IFPO, 2E:* Chapter 2, page 33.

Question #32

Steering wheel inspection should be limited to determining excessive play. Damage to the steering system can occur when the steering wheel is turned so that the wheels turn. NFPA 1002 4.2.1. *IFPO, 2E:* Chapter 2, pages 35 – 36.

Question #33

NFPA 1500 suggests that the following safety-related components be inspected on a routine basis:

- Tires
- Brake
- Warning systems
- Windshield wipers
- Headlights and clearance lights
- Mirrors

Typically, manufacturers recommend the following components be inspected:

- Engine oil
- Coolant level
- Transmission oil
- Brake system
- Belts

NFPA 1002 4.2.1, 5.1.1. *IFPO, 2E:* Chapter 2, pages 28 – 29.

Question #34

It is important that manufacturer recommendations be followed when conducting vehicle preventive maintenance inspections. NFPA 1002 4.2.1, 5.1.1. *IFPO, 2E:* Chapter 2, pages 28 – 29.

Question #35

Check to ensure that no one is in close proximity to audible warning devices prior to conducting the operation check. NFPA 1002 4.2.1. *IFPO, 2E:* Chapter 2, page 35.

Question #36

Maintaining emergency vehicles clean at all times is important because the practice:

- Helps maintain a good public image (municipal apparatus usually belong to tax-paying public).
- Helps ensure the pump, systems, and equipment operate as intended.
- Helps ensure vehicles can be inspected properly (dirt and grime could cover defects or potential problems).

NFPA 1002 4.2.1. *IFPO, 2E:* Chapter 2, page 37.

Question #37

According to annual reports published in the *NFPA Journal*, over 11,000 emergency-vehicle accidents occurred each year from 1990 to 2000. NFPA 1002 4.3.1. *IFPO, 2E:* Chapter 3, pages 42 – 43.

Question #38
Most drivers possess a license, as required by NFPA 1500 and most state laws. Having driver certification is not a factor, per se, but can certainly help in understanding, avoiding, and/or adequately compensating for the common factors. NFPA 1002 4.3.2, 4.3.3, 4.3.5, and 4.3.6. *IFPO, 2E:* Chapter 3, pages 43 – 45.

Question #39
- Chapter 6 of NFPA 1500 includes specific requirements for fire apparatus, equipment, and driver/operators.
- State laws set the overall rules and standards for the state.
- Local ordinances can be more restrictive and provide specific detail for operating within the city/county.
- The Federal Fire Act does not address safe emergency-vehicle operations.

NFPA 1002 4.3.1| 4.3.6. *IFPO, 2E:* Chapter 3, pages 44 – 50.

Question #40
Laws are rules that are legally binding and enforceable. Standards are guidelines that are not legally binding or enforceable. NFPA 1002 4.3. *IFPO, 2E:* Chapter 3, pages 44 – 45.

Question #41
Total stopping distance is measured from the time a hazard is detected until the vehicle comes to a complete stop. Total stopping distance consists of:
- *Perception distance* (distance apparatus travels from the time the hazard is seen until the brain recognizes it as a hazard)
- *Reaction distance* (distance apparatus travels from the time the brain sends the message to depress the brakes until the brakes are depressed)
- *Braking distance* (distance of travel from the time the brake is depressed until the vehicle comes to a complete stop)

NFPA 1002 4.3. *IFPO, 2E:* Chapter 3, pages 55 – 65.

Question #42
Traction is the friction between the tires and the road surface, and it is essential for steering. NFPA 1002 4.3. *IFPO, 2E:* Chapter 3, pages 55 – 56.

Question #43
One method to help ensure the safe control of a stationary apparatus is to properly align the front wheels as follows:
- When parked next to curb – rotate the front wheels so that they point toward the curb.
- When no curb is present – the front wheels should be positioned to roll the apparatus away from the road.

NFPA 1002 4.3. *IFPO, 2E:* Chapter 3, pages 58 – 59.

Question #44
Hydrodynamics refers to a branch of hydraulics that deals with liquids in motion. The other two answer selections are types of pumps (remove the term hydro) that are based on hydrostatic and hydrodynamic principles. NFPA 1002 5.2.1, 5.2.2. *IFPO, 2E:* Chapter 4, page 76.

Question #45
Piston pumps are positive-displacement pumps, but are categorized as reciprocating pumps. NFPA 1002 5.2.1, 5.2.2. *IFPO, 2E:* Chapter 4, pages 80 – 84.

Question #46
The split-shaft transfers power from the rear axle to the pump. NFPA 1002 5.2.1, 5.2.2. *IFPO, 2E:* Chapter 4, pages 92 – 93.

Question #47
Priming is the process of getting water into the pump through the use of atmospheric pressure, not a suctioning process. NFPA 1002, 5.2.1, 5.2.2. *IFPO, 2E:* Chapter 4, pages 75 – 76.

Question #48
The volume mode of a two-stage centrifugal pump moves water from a common source to both impellers at the same time, and then to a common discharge. The volume of both impellers is added together for total flow, while the pressure remains constant. NFPA 1002 5.2.1, 5.2.2. *IFPO, 2E:* Chapter 4, pages 89 – 90.

Question #49
According to NFPA 1901, pumps must have a rated capacity as follows:

- 100% of its rated capacity at 150 psi
- 70% of its rated capacity at 200 psi
- 50% of its rated capacity at 250 psi

NFPA 1002 5.2.1, 5.2.2. *IFPO, 2E:* Chapter 4, page 91.

Question #50
The intake side of the pump is the point at which water enters the pump; it also is referred to as the *supply side* and the *suction side* of the pump.
The discharge side of the pump is the location at which water leaves the pump. NFPA 1002 5.2. *IFPO, 2E:* Chapter 4, page 75.

Question #51
Basic pump terms include:

- *Flow* refers to the rate and quantity of water delivered by the pump and is expressed in gallons per minute (gpm).
- *Pressure* refers to the amount of force generated by the pump or the resistance encountered on the discharge side of the pump; it is expressed in pounds per square inch (psi).
- *Speed* refers to the rate at which the pump is operating and is typically expressed in revolutions per minute (rpm).
- *Slippage* is the term used to describe the leaking of water between the surface of the internal moving parts of a pump.

NFPA 1002 5.2. *IFPO, 2E:* Chapter 4, page 75.

Question #52
This is an example of a two-stage centrifugal pump. Note the two impellers on the same shaft. NFPA 1002 5.2.1, 5.2.2. *IFPO, 2E:* Chapter 4, pages 87 – 89.

Question #53
The two largest gauges on a pump panel are typically the main pump gauges, called pump intake gauge and pump discharge gauge; they are most often of Bourdon-tube construction. NFPA 1002 5.2.1, 5.2.2. *IFPO, 2E:* Chapter 5, pages 110 – 112.

Question #54
Flow meters measure gpm (quantity and rate) and, because the flow is known, normal friction loss calculations are not required. Most flow meters use paddle wheels to measure the flow. NFPA 1002 5.2.1, 5.2.2. *IFPO, 2E:* Chapter 5, pages 113 – 114.

Question #55
Pneumatically activated control valve operation is not considered a manual operation. NFPA 1002 5.2.1, 5.2.2. *IFPO, 2E:* Chapter 5, pages 114–115.

Question #56

A pressure governor controls pressure by increasing or decreasing engine speed to maintain the desired pressure. NFPA 1002 5.2.1, 5.2.2, 5.2.4. *IFPO, 2E:* Chapter 5, pages 122–125.

Question #57

The radiator-fill system pumps water directly into the radiator. NFPA 1002 5.2.1, 5.2.2, 5.2.4. *IFPO, 2E:* Chapter 5, pages 125 – 126.

Question #58

Foam systems found on pumping apparatus include:

- Pre-mixed
- In-line eductor
- Around-the-pump proportioning
- Balanced pressure
- Direct injection/compressed-air

NFPA 1002 5.2.3. *IFPO, 2E:* Chapter 5, pages 126 – 130.

Question #59

Because priming pumps rely on close fitting parts, oil helps to both lubricate the moving parts as well as provide a tighter seal. NFPA 1002 5.2.1, 5.2.2, 5.2.4. *IFPO, 2E:* Chapter 5, pages 120 – 122.

Question #60

Instrumentation, such as gauges, flow meters, and indicators, is used to ensure that the pump is operating efficiently while providing appropriate pressure and flows. A foam system is not considered an instrumentation device. NFPA 1002 5.2.1, 5.2.2. *IFPO, 2E:* Chapter 5, page 99.

Question #61

- *Tank-to-pump* control valves allows water to flow from the on-board water supply to the intake side of the pump.
- *Pump-to-tank* (tank fill) control valves allow water to flow from the discharge side of the pump to the tank.
- *Transfer valve* control valves are found on multistage pumps and redirect water from the pump between the pressure mode and the volume mode.

NFPA 1002 5.2.1, 5.2.2. *IFPO, 2E:* Chapter 5, page 119.

Question #62

- *NFPA 1500* is the standard for occupational safety and health programs.
- *NFPA 1962* is the standard for the inspection, care, and use of fire hose, couplings, and nozzles, and the service testing of hose.
- *NFPA 1965* is the standard for fire hose appliances.

NFPA 1002 5.2.1, 5.2.2, 5.2.4. *IFPO, 2E:* Chapter 6, pages 138 – 139.

Question #63

Supply lines should not be operated at pressures exceeding 185 psi, and attack lines should have a highest normal operating pressure of 275 psi. NFPA 1002 5.2.1, 5.2.2, 5.2.4. *IFPO, 2E:* Chapter 6, page 140.

Question #64

Although NFPA 1961 no longer specifies the required length of hose, a section of hose is commonly 50 feet. NFPA 1002 5.2.1, 5.2.2, 5.2.4. *IFPO, 2E:* Chapter 6, page 140.

Question #65

NFPA 1963 refers to the threading coupling construction as American National Fire Hose Connection Screw Thread; it is abbreviated with the thread symbol NH. Storz is not a threaded coupling, and NTS is not a type of thread used in the fire service. Occasionally, threads are referred to as National Standard Thread (NST). NFPA 1002 5.2.1, 5.2.2, 5.2.4. *IFPO, 2E:* Chapter 6, pages 141 – 152.

Question #66

- A *wye* has a single female inlet connection and two or more male outlet connections.

- A *siamese* has two or more female inlet connections and one male outlet connection.

Although the wye in this picture can be used as a siamese, it is still considered a wye and requires double male and female adapters. NFPA 1002 5.2.1, 5.2.2, 5.2.4. *IFPO, 2E:* Chapter 6, pages 148 – 149.

Question #67

Double female adapters have a female coupling on both sides. NFPA 1002 5.2.1, 5.2.2, 5.2.4. *IFPO, 2E:* Chapter 6, page 150.

Question #68

- 50 psi is the operating pressure for smooth-bore hand-lines.

- 80 psi is the operating pressure for smooth-bore master streams.

- 100 psi is the operating pressure for combination nozzles.

- 125 psi is not a normal operating pressure for nozzles used in the fire service.

NFPA 1002 5.2.1, 5.2.2, 5.2.4. *IFPO, 2E:* Chapter 6, pages 153 – 157.

Question #69

- The *fixed-flow* or *constant-flow* nozzle provides a constant flow regardless of stream pattern.

- A *selectable-flow* nozzle allows the ability to adjust flow at the nozzle.

- An *automatic* nozzle maintains a constant nozzle pressure over a wide variety of flows.

- A *smooth-bore* nozzle is not a type of combination nozzle.

NFPA 1002 5.2.1, 5.2.2, 5.2.3. *IFPO, 2E:* Chapter 6, pages 155 – 157.

Question #70

A pump operator may use each of the items listed, as well as hose jackets. NFPA 1002 5.2.1, 5.2.2, 5.2.4. *IFPO, 2E:* Chapter 6, page 145.

Question #71

The basic parts of a hose include:

- Reinforced inner liner

- Outer protective shell

- Couplings attached to both ends

NPFA 1002 5.2.1, 5.2.2, 5.2.4. *IFPO, 2E:* Chapter 6, pages 140 – 141.

Question #72

The eductor uses the Venturi principle to draw foam into the water stream. NFPA 1002 5.2.3. *IFPO, 2E:* Chapter 6, page 151.

Question #73

- *Nozzle pressure* is the designed operating pressure for a particular nozzle.

- *Nozzle flow* is the amount of water flowing from a nozzle.

- *Nozzle reach* is the distance water travels after leaving a nozzle.

- *Nozzle reaction* is the tendency of a nozzle to move in opposite direction of water flow.

NFPA 1002 5.2.1. *IFPO, 2E:* Chapter 6, pages 153 – 154.

Question #74

NFPA specifies minimum tank capacity for apparatus as follows:

- 200 gallons – initial attack

- 300 gallons – pumper

- 1,000 gallons – mobile water apparatus/tanker

NFPA 1002 5.2.1. *IFPO, 2E:* Chapter 7, page 167.

Question #75

Hydrants located on a main that is suppled from one direction (dead-end main) are called dead-end hydrants. NFPA 1002 5.2.1, 5.2.2. *IFPO, 2E:* Chapter 7, page 173.

Question #76

This hydrant is dry and requires the stem nut located on the bonnet to be turned before water can enter the hydrant barrel.

- *Wet-barrel hydrants* have stem nuts to control each discharge outlet.

- *Static-source hydrants* typically do not have control valves, and *dead-end hydrants* could be either wet-or dry-barrel.

NFPA 1002 5.2.1. *IFPO, 2E:* Chapter 7, pages 175 – 176.

Question #77

A straight or forward lay would cause the pump to be located near the incident. Each of the other answer choices would cause the pump to be located near the hydrant. NPFA 1002 5.2.1, 5.2.2, 5.2.4. *IFPO, 2E:* Chapter 7, pages 182 – 183.

Question #78

Water will rise approximately 2.3 feet for each 1 psi of pressure. With a perfect vacuum, atmospheric pressure (14.7) will force water to a height of 33.81 feet (2.3 × 14.7 = 33.81). NFPA 1002 5.2.1, 5.2.2, 5.2.4. *IFPO, 2E:* Chapter 7, pages 185 – 186.

Question #79

Intake pressure for pumps in a relay should not fall below 20 psi, to reduce the chances of the pump cavitating. NFPA 1002 5.2.2. *IFPO, 2E:* Chapter 7, pages 189 – 194.

Question #80

Should the pump not properly engage, the apparatus may accidently move. Chocking the wheels will help guard against accidently movement. NFPA 1901 does not require an apparatus to be chocked. Apparatus should be chocked immediately upon leaving the driver seat and especially before engaging and operating the pump. NFPA 1002 5.2.1, 5.2.2. *IFPO, 2E:* Chapter 8, page 213.

Question #81

- The *vacuum test* is used to determine the ability of the pump to maintain a vacuum.

- The *priming device test* ensures the primer can develop sufficient vacuum to draft.

- The *pressure control test* is conducted to ensure the pressure control can adequately maintain safe discharge pressures.

- The *draft test* is a non-specific test.

NFPA 1002 5.1.1. *IFPO, 2E:* Chapter 8, pages 226 – 231.

Question #82

Each of the answer selections can cause a water hammer except cavitation, which is caused by the pump attempting to pump more water than is available. If anything, cavitation will cause a decrease in pressure, as opposed to a sudden surge of pressure indicative of a water hammer. NFPA 1002 5.1.1. *IFPO, 2E:* Chapter 9, pages 244 – 245.

Question #83

Cavitation is the process that explains the formation and collapse of vapor pockets when certain conditions exists during pumping operations. NFPA 1002 5.1.1. *IFPO, 2E:* Chapter 9, pages 245 – 247.

Question #84

Spraying water on the engine block or placing ice packs on the radiator may cause serious damage, and is a dangerous practice. Recirculating the water between the pump and the tank will help keep the pump temperature cool, but not the engine. Increasing ventilation by opening the engine compartment may help keep the temperature within normal operating parameters. NFPA 1002 5.2.1, 5.2.2. *IFPO, 2E:* Chapter 9, page 242.

Question #85

- *Dry-pipe systems* maintain air or compressed gas under pressure within the system. When a head fuses, water enters the system and discharges through any fused heads.

- *Pre-action systems* are similar to dry-pipe systems in that air or compressed gas is maintained in the system. However, an automatic detection system (smoke, heat, flame, etc.) or manual system (pull box) must operate to allow water to enter the system. At this point, the dry-pipe system is similar to a wet-pipe system, and when a sprinkler head fuses, water is discharged.

- A *deluge system* maintains all sprinkler heads in the open position. When a detection system indicates, water enters the system and is discharged through all the open heads.

- A *wet-pipe system* maintains water in the system at all times. When a sprinkler head is fused, water is immediately discharged.

NFPA 1002 5.2.4. *IFPO, 2E:* Chapter 9, pages 148 – 253.

Question #86

The outside stem and yoke (OS&Y) valve allows for the quick determination of valve position. If the stem is out and exposed, the valve is open. If the stem is in or only protruding a short distance, the valve is closed. NFPA 1002 5.2.4. *IFPO, 2E:* Chapter 9, page 249.

Question #87

The fire department connection is similar to a siamese in that it combines two or more lines into one line, and the inlet connections are female. NFPA 1002 5.2.4. *IFPO, 2E:* Chapter 9, page 252.

Question #88

- *Class 1* standpipes provide 2 ½-inch hose connections and are intended for use by firefighters or fire brigade members.

- *Class 2* standpipes provide 1 ½-inch hose station and are intended primarily for trained personnel during initial attack efforts.

- *Class 3* standpipes provide 1 ½-inch and 2 ½-inch hose connections for use by firefighters or fire brigade members.

NFPA 1002 5.2.4. *IFPO, 2E:* Chapter 9, page 253.

Question #89

The basic characteristics and properties of water are listed below. These are considered approximate in that water purity, atmospheric conditions, and rounding can effect the values:

- Virtually incompressible
- Freezes at 32°F
- Expands when frozen
- Boils at 212°F (considered water boiling point)
- 1 gallon weighs 8.34 lbs
- Density is 62.4 lb/ft^3
- Number of gallons in 1 cubic foot is 7.38 gallons/ft^3

NFPA 1002 5.2.1, 5.2.2. *IFPO, 2E:* Chapter 10, pages 262 – 263.

Question #90

- *Evaporation* is the physical change of state from a liquid to a vapor.
- *Vapor pressure* (VP) is the pressure exerted on the atmosphere by molecules as they evaporate from the surface of the liquid.
- *Boiling point* (BP) is the temperature at which the vapor pressure of a liquid equals the surrounding pressure.

NFPA 1002 5.2.1, 5.2.2. *IFPO, 2E:* Chapter 10, pages 262 – 263.

Question #91

- *Weight* is the downward force exerted on an object by the Earth's gravity, typically expressed in pounds.
- *Density* is the weight of a substance expressed in units of mass per volume.
- *Volume* is the amount of space occupied by an object.
- *Pressure* is the force exerted by a substance in units of weight per area, typically expressed in pounds per square inch (psi).

NFPA 1002 5.2.1, 5.2.2. *IFPO, 2E:* Chapter 10, pages 264 – 269.

Question #92

Pressure is the force exerted by a substance in *units of weight per area*, typically expressed in *pounds per square inch* (psi or lb/in^2).
The unit lb/ft^3 is used with density. NFPA 1002 5.2.1, 5.2.2. *IFPO, 2E:* Chapter 10, pages 269 – 270.

Question #93

- *Gauge pressure* is the measurement of pressure that does not include atmospheric pressure, typically expressed as psig.
- *Absolute pressure* is the measurement of pressure that include atmospheric pressure, typically expressed as psia.
- *Vacuum* is the measurement of pressure that is less than atmospheric pressure, typically expressed in inches of mercury (in. Hg).
- *Head pressure* is the pressure exerted by the vertical height of a column of liquid expressed in feet.

NFPA 1002 5.2.1, 5.2.2. *IFPO, 2E:* Chapter 10, pages 275 – 277.

Question #94

- *Atmospheric pressure* is the pressure exerted by the atmosphere on the Earth.
- *Gauge pressure* is the measurement of pressure that does not include atmospheric pressure, typically expressed as psig.
- *Absolute pressure* is the measurement of pressure that include atmospheric pressure, typically expressed as psia.
- *Vacuum* is the measurement of pressure that is less than atmospheric pressure, typically expressed in inches of mercury (in. Hg).
- *Head pressure* is the pressure exerted by the vertical height of a column of liquid expressed in feet.

NFPA 1002 5.2.1, 5.2.2. *IFPO, 2E:* Chapter 10, pages 275 – 277.

Question #95

- *Static pressure* is the pressure in a system when no water is flowing.
- *Residual pressure* is the pressure remaining in the system after water has been flowing.
- *Pressure drop* is the difference between the static pressure and the residual pressure when measured at the same location.
- *Normal pressure* is the water flow pressure found in a municipal water supply during normal consumption demands.

NFPA 1002 5.2.1, 5.2.2. *IFPO, 2E:* Chapter 10, pages 276 – 277.

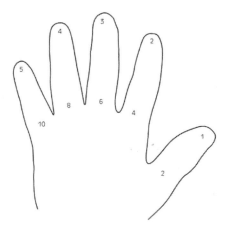

Question #96

The hand method is a fireground method used to estimate friction loss in 100-foot sections of 2 ½-inch hose. Simply multiply the two figures on a finger for the approximate friction loss pressure for each 100-section of 2 ½-inch hose. NFPA 1002 5.2.1, 5.2.2. *IFPO, 2E:* Chapter 10, pages 302 – 303.

Question #97

Pressure is the force exerted by a substance in *units of weight per area*, typically expressed in *pounds per square inch* (psi or lb/in^2)
P = Pressure
F = Force (weight in pounds)
A = Area (square inches

NFPA 1002 5.2.1, 5.2.2. *IFPO, 2E:* Chapter 10, pages 269 – 270.

Question #98

One gallon of water weighs 8.34 pounds and is expressed as 8.34 lb/gal.
The density of water is 62.4 pounds and is expressed as 62.34 lb/ft^3 (cubic feet). NFPA 1002 5.2.
IFPO, 2E: Chapter 10, pages 264 – 265.

Question #99

This friction loss formula was derived from a combination of Bernoulli's equation, the Darcy-Weibach equation, and the Continuity equation. It is both simple and reasonably accurate for varying hose sizes, and *it can be use for both preplanning and fireground calculations:*

$$FL = c \times q^2 \times L$$

where   FL = friction loss
          c = constant for a specific hose diameter
          q = gm ÷ 100 (flow in hundreds of gpm)
          L = length of hose in hundreds of feet

NFPA 1002 5.2.1, 5.2.2. *IFPO, 2E:* Chapter 11, pages 306 – 309.

Question #100

Maintaining emergency vehicles clean at all times is important because it:

- Helps maintain a good public image (municipal apparatus usually belong to tax-paying public)
- Helps ensure the pump, systems, and equipment operate as intended
- Helps ensure vehicles can be inspected properly (dirt and grime could cover defects or potential problems)

NFPA 1002 4.2.1. *IFPO, 2E:* Chapter 2, page 37.

Phase One, Exam Two

1. Pump operators are subject to periodic medical examinations, as required by NFPA 1901, to help ensure they are medically fit to perform the duties of a driver pump operator.
   a. True
   b. False

2. Driver pump operators must posses a valid driver's license and be licensed to drive the largest apparatus within their department.
   a. True
   b. False

3. Some tires have wear bars – grooves in the tire treads to help determine wear. These should be disregarded for tires on emergency vehicles.
   a. True
   b. False

4. Fire Department mechanics must be certified and are responsible for inspecting, servicing, and testing apparatus.
   a. True
   b. False

5. Doubling the speed of a vehicle decreases the stopping distance an estimated four times.
   a. True
   b. False

6. A pump that receives its power from a PTO on the transmission can be used for both stationary and mobile pump operations.
   a. True
   b. False

7. Posted speed limits are based on ideal driving conditions. Even when responding to an incident, speeds should be reduced when driving in adverse weather conditions.
   a. True
   b. False

8. When pressure is applied to a confined liquid, the same pressure is transmitted within the liquid equally.
   a. True
   b. False

9. The purpose of flow meters is to measure the quantity and rate of water flow in gallons per minute (gpm).
   a. True
   b. False

10. The in-line eductor foam system mixes foam with water by means of an eductor and separate foam pump.
    a. True
    b. False

11. The two pressure regulators commonly found on pumps are pressure-relief devices and pressure governors.
    a. True
    b. False

12. The two main types of hose construction are woven jacket and rubber covered.
    a. True
    b. False

13. Securing a water supply from static sources often requires the apparatus to position fairly close to the source. A rule of thumb is to park the apparatus no further than about 22 feet from the source.
    a. True
    b. False

14. Newer braking systems eliminate the need to chock or block the wheels of emergency vehicles when parked.
    a. True
    b. False

15. Most main pumps on apparatus are powered by the drive engine, as opposed to a separate engine.
    a. True
    b. False

16. Feathering or gating is a process of partially opening or closing control valves to regulate pressure and flow for individual lines.
    a. True
    b. False

17. Today, most sprinkler system water supplies are designed to provide adequate flow and pressure to most of the sprinkler heads on the system. Because of this, fire departments rarely need to provide additional support to sprinkler systems.
    a. True
    b. False

18. Standpipe systems can be connected to a water supply in the same manner as automatic sprinkler systems, and can also be directly connected and solely supported by a fire department connection.
    a. True
    b. False

19. Water is virtually incompressible and does not expand when frozen.
    a. True
    b. False

20. The pressure of a fluid acting on a surface is perpendicular to that surface.
    a. True
    b. False

21. When the flow remains constant, friction loss varies inversely with hose diameter.
    a. True
    b. False

22. The drop-ten method is a fireground method to estimate friction loss only in 1 ½-inch hose.
    a. True
    b. False

23. Appliance friction loss is the reduction in pressure resulting from increased turbulence caused by the appliance. The appliance pressure loss within standpipe systems should be estimated as 25 psi, for sprinkler systems 150 psi, and for combination systems 175 psi.
    a. True
    b. False

24. Hydrodynamics is a branch of science dealing with the principles of fluid at rest or in motion.
    a. True
    b. False

25. NFPA 1002 requires pump operators to meet the requirements of NFPA _____ prior to being certified.
    a. NFPA 1001, Fire Fighter 1
    b. NFPA 1001, Fire Fighter 2
    c. NFPA 1901
    d. both a and b are correct

26. The correct title for NFPA 1002 is
    a. Professional Qualifications for Fire Pump Operations.
    b. Fire Apparatus Driver/Operator Professional Qualifications.
    c. Fire Pump Operator Professional Qualifications.
    d. Professional Qualifications for Drivers and Engineers.

27. The title used to describe the position responsible for driving and operating pumping apparatus is _____?
    a. driver
    b. engineer
    c. chauffeur
    d. all are correct

28. _____ are guidelines that are not legally binding and enforceable.
    a. Standards
    b. Laws
    c. Regulations
    d. Codes

29. When conducting a vehicle inspection, the brake pedal should be depressed and should
    a. stop when it reaches the floorboard.
    b. have 2 inches of play.
    c. stop prior to reaching the floorboard.
    d. have 2 inches of free play.

30. Keeping emergency vehicles clean at all times is important because it helps
    a. maintain a good public image.
    b. ensure the pump, systems, and equipment operate as intended.
    c. ensure vehicles can be inspected properly.
    d. all the answers are correct.

31. Scheduling of preventive maintenance activities should be based on each of the following NFPA standards *except*
    a. 1002.
    b. 1500.
    c. 1901.
    d. 1911.

32. Engine oils are classified using a rating system developed by API and SAE. SAE stands for
    a. Safety Aspects of Engineering.
    b. Society of American Engineers.
    c. Safety and Automotive Engineers.
    d. Society of Automotive Engineers.

33. Inspection of coolant systems on emergency vehicles should include which of the following?
    a. pH level of coolant fluid
    b. coolant level when hot
    c. coolant pressure within the system
    d. obvious signs of damage such as dents and leaks

34. Documenting preventive maintenance activities is important for all the following reasons *except*
    a. tracking needed maintenance.
    b. tracking needed repairs.
    c. tracking inspection duration.
    d. determine maintenance trends.

35. NFPA 1500 suggests that all the following should be inspected on a routine basis *except*
    a. warning systems.
    b. wipers.
    c. engine belts.
    d. mirrors.

36. Batteries can produce _____ gas while charging.
    a. hydrogen
    b. acetylene
    c. combustible
    d. phosgene

37. Each year, approximately _____ firefighter fatalities occurred while responding to or returning from an incident.
    a. 21
    b. 75
    c. 100
    d. 500

38. The cause of emergency-vehicle accidents can be attributed to several common factors. Each of the following are common factors that appear in most vehicle accidents *except*
    a. lacking driver licensure and certification.
    b. not fully being aware of or appreciating apparatus limitations.
    c. lack of appreciation for weather and/or traffic conditions.
    d. not obeying and following state laws and national standards.

39. Safe emergency-vehicle operations and requirements can be found in each of the following *except*
    a. NFPA 1002.
    b. NFPA 1500.
    c. state statutes.
    d. NFPA 1403.

40. The distance of travel from the time the hazard is seen until the brain recognizes it as a hazard is the
    a. perception distance.
    b. reaction distance.
    c. braking distance.
    d. total stopping distance.

41. _____ provide suggested guidelines for drivers.
    a. Laws
    b. Standards
    c. Statutes
    d. Ordinances

42. The time between when a hazard is detected until the vehicle comes to a complete stop is known as
    a. perception distance.
    b. reaction distance.
    c. braking distance.
    d. total stopping distance.

43. The tendency to move outward from the center is known as
    a. centrifugal force.
    b. pressure surge.
    c. cavitation.
    d. friction.

44. Which of the following NFPA standards requires drivers to demonstrate the ability to safely maneuver the apparatus over a predetermined route in compliance with all applicable state, local, and departmental rules and regulations?
    a. NFPA 1001
    b. NFPA 1002
    c. NFPA 1500
    d. NFPA 1901

45. When parking an apparatus with no curb present, the rotation of the front tire should be positioned
    a. toward the curb.
    b. away from the curb.
    c. straight ahead.
    d. to roll the apparatus away from the road.

46. Piston and rotary pumps generally provide
    a. higher flows with lower pressures.
    b. higher flows with higher pressures.
    c. lower flows with lower pressures.
    d. lower flows with higher pressures.

47. Centrifugal pump construction includes all of the following *except*
    a. impeller.
    b. shroud.
    c. volute.
    d. lobe.

48. Which of the following is true concerning pump priming?
    a. Centrifugal pumps are self-priming.
    b. Positive-displacement pumps are used to prime centrifugal pumps.
    c. Priming is a suction process that moves water from a static source to the suction side of a pump.
    d. Centrifugal pumps are used to prime-positive displacement pumps.

49. For a two-stage centrifugal pump operating in pressure mode
    a. water enters both impellers at the same time.
    b. water enters both impellers and then reenters the same impeller for increased pressure.
    c. pressure from one impeller is added to the pressure generated in the second impeller.
    d. each impeller delivers half the total flow.

50. According to NFPA 1901, a pump must be able to deliver 70% of its rated capacity at _____ psi.
    a. 100 psi
    b. 150 psi
    c. 200 psi
    d. 250 psi

51. The _____ side of the pump is the location where water leaves the pump.
    a. intake
    b. discharge
    c. draft
    d. all are correct

52. The amount of force generated by the pump is called
    a. pressure.
    b. slippage.
    c. speed.
    d. flow.

53. NFPA 1901 requires all of the following *except*
    a. pump panels with 2 ½-inch discharges or smaller.
    b. pump panels located in the crew cab.
    c. relief valves on intakes.
    d. standard color code to match discharges with their gauges.

54. Which of the following is correct concerning pressure gauges on pump panels?
    a. most are Bourdon tube gauges.
    b. most measure positive pressure.
    c. heavy liquids or external dampening devices are used to reduce needle fluctuations.
    d. all are correct.

55. Electrically activated control valves are increasingly being installed on pumps because
    a. they increase pump panel design and location opportunities.
    b. they reduce potential water hammer because electric motors control the speed of operation.
    c. they are easier to operate and control under higher pressure.
    d. all are correct.

56. To determine the size of the pump, count the number of 2 ½-inch discharges and multiply by

    _____ .
    a. 250 gpm
    b. 100 gpm
    c. a factor of 2
    d. a factor of 4

57. A _____ device controls pressure buildup by sending excess water pressure back to the intake side of the pump.
    a. pressure-relief
    b. pressure governor
    c. auxiliary pressure
    d. automatic pressure

58. The _____ cooling system consists of tubing running from the pump discharge through a heat exchanger and back to the intake side of the pump.
    a. auxiliary
    b. radiator
    c. primary
    d. pump

59. Foam systems found on pumping apparatus include all *except*
    a. in-line pump.
    b. balanced pressure.
    c. direct injection/compressed-air.
    d. pre-mixed.

60. During the priming process, an indication that the pump is primed is
    a. a positive reading on the pressure gauge.
    b. the priming motor will stop.
    c. the main pump will sound as if it is under load.
    d. both a and c are correct.

61. Components directly or indirectly attached to the pump that are used to control and monitor the pump and the engine are called pump
    a. gauges.
    b. flow meters.
    c. peripherals.
    d. control valves.

62. The _____ control valve allows water to flow from the discharge side of the pump to the tank.
    a. tank-to-pump
    b. pump-to-tank
    c. transfer valve
    d. discharge

63. NFPA 1962 requires that supply hose not be operated at pressures exceeding
    a. 80 psi.
    b. 100 psi.
    c. 185 psi.
    d. 275 psi.

64. NFPA 1961 and 1962 refer to discharge lines as
    a. attack hose.
    b. supply hose.
    c. 2 ½-inch hose.
    d. 4-inch hose.

65. NFPA 1961 requires that hose section lengths not exceed
    a. 50 feet.
    b. 100 feet.
    c. 150 feet.
    d. none is correct.

66. The type of coupling in this picture is called a
    a. Storz.
    b. NH or NST.
    c. sexless.
    d. both a and c are correct.

67. The appliance in this picture
    a. is a wye.
    b. is a siamese.
    c. can be either a wye or siamese.
    d. no correct answer provided.

68. Which of the following is an example of a water thief?
    a.

    b.

    c.

    d.

69. The recommended operating pressure for the nozzle is this picture is
    a. 50 psi.
    b. 80 psi.
    c. 100 psi.
    d. 125 psi.

70. The effect of making changes to a selectable-flow nozzle is that
    a. the nozzle will automatically flow the selected rate without the pump operator making any changes.
    b. the nozzle automatically adjusts to provide the correct nozzle operating pressure.
    c. the pump operator must also make changes to maintain correct flow and nozzle operating pressure.
    d. all are correct.

71. Which of the following is considered the most common sexless coupling used in the fire service?
    a. Storz
    b. NH or NST
    c. lug
    d. rocker

72. A _____ is used to divide one hose line into two or more lines.
    a. wye
    b. siamese
    c. adapter
    d. increaser

73. The distance water travels after leaving a nozzle is called
    a. nozzle pressure.
    b. nozzle reaction.
    c. nozzle flow.
    d. nozzle reach.

74. An apparatus with a 500-gallon tank is classified as
    a. an initial attach apparatus.
    b. a pumper fire apparatus.
    c. a mobile water apparatus.
    d. a tanker apparatus.

75. A potential problem with dead-end mains is that
    a. if the main is damaged all subsequent hydrants on the main become inoperable.
    b. large quantities of water pumped from one hydrant will significantly reduce water pressure and flow to other hydrants on the main.
    c. pressure tends to increase toward the end of the main.
    d. both a and b are correct.

76. The hydrant in this picture is a
    a. dry-barrel hydrant.
    b. wet-barrel hydrant.
    c. dead-end hydrant.
    d. static hydrant.

77. A hydrant with a red bonnet and discharge caps will flow
    a. over 1,500 gpm.
    b. 1,000 to 1,499 gpm.
    c. 500 to 999 gpm.
    d. less than 500 gpm.

78. Priming systems are not able to create a perfect vacuum. Because of this, a general rule of thumb is to pump no more than two-thirds of the theoretical lift, which is
    a. 2.35.
    b. 14.7.
    c. 22.5.
    d. 33.81.

79. When designing a relay operation, the largest pump should be placed
    a. at the end of the relay as close to the scene as safely possible.
    b. at or near the water source.
    c. in the middle of the relay.
    d. within 500 feet of the water source.

80. The device in this picture is
    a. an eductor used to mix foam and water directly into a nozzle stream.
    b. a jet siphon used to help move water from one dump tank to another.
    c. a pressure nozzle that helps improve stream reach.
    d. a drafting device that helps improve drafting operations.

81. The _____ flow is the estimated flow of water needed for a specific incident.
    a. critical
    b. available
    c. required
    d. incident

82. The maximum time allowed to achieve a prime for a pump rated at 1,500 gpm or higher is
    a. 15 seconds.
    b. 30 seconds.
    c. 45 seconds.
    d. 60 seconds.

83. The test conducted to ensure the priming device is able to develop a sufficient vacuum draft is called the
    a. priming device test.
    b. draft test.
    c. vacuum test.
    d. pressure control test.

84. Which of the following is not a typical weekly or monthly test required by manufacturers?
    a. regulating device test
    b. priming system test
    c. engine speed check
    d. transfer valve operation check

85. The process that explains the formation and collapse of vapor pockets when certain conditions exist during pumping operations is called
    a. cavitation.
    b. water hammer.
    c. vapor pressure.
    d. pressure fluctuation.

86. The _____ sprinkler system maintains water within the pipes. When a sprinkler head is fused, water immediately discharges.
    a. dry-pipe
    b. pre-action
    c. deluge
    d. wet-pipe

87. Which of the following is *not* correct when supporting sprinkler or standpipe systems?
    a. Check for debris in the fire department connection before connecting hose lines.
    b. Start pumping immediately if smoke or fire is evident.
    c. Place the transfer valve into the pressure position.
    d. Pump the fire department connection at a minimum of 200 psi.

88. According to NFPA 14, a standpipe that provides 1 ½-inch and 2 ½-inch hose connections for use by firefighters or fire brigade members is a
    a. Class 1 standpipe.
    b. Class 2 standpipe.
    c. Class 3 standpipe.
    d. Class 4 standpipe.

89. Water starts to change from a liquid state to a vapor state at _____ °F.
    a. 0
    b. 32
    c. 100
    d. 212

90. The pressure exerted on the atmosphere by molecules as they evaporate from the surface of the liquid is called
    a. boiling point.
    b. vapor pressure.
    c. evaporation.
    d. latent heat of vaporization.

91. The weight of a substance expressed in units of mass per volume is the definition of
    a. density.
    b. weight.
    c. volume.
    d. pressure.

92. In the formula $P = \dfrac{F}{A}$, the unit A is
    a. area.
    b. atmospheric pressure.
    c. absolute pressure.
    d. both a and b are correct.

93. Vacuum pressure is typically expressed as
    a. in. Hg.
    b. psia.
    c. psi.
    d. psig.

94. The measurement of pressure that includes atmospheric pressure is known as
    a. vacuum.
    b. gauge pressure.
    c. absolute pressure.
    d. head pressure.

95. The water flow pressure found in a municipal water supply during normal consumption demands is referred to as
    a. static pressure.
    b. residual pressure.
    c. pressure drop.
    d. normal pressure.

96. When using the hand method to calculate friction loss in 100-foot sections of 2 ½-inch hose, the pinky finger has a value of _____ at the time and _____ at the base.
    a. 1 (100 gpm) and 2 (200 gpm)
    b. 2 (200 gpm) and 4 (400 gpm)
    c. 4 (400 gpm) and 8 (800 gpm)
    d. 5 (500 gpm) and 10 (1,000 gpm)

97. One gallon of water weighs _____, whereas the density of water is _____.
    a. 8.34 lb/gal, 62.4 lb/ft^3
    b. 8.34 lb/ft^3, 62.4 lb/gal
    c. 1 lb, 8.34 lb/gal
    d. 8.34 lb/gal, 1 lb

98. The volume of a tank that is 30 feet high, by 10 feet wide, by 20 feet deep is
    a. 200 ft^3.
    b. 600 ft^3.
    c. 6,000 ft^3.
    d. 60,000 ft^3.

99. Which of the following is not true concerning the formula $FL = 2q^2 + q$?
    a. q = flow in hundreds of gallons $\dfrac{gpm}{100}$ per minute
    b. It was designed for use with cotton jacketed hose but provides accurate results for newer hose.
    c. It was designed for flows greater than 100 gpm.
    d. All are correct.

100. For the formula cq^2L, all of the following are correct *except*

   a. c = constant or coefficient for a specific hose diameter.

   b. $q = \dfrac{GPM}{100}$ (flow in hundreds of gpm).

   c. L = length of hose in hundreds of feet.

   d. it should only be used for preplanning and not for fireground calculations.

Phase I, Exam II: Answers to Questions

1.	F	26.	B	51.	B	76.	B
2.	F	27.	D	52.	A	77.	D
3.	F	28.	A	53.	B	78.	D
4.	F	29.	C	54.	D	79.	B
5.	F	30.	D	55.	D	80.	B
6.	T	31.	A	56.	A	81.	C
7.	T	32.	D	57.	A	82.	C
8.	T	33.	D	58.	A	83.	A
9.	T	34.	C	59.	A	84.	C
10.	F	35.	C	60.	D	85.	A
11.	T	36.	A	61.	C	86.	D
12.	T	37.	A	62.	B	87.	D
13.	T	38.	A	63.	C	88.	C
14.	F	39.	D	64.	A	89.	D
15.	T	40.	A	65.	D	90.	B
16.	T	41.	B	66.	A	91.	A
17.	F	42.	D	67.	B	92.	A
18.	T	43.	A	68.	B	93.	A
19.	F	44.	B	69.	C	94.	C
20.	T	45.	D	70.	C	95.	D
21.	T	46.	D	71.	A	96.	D
22.	F	47.	D	72.	A	97.	A
23.	T	48.	B	73.	D	98.	C
24.	F	49.	C	74.	B	99.	B
25.	A	50.	C	75.	D	100.	D

Phase I, Exam II:
Rationale & References for Questions

Question #1
NFPA 1002 requires that pump operators be subject to periodic medical evaluations as required by NFPA 1500. *NFPA 1901* is the standard on automotive fire apparatus and as such does not contain medical requirements for pump operator. NFPA 1002 1.4.2. *IFPO, 2E:* Chapter 1, pages 15 – 17.

Question #2
According to NFPA1002, pump operators must be licensed to drive all the vehicles they are expected to operate, not just the largest. NFPA 1002 1.4.1. *IFPO, 2E:* Chapter 1, page 12.

Question #3
Wear bars should be used and can help determine if tire tread wear is excessive. NFPA 1002 1.4.2. *IFPO, 2E:* Chapter 2, page 30.

Question #4
Although mechanics share some responsibility for preventive maintenance, they usually complete work that:

- Requires apparatus to be taken out of service
- Requires several hours to complete
- Involves detailed and complicated repairs

NFPA 1002 4.2.1, 5.1.1. *IFPO, 2E:* Chapter 2, page 23.

Question #5
The statement states that if you increase speed, you can actually stop the vehicle in a shorter distance by four times. This is incorrect. The correct concept is that the faster the vehicle travels, the greater the distance it takes to stop. NFPA 1002 4.3.1, 4.3.6. *IFPO, 2E:* Chapter 3, page 55.

Question #6
Pumps receiving power from a PTO (power take-off) can operate while stationary or while the apparatus is in motion. NFPA 1002 5.2.1, 5.2.2. *IFPO, 2E:* Chapter 4, pages 92 – 93.

Question #7
Laws governing the rules of the road are established for ideal conditions such as dry roads and good visibility. Appropriate adjustments when adverse environmental conditions are encountered must be made, even when responding to incidents. NFPA 1002 4.3. *IFPO, 2E:* Chapter 3, pages 50 – 60.

Question #8
This is a basic pressure principle. Another way of stating the principle is that when no water is flowing, the pressure at any point in a hose line will be the same. NPFA 1002 5.2.1, 5.2.2. *IFPO, 2E:* Chapter 4, page 77.

Question #9
Flow meters use a paddle-wheel to measure the quantity (gallons) and rate (per minute) of water flow. NFPA 1002 5.2.1, 5.2.2. *IFPO, 2E:* Chapter 5, pages 113 – 114.

Question #10
This system only uses an eductor, which draws foam into the system. NFPA 1002 5.2.3. *IFPO, 2E:* Chapter 5, pages 126 – 130.

Question #11
Both pressure-relief devices and pressure governors are commonly found on pumping apparatus. NFPA 1002 5.2.1, 5.2.2. *IFPO, 2E:* Chapter 5, pages 122 – 125.

Question #12
The two main types of hose construction used today are woven jacket and rubber covered.
NFPA 1002 5.2.1, 5.2.2, 5.2.4. *IFPO, 2E:* Chapter 6, pages 140 – 141.

Question #13
The rule of thumb is to park no further than two-thirds of the theoretical lift ($33.81 \times \frac{2}{3} = 22.54$)
when drafting. NFPA 1002 5.2., 5.2.2, 5.2.4. *IFPO, 2E:* Chapter 7, pages 185 – 186.

Question #14
Chocking or blocking the wheels helps prevent inadvertent movement of apparatus while parked and
engaged in pumping operations. NFPA 1002 5.2.1, 5.2.2. *IFPO, 2E:* Chapter 8, page 213.

Question #15
The two basic methods for providing power to a pump is either a separate engine or from the drive
engine. NFPA 1002 5.2.1, 5.2.2, 5.2.4. *IFPO, 2E:* Chapter 8, page 210.

Question #16
Feathering or gating is most often used when initiating, changing, or shutting down a discharge line
while other discharge lines are in operation. NFPA 1002 5.2.1. *IFPO, 2E:* Chapter 9, page 237.

Question #17
Sprinkler system water supplies are most often designed to support a small number of sprinkler
heads. Fire departments must support sprinkler systems when a large number of heads fuse or when
pipes break. NFPA 1002 5.2.4. *IFPO, 2E:* Chapter 9, pages 248 – 253.

Question #18
A standpipe can be supplied in a similar manner to automatic sprinkler systems and through a direct
connection to a fire department connection. NFPA 1002 5.2.4. *IFPO, 2E:* Chapter 9, page 253.

Question #19
Although water is virtually incompressible, it does expand when frozen. NFPA 1002 5.2.1, 5.2.2.
IFPO, 2E: Chapter 10, pages 262 – 263.

Question #20
This concept is one of five pressure principles related to the manner in which liquids behave while
under pressure.
A hose line is flat before it is charged, and becomes round after charging because the pressure is
perpendicular to the internal surface of the hose. NFPA 1002 5.2.1, 5.2.2. *IFPO, 2E:* Chapter 10,
page 271.

Question #21
This is one of four fundamental friction loss principles.
Another way of saying this is that when hose diameter increases, friction loss will decrease; when
hose diameter decreases, friction loss increases when all other variables are held constant.
NFPA 1002 5.2.1, 5.2.2. *IFPO, 2E:* Chapter 10, pages 280 – 282.

Question #22
The drop-ten method is a fireground method used to estimate friction loss in 2 ½-inch hose. It can
also be used for other size hose, although it is less accurate.
For this method, simply subtract 10 from the first two numbers of the gpm flow. For example, the
friction loss in 100 feet of 2 ½-inch hose flowing 250 gpm is 15 psi ($25 – 10 = 15$).
NFPA 1002 5.2.1, 5.2.2. *IFPO, 2E:* Chapter 11, pages 303 – 304.

Question #23
These are estimated friction losses for the systems and can be used for fireground calculations.
NFPA 1002 5.2.1, 5.2.2. *IFPO, 2E:* Chapter 11, pages 309 – 310.

Question #24
- *Hydraulics* is that branch of science dealing with the principles of fluid at rest or in motion.
- *Hydrodynamics* is that branch of hydraulics that deals with the principles and laws of fluids in motion.
- *Hydrostatics* is that branch of hydraulics that deals with the principles and laws of fluids at rest and the pressures they exert or transmit.

NFPA 2001 5.2. *IFPO, 2E:* Chapter 10, page 261.

Question #25
NFPA 1002 requires that pump operators meet the requirements of NFPA 1001, Fire Fighter 1, before being certified as a driver pump operator. In addition, NFPA 1002 requires that pump operators be licensed to drive those vehicles they are expected to operate. NFPA 1002 5.1. *IFPO, 2E:* Chapter 1, page 15.

Question #26
The title for NFPA 1002 is Fire Apparatus Driver/Operator Professional Qualifications. NFPA 1002 1.4.1. *IFPO, 2E:* Chapter 1, page 15.

Question #27
Over the years, several names or titles have been used for the position responsible for driving and operating pumping apparatus, including:
- Wagon driver or tender
- Chauffeur
- Apparatus operator
- Motorized pump operator
- Driver/pump operator
- Apparatus engineer
- Engineer
- Pump operator
- Driver

NFPA 1002 1.4. *IFPO, 2E:* Chapter 1, pages 5 – 6.

Question #28
Laws are rules that are legally binding and enforceable. *Standards* are guidelines that are not legally binding or enforceable. NFPA 1002 4.2, 4.3, 5.1. *IFPO, 2E:* Chapter 1, page 15.

Question #29
Brake pedals should be tested during preventative maintenance inspections. When depressed, brake pedals should stop prior to reaching the floorboard. NFPA 1002 4.2.1. *IFPO, 2E:* Chapter 2, page 35.

Question #30
Maintaining emergency vehicles clean at all times is important because this practice:
- Helps maintain a good public image (municipal apparatus usually belong to tax-paying public).
- Helps ensure the pump, systems, and equipment operate as intended.
- Helps ensure vehicles can be inspected properly (dirt and grime could cover defects or potential problems).

NFPA 1002 4.2.1. *IFPO, 2E:* Chapter 2, page 37.

Question #31

NFPA 1002 requires that the pump operator be able to inspect and conduct preventative maintenance activities in accordance with standards and manufacturer's requirements. Specific requirements related to the schedule of preventative maintenance activities can be found in each of the other NFPA standards. NFPA 1002 4.2.1, 5.1.1. *IFPO, 2E:* Chapter 2, page 25.

Question #32

SAE stands for the Society of Automotive Engineers. SAE uses a numbering system to grade or rate engine oil viscosity. NFPA 1002 4.2.1. *IFPO, 2E:* Chapter 2, pages 32 – 33.

Question #33

Coolant pH and pressure are usually not included in vehicle inspections conducted by pump operators. Also, the coolant level should only be inspected when the engine is cool. NFPA 1002 4.2.1. *IFPO, 2E:* Chapter 2, pages 33 – 34.

Question #34

All the answers listed are important reasons for documenting preventive maintenance activities except that of tracking inspection duration. The time it takes to conduct an inspection is not a preventative maintenance issue as much as a personnel productivity issue. NFPA 1002 4.2.2. *IFPO, 2E:* Chapter 2, page 25.

Question #35

NFPA 1500 suggests that the following safety-related components be inspected on a routine basis:

- Tires
- Brake
- Warning systems
- Windshield wipers
- Headlights and clearance lights
- Mirrors

Typically, manufacturers recommend the following components be inspected:

- Engine oil
- Coolant level
- Transmission oil
- Brake system
- Belts

NFPA 1002 4.2.1, 5.1.1. *IFPO, 2E:* Chapter 2, pages 28 – 29.

Question #36

While charging vehicle batteries, hydrogen gas can be generated. Ventilation is important to reduce the accumulation of this combustible gas. NFPA 1002 4.2.1. *IFPO, 2E:* Chapter 2, page 33.

Question #37

According to reports published in the *NFPA Journal*, an average of 21 firefighter fatalities occur each year while responding to or returning from an incident. NFPA 1002 4.3.1, 4.3.2, 4.3.3, 4.3.4, 4.3.5, 4.3.6. *IFPO, 2E:* Chapter 3, page 42.

Question #38

Having driver license or certification is not a factor, per se, but can certainly help in understanding, avoiding, and/or adequately compensating for the common factors. Also, NFPA 1500 requires that all drivers posses a valid driver license prior to operating emergency vehicles. NFPA 1002 4.3.2, 4.3.3, 4.3.5, and 4.3.6. *IFPO, 2E:* Chapter 3, pages 43 – 45.

Question #39

Safe emergency-vehicle operations and requirements can be found in each of items listed except NFPA 1403. *NFPA 1002* includes several knowledge and skill requirements that will help drivers operate emergency vehicles safety. Chapter 6 of *NFPA 1500* incudes specific requirements for fire apparatus, equipment, and driver/operators. *State laws* set the overall rules and standards for emergency-vehicle drivers in the state. NFPA 1002 4.3.1 and 4.3.6. *IFPO, 2E:* Chapter 3, pages 44 – 50.

Question #40

Total stopping distance is measured from the time a hazard is detected until the vehicle comes to a complete stop. Total stopping distance consists of:

- *Perception distance* (distance apparatus travels from the time the hazard is seen until the brain recognizes it as a hazard)
- *Reaction distance* (distance apparatus travels from the time the brain sends the message to depress the brakes until the brakes are depressed)
- *Braking distance* (distance of travel from the time the brake is depressed until the vehicle comes to a complete stop)

NFPA 1002 4.3.1, 4.3.6. *IFPO, 2E:* Chapter 3, page 5.

Question #41

Laws (statutes, ordinances, legislation) are rules that are legally binding and enforceable. Standards are guidelines that are not legally binding or enforceable. NFPA 1002 4.3. *IFPO, 2E:* Chapter 3, pages 44 – 45.

Question #42

Total stopping distance is measured from the time a hazard is detected until the vehicle comes to a complete stop. Total stopping distance consists of:

- *Perception distance* (distance apparatus travels from the time the hazard is seen until the brain recognizes it as a hazard)
- *Reaction distance* (distance apparatus travels from the time the brain sends the message to depress the brakes until the brakes are depressed)
- *Braking distance* (distance of travel from the time the brake is depressed until the vehicle comes to a complete stop)

NFPA 1002 4.3. *IFPO, 2E:* Chapter 3, pages 55 – 65.

Question #43

Centrifugal force is the tendency of an object to move outward from the center. NFPA 1002 4.3. *IFPO, 2E:* Chapter 3, pages 56 – 57.

Question #44

NFPA 1002 requires drivers to demonstrate the ability to safely maneuver the apparatus over a predetermined route in compliance with all applicable state, local, and departmental rules and regulations. NFPA 1002 4.3. *IFPO, 2E:* Chapter 3, pages 49 – 50.

Question #45

One method to help ensure the safe control of a stationary apparatus is to properly align the front wheels as follows:

- When parked next to curb – rotate the front wheels so that they point toward the curb.
- When no curb is present – the front wheels should be positioned to roll the apparatus away from the road.

NFPA 1002 4.3. *IFPO, 2E:* Chapter 3, pages 58 – 59.

Question #46
Piston and rotary pumps are classified as positive-displacement pumps that tend to produce higher pressures and lower flows based on their operating characteristics. NFPA 1002 5.2.1, 5.2.2. *IFPO, 2E:* Chapter 4, page 76.

Question #47
A lobe is a component of a positive-displacement pump, specifically a rotary-lobe pump. NFPA 1002 5.2.1, 5.2.2. *IFPO, 2E:* Chapter 4, pages 83 – 87.

Question #48
Centrifugal pumps must be primed in order to pump water. Positive-displacement pumps are used to prime centrifugal pumps. NFPA 1002 5.2.1, 5.2.2. *IFPO, 2E:* Chapter 4, pages 75 – 76.

Question #49
The pressure mode of a two-stage centrifugal pump moves water from one impeller to the second impeller. The pressure from one impeller is added to the pressure generated by the second impeller. The volume of the first impeller is delivered to the second impeller, which equals the total flow. NFPA 1002 5.2.1, 5.2.2. *IFPO, 2E:* Chapter 4, pages 89 – 90.

Question #50
According to NFPA 1901, a pump must have a rated capacity as follows:

- 100% of its rated capacity at 150 psi

- 70% of its rated capacity at 200 psi

- 50% of its rated capacity at 250 psi

NFPA 1002 5.2.1, 5.2.2. *IFPO, 2E:* Chapter 4, page 91.

Question #51
The intake side of the pump is the point at which water enters the pump; it is also referred to as the *supply side* and the *suction side* of the pump.
The discharge side of the pump is the location where water leaves the pump. NFPA 1002 5.2. *IFPO, 2E:* Chapter 4, page 75.

Question #52
Basic pump terms include:

- *Flow* refers to the rate and quantity of water delivered by the pump and is expressed in gallons per minute (gpm).

- *Pressure* refers to the amount of force generated by the pump or the resistance encountered on the discharge side of the pump and is expressed in pounds per square inch (psi).

- *Speed* refers to the rate at which the pump is operating and is typically expressed in revolutions per minute (rpm).

- *Slippage* is the term used to describe the leaking of water between the surfaces of the internal moving parts of a pump.

NFPA 1002 5.2. *IFPO, 2E:* Chapter 4, page 75.

Question #53
Although pump panels located in the crew cab provide a margin of safety and comfort to the pump operator, it is not required by NFPA 1901. NFPA 1002 5.2.1, 5.2.2. *IFPO, 2E:* Chapter 5, pages 101 – 105.

Question #54
Each of the answer selection options is correct concerning pressure gauges located on pump panels. NFPA 1002 5.2.1, 5.2.2, 5.2.4. *IFPO, 2E:* Chapter 5, pages 110 – 105.

Question #55
All the answer selections are reasons why electrically controlled valves are gaining popularity. NFPA 1002 5.2.1, 5.2.2. *IFPO, 2E:* Chapter 5, page 115.

Question #56
As a rule of thumb, a pump will have one 2 ½-inch discharge for each 250 gpm of rated capacity. A 1,250 gpm pumper will usually have five 2 ½-inch discharges, $5 \times 250 = 1,250$. NFPA 1002 5.2.1, 5.2.2. *IFPO, 2E:* Chapter 5, pages 118 – 119.

Question #57
Pressure-relief devices control pressure buildup by automatically opening or closing a relief valve that diverts excess pressure from the discharge side of the pump. NFPA 1002 5.2.1, 5.2.2, 5.2.4. *IFPO, 2E:* Chapter 5, pages 122 – 125.

Question #58
The auxiliary pump uses the heat exchanger to help cool the engine. NFPA 1002 5.2.1, 5.2.2, 5.2.4. *IFPO, 2E:* Chapter 5, pages 125 – 126.

Question #59
Foam systems found on pumping apparatus include:

- Pre-mixed
- In-line eductor
- Around-the-pump proportioning
- Balanced pressure
- Direct injection/compressed-air

NFPA 1002 5.2.3. *IFPO, 2E:* Chapter 5, pages 126 – 130.

Question #60
When the pump is primed, it starts pumping water. The result will be a positive pressure reading, and the main pump will sound differently as it begins to pump water. The priming motor will sound as if it is slowing and will only stop when deactivated by the pump operator. NFPA 1002 5.2.1, 5.2.2. *IFPO, 2E:* Chapter 5, pages 120 – 122.

Question #61
Pump peripherals are those components directly or indirectly attached to the pump that are used to control and monitor the pump and the engine. NFPA 1002 5.2.1, 5.2.2. *IFPO, 2E:* Chapter 5, page 99.

Question #62
- *Tank-to-pump* control valves allow water to flow from the on-board water supply to the intake side of the pump.
- *Pump-to-tank* (tank fill) control valves allow water to flow from the discharge side of the pump to the tank.
- *Transfer valve* control valves found on multistage pumps redirect water from the pump between the pressure mode and the volume mode

NFPA 1002 5.2.1, 5.2.2. *IFPO, 2E:* Chapter 5, page 119.

Question #63
According to NFPA 1962, supply hose should not be operated at pressures exceeding 185 psi.

- 80 psi is the nozzle pressure for smooth-bore master streams.
- 100 psi is the nozzle pressure for combination nozzles.
- 275 psi is the highest recommended normal operating pressure for attack lines.

NFPA 1002 5.2.1, 5.2.2, 5.2.4. *IFPO, 2E:* Chapter 6, pages 138 – 140.

Question #64
Both the standards refer to discharge lines as attack hose. NFPA 1002 5.2.1, 5.2.2, 5.2.4. *IFPO, 2E:* Chapter 6, pages 138 – 142.

Question #65
NFPA 1961 no longer specifies the required length of hose. NFPA 1002 5.2.1, 5.2.2, 5.2.4. *IFPO, 2E:* Chapter 6, page 140.

Question #66
The coupling illustrated is a Storz couple, which is considered a sexless couple. NFPA 1002 5.2.1, 5.2.2, 5.2.4. *IFPO, 2E:* Chapter 6, pages 141 – 142.

Question #67
- A *wye* has a single female inlet connection and two or more male outlet connections.

- A *siamese* has two or more female inlet connections and one male outlet connection.

Although the siamese in this picture can be used as a wye, it is still considered a siamese and would require double male and female adapters. NFPA 1002 5.2.1, 5.2.2, 5.2.4. *IFPO, 2E:* Chapter 6, pages 148 – 149.

Question #68
The appliance in answer selection c is a four-way hydrant valve.
The appliance in answer selection a is a large-gated wye.
The appliance in answer selection d is a reducer. NFPA 1002 5.2.1, 5.2.2, 5.2.4. *IFPO, 2E:* Chapter 6, page 151.

Question #69
The nozzle in the picture is a combination nozzle, and 100 psi is the operating pressure for combination nozzles.

- 50 psi is the operating pressure for smooth-bore hand-lines.

- 80 psi is the operating pressure for smooth-bore master streams.

- 125 psi is not a normal operating pressure for nozzles used in the fire service.

NFPA 1002 5.2.1, 5.2.2, 5.2.4. *IFPO, 2E:* Chapter 6, pages 153 – 157.

Question #70
Selectable-flow nozzles allow for changes in flow rate to be made at the nozzle. However, the pump operator must change pump pressure to ensure proper nozzle operating pressure is maintained and the selected flow rate is delivered. NFPA 1002 5.2.1, 5.2.2, 5.2.4. *IFPO, 2E:* Chapter 6, pages 154 – 157.

Question #71
The most common sexless coupling a called a Storz. NFPA 1002 5.2.1, 5.2.2, 5.2.4. *IFPO, 2E:* Chapter 6, pages 141 – 142.

Question #72
- A *siamese* is used to combine two or more lines into a single line.

- A *wye* is used to divide one hose line into two or more lines.

NFPA 1002 5.2.1, 5.2.2, 5.2.4. *IFPO, 2E:* Chapter 6, pages 148 – 149.

Question #73
- *Nozzle pressure* is the designed operating pressure for a particular nozzle.

- *Nozzle flow* is the amount of water flowing from a nozzle.

- *Nozzle reach* is the distance water travels after leaving a nozzle.

- *Nozzle reaction* is the tendency of a nozzle to move in the direction opposite to water flow.

NFPA 1002 5.2.1. *IFPO, 2E:* Chapter 6, pages 153 – 154.

Question #74

NFPA specifies *minimum* tank capacity for apparatus as follows:

- 200 gallons – initial attack
- 300 gallons – pumper (500 gallons being the most common)
- 1,000 gallons – mobile water apparatus/tanker

NFPA 1002 5.2.1. *IFPO, 2E:* Chapter 7, page 167.

Question #75

Dead-end water mains lack the ability to receive water from different directions, as in a grid system. Therefore, one or more hydrants can be come inoperable due to a damaged main, and hydrants can experience significant reduction in pressure and volume based on flow from other hydrants on the dead-end main. NFPA 1002 5.2.1, 5.2.2. *IFPO, 2E:* Chapter 7, pages 172 – 173.

Question #76

This is a *wet-barrel hydrant* because it has stem nuts to control each discharge outlet. A dry-barrel hydrant has a stem nut located on the bonnet, which must be turned before water can enter the hydrant barrel. *Static-source hydrants* typically do not have control valves, and *dead-end hydrants* could be either wet- or dry-barrel. NFPA 1002 5.2.1. *IFPO, 2E:* Chapter 7, pages 175 – 176.

Question #77

NFPA 291 suggest the following classification and color coding for hydrants:

Class	Rated Capacity	Color of Bonnets and Caps
AA	1,500 gpm or greater	Light blue
A	1,000 to 1,499 gpm	Green
B	500 to 999 gpm	Orange
C	Less than 500 gpm	Red

NFPA 1002 5.2.1, 5.2.2. *IFPO, 2E:* Chapter 7, page 178.

Question #78

Water will rise approximately 2.3 feet for each 1 psi of pressure. With a perfect vacuum, atmospheric pressure (14.7) will force water to a height of 33.81 feet (2.3 × 14.7 = 33.81).

$33.8 \times \frac{2}{3} = 22.5$

NFPA 1002 5.2.1, 5.2.2, 5.2.4. *IFPO, 2E:* Chapter 7, pages 185 – 186.

Question #79

When possible, the largest tanker should be placed at the water source. NFPA 1002 5.2.2. Chapter 7, pages 189 – 194.

Question #80

The device shown in the picture is a jet siphon that helps move water from one dump tank to another. NFPA 1002 5.2.1. *IFPO, 2E:* Chapter 7, pages 197 – 198.

Question #81

Required flow is the estimated flow of water needed for a specific incident. NFPA 1002 5.2.1. *IFPO, 2E:* Chapter 7, page 164.

Question #82

NFPA 1901 requires that pumps rated at 1,500 gpm or higher achieve a prime within 45 seconds. NFPA 1002 5.1.1. *IFPO, 2E:* Chapter 8, page 214.

Question #83

- The *priming device test* ensures the primer can develop sufficient vacuum to draft.

- The *vacuum test* is use to determine the ability of the pump to maintain a vacuum.

- The *pressure control test* is conducted to ensure the pressure control can adequately maintain safe discharge pressures.

- The *draft test* is a nonspecific test.

NFPA 1002 5.1.1. *IFPO, 2E:* Chapter 8, pages 226 – 231.

Question #84

Each of the tests or checks listed are often required by manufacturers on a weekly and/or monthly basis. The engine speed check is usually performed on an annual basis. NFPA 1002 5.1.1. *IFPO, 2E:* Chapter 8, pages 226 – 232.

Question #85

Cavitation is the process that explains the formation and collapse of vapor pockets when certain conditions exist during pumping operations. NFPA 1002 5.1.1. *IFPO, 2E:* Chapter 9, pages 245 – 247.

Question #86

- *Wet-pipe systems* maintain water in the system at all times. When a sprinkler head is fused, water immediately discharges.

- *Dry-pipe systems* maintain air or compressed gas under pressure within the system. When a head fuses, water enters the system and discharges through any fused heads.

- *Pre-action systems* are similar to dry-pipe systems in that air or compressed gas is maintained in the system. However, an automatic detection system (smoke, heat, flame, etc.) or manual system (pull box) must operate to allow water to enter the system. At this point, the system is similar to a wet-pipe system; when a sprinkler head fuses, water discharges.

- A *deluge system* maintains all sprinkler heads open. When a detection system operates, water enters the system and is discharged through all the open heads.

NFPA 1002 5.2.4. *IFPO, 2E:* Chapter 9, pages 148 – 253.

Question #87

A general rule of thumb is to pump a fire department connection at 150 psi. NFPA 1002 5.2.4. *IFPO, 2E:* Chapter 9, pages 251 – 252.

Question #88

- *Class 1* connections provide 2 ½-inch hose connections and are intended for use by firefighters or fire brigade members.

- *Class 2* connections provide 1 ½-inch hose station and are intended primarily for trained personnel during initial attack efforts.

- *Class 3* connections provide 1 ½-inch and 2 ½-inch hose connections for use by firefighters or fire brigade members.

NFPA 1002 5.2.4. *IFPO, 2E:* Chapter 9, page 253.

Question #89

The basic characteristics and properties of water are listed below. These are considered approximate in that water purity, atmospheric conditions, and rounding can effect the values.

- Virtually incompressible
- Freezes at 32°F
- Expands when frozen
- Boils at 212°F (considered water's boiling point)
- 1 gallon weighs 8.34 lbs
- Density is 62.4 lb/ft^3
- Number of gallons in 1 cubic foot is 7.38 gallons/ft^3

NFPA 1002 5.2.1, 5.2.2. *IFPO, 2E:* Chapter 10, pages 262 – 263.

Question #90

- *Evaporation* is the physical change of state from a liquid to a vapor.
- *Vapor pressure* (VP) is the pressure exerted on the atmosphere by molecules as they evaporate from the surface of the liquid.
- *Boiling point* (BP) is the temperature at which the vapor pressure of a liquid equals the surrounding pressure.

NFPA 1002 5.2.1, 5.2.2. *IFPO, 2E:* Chapter 10, pages 262 – 263.

Question #91

- *Density* is the weight of a substance expressed in units of mass per volume.
- *Weight* is the downward force exerted on an object by the Earth's gravity, typically expressed in pounds.
- *Volume* is the amount of space occupied by an object.
- *Pressure* is the force exerted by a substance in units of weight per area, typically expressed in pounds per square inch (psi).

NFPA 1002 5.2.1, 5.2.2. *IFPO, 2E:* Chapter 10, pages 264 – 269.

Question #92

Pressure is the force exerted by a substance in *units of weight per area*, typically expressed in *pounds per square inch* (psi or lb/in^2)

P = Pressure

F = Force (weight in pounds)

A = Area (square inches)

NFPA 1002 5.2.1, 5.2.2. *IFPO, 2E:* Chapter 10, pages 269 – 270.

Question #93

- *Head pressure* is the pressure exerted by the vertical height of a column of liquid expressed in feet.
- *Gauge pressure* is the measurement of pressure that does not include atmospheric pressure, typically expressed as psig.
- *Absolute pressure* is the measurement of pressure that includes atmospheric pressure and is typically expressed as psia.
- *Vacuum* pressure is the measurement of pressure that is less than atmospheric pressure, typically expressed in inches of mercury (in. Hg).

NFPA 1002 5.2.1, 5.2.2. *IFPO, 2E:* Chapter 10, pages 275 – 277.

Question #94

- *Absolute pressure* is the measurement of pressure that includes atmospheric pressure and is typically expressed as psia.
- *Atmospheric pressure* is the pressure exerted by the atmosphere on the Earth.
- *Gauge pressure* is the measurement of pressure that does not include atmospheric pressure, typically expressed as psig.
- *Vacuum pressure* is the measurement of pressure that is less than atmospheric pressure, typically expressed in inches of mercury (in. Hg).
- *Head pressure* is the pressure exerted by the vertical height of a column of liquid expressed in feet.

NFPA 1002 5.2.1, 5.2.2. *IFPO, 2E:* Chapter 10, pages 275 – 277.

Question #95

- *Normal pressure* is the water flow pressure found in a municipal water supply during normal consumption demands.
- *Static pressure* is the pressure in a system when no water is flowing.
- *Residual pressure* is the pressure remaining in the system after water has been flowing.
- *Pressure drop* is the difference between the static pressure and the residual pressure when measured at the same location.

NFPA 1002 5.2.1, 5.2.2. *IFPO, 2E:* Chapter 10, pages 276 – 277.

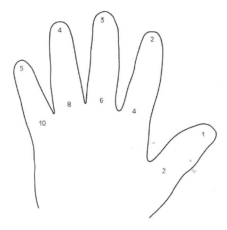

Question #96

The hand method is a fireground method used to estimate friction loss in 100-foot sections of 2 ½-inch hose. Simply multiply the two figures on a finger for the approximate friction loss pressure for each 100-section of 2 ½-inch hose. NFPA 1002 5.2.1, 5.2.2. *IFPO, 2E:* Chapter 10, pages 302 – 303.

Question #97

One gallon of water weighs 8.34 pounds and is expressed 8.34 lb/gal.
The density of water is 62.4 pounds and is expressed 62.34 lb/ft^3 (cubic feet)
NFPA 1002 5.2. *IFPO, 2E:* Chapter 10, pages 264 – 265.

Question #98

V = L × W × H
V = 20 feet × 10 feet × 30 feet
V = 6,000

NFPA 1002 5.2. *IFPO, 2E:* Chapter 10, page 266.

Question #99

Improvements in hose construction has reduced friction loss within the hose; this requires the use of a more accurate formula and $FL = 2q^2 + q$. NFPA 1002 5.2.1, 5.2.2. *IFPO, 2E:* Chapter 11, pages 304 – 305.

Question #100

This friction loss formula was derived from a combination of Bernoulli's equation, the Darcy-Weibach equation, and the Continuity equation. It is both simple and reasonably accurate for varying hose sizes, and *it can be use for both preplanning and fireground calculations.*

$$FL = c \times q^2 \times L$$

where   FL = friction loss

c = constant for a specific hose diameter

q = gm ÷ 100 (flow in hundreds of gpm)

L = length of hose in hundreds of feet

NFPA 1002 5.2.1, 5.2.2. *IFPO, 2E:* Chapter 11, pages 306 – 309.

Phase One, Exam Three

1. NFPA 1002 is the standard on Fire Apparatus Driver/Operator Professional Qualifications.
   a. True
   b. False

2. Pump operators are subject to periodic medical examinations, as required by NFPA 1500, to help ensure they are medically fit to perform the duties of a driver pump operator.
   a. True
   b. False

3. Engine oil should only be checked when the engine is cold.
   a. True
   b. False

4. When inspecting the steering, turn the steering wheel so that the wheels turn at least 45 degrees in both directions.
   a. True
   b. False

5. For each year between 1990 and 2000, emergency vehicles were involved in over 11,000 accidents while returning from incidents.
   a. True
   b. False

6. A pump that receives its power from a split shaft can be used for both stationary and mobile pump operations.
   a. True
   b. False

7. Positive-displacement pumps theoretically discharge (displace) a specific quantity of water for each revolution or cycle of the pump.
   a. True
   b. False

8. Individual discharge gauges located on pump panels can typically read both positive and negative pressure.
   a. True
   b. False

9. The two most common types of pressure-regulating devices found on pumpers are intake and discharge reliefs.
   a. True
   b. False

10. Hydrants located on a dead-end main will experience an increase in pressure and flow toward the end of the main.
    a. True
    b. False

11. The red button on most throttle controls is intended to allow for rapid decrease in pump pressure in case of an emergency. Depressing the button will also allow for a rapid increase in pump pressure but may cause high-pressure surges.
    a. True
    b. False

12. During the tank-to-pump piping flow test, either a flow meter or a pitot tube reading can be used.
    a. True
    b. False

13. During pump operations, the auxiliary cooling system should be used to help keep the water temperature from rising within the pump as the result of reduced flows or discharge lines temporarily shutting down.
    a. True
    b. False

14. Standpipe systems can be connected and solely supported by a fire department connection.
    a. True
    b. False

15. Because water used for fire suppression is not pure, the properties of domestically prepared water are used for fire-protection hydraulic calculation purposes.
    a. True
    b. False

16. The pressure at any point beneath the surface of a liquid in an open container is directly proportional to its depth.
    a. True
    b. False

17. For any given velocity, the friction loss will be about the same regardless of water pressure.
    a. True
    b. False

18. Friction loss in common appliances used in pumping operations is usually minimal and is usually not included in pump discharge pressure calculations.
    a. True
    b. False

19. Hydrostatics is a branch of science dealing with the principles of fluid at rest or in motion.
    a. True
    b. False

20. Which of the following is correct concerning driver pump operator requirements contained in NFPA 1002?
    a. must be 21 years of age
    b. must meet the requirements of NFPA 1001, Fire Fighter II
    c. must be licensed to drive the vehicles they are expected to operate
    d. both b and c are correct

21. The titles used to describe the position responsible for driving and operating pumping apparatus include which of the following?
    a. driver and pump operator
    b. engineer and apparatus operator
    c. chauffeur and wagon driver
    d. all are correct

22. Laws are _____ that are legally binding and enforceable.
    a. rules
    b. guidelines
    c. codes
    d. ordinances

23. NFPA 1500 suggests that fire department vehicles used on a daily basis should be inspected
    _____.
    a. daily
    b. weekly
    c. monthly
    d. annually

24. The inspection of emergency vehicles includes three basic steps. Which of the following is *not* one of the three steps?
    a. reviewing previous inspection reports
    b. inspecting the vehicle
    c. topping of fluid levels as needed
    d. documenting and reporting inspection results

25. Which of the following is *not* correct concerning the inspection of belts and wires in the engine compartment?
    a. Both should be free from dirt and debris.
    b. Belts should be checked for proper alignment and tension while the vehicle is running.
    c. Wires should be inspected for loose connections, worn or frayed insulation, and exposed metal.
    d. Both should be visually inspected for obvious signs of damage.

26. According to some manufacturers' recommendations, which of the following fluid levels can be checked while the engine is running?
    a. oil level
    b. fuel level
    c. transmission level
    d. coolant level

27. When conducting a vehicle inspection, a penny test can be use used to assess which of the following?
    a. tire tread wear
    b. fan belt slack
    c. distance of brake pedal travel
    d. tandem tire distance

28. Inspections are conducted to verify the _____ of components.
    a. condition
    b. status
    c. performance
    d. serviceability

29. According to NFPA 1500, the following safety-related components should be inspected on a routine basis *except*
    a. brakes.
    b. warning systems.
    c. headlights.
    d. coolant level.

30. Caution should be exercised and adequate ventilation provided when charging batteries because
    a. excessive heat can build up.
    b. the batteries may explode.
    c. the acid in the battery may overflow.
    d. hydrogen gas can be produced.

31. Air tanks on emergency vehicles should
    a. be drained slightly to remove moisture.
    b. be drained fully every day.
    c. never be drained.
    d. be drained according to manufacturer's recommendations.

32. On newer vehicles, the fluid level for each of the following components is checked during preventive maintenance inspections *except*
    a. power steering.
    b. brake.
    c. engine.
    d. battery.

33. As part of a vehicle's inspection, the brake pedal should be depressed and should
    a. have at least 2 inches of free play.
    b. stop prior to reaching the floorboard.
    c. stop when the pedal reaches the floorboard.
    d. have virtually no resistance until it stops.

34. Doubling the speed of an apparatus will
    a. decrease stopping distance by two times.
    b. decrease stopping distance by four times.
    c. increase stopping distance by six times.
    d. increase stopping distance by four times.

35. Most state laws identify or define fire department vehicles with appropriate identification and warning devices as
    a. state-authorized emergency-response vehicles.
    b. emergency-response vehicles.
    c. exempt emergency-response vehicles.
    d. authorized emergency vehicles.

36. A vehicle navigating a curve at a high rate of speed will experience
    a. loss of traction.
    b. pressure surge.
    c. centrifugal force.
    d. friction.

37. Steering is a function of
    a. speed.
    b. traction.
    c. road surface.
    d. good tires.

38. Which of the following driving conditions might affect vision and traction?
    a. night driving
    b. fog or rain
    c. snow
    d. All of these conditions may affect vision and traction.

39. The requirement that emergency-vehicle drivers bring the apparatus to a complete stop at traffic red lights or stop signs is contained in which of the following NFPA standards?
    a. 1001
    b. 1002
    c. 1500
    d. 1901

40. The rotation of the front tire when parking an apparatus with no curb present should be
    a. toward the curb.
    b. away from the curb.
    c. straight ahead.
    d. positioned to roll the apparatus away from the road.

41. Centrifugal pumps generally provide
    a. higher flows with lower pressures.
    b. higher flows with higher pressures.
    c. lower flows with lower pressures.
    d. lower flows with higher pressures.

42. Positive-displacement pump construction includes which of the following?
    a. volute
    b. vane
    c. impeller
    d. shroud

43. Which is correct concerning pump priming?
    a. Positive-displacement pumps are self-priming.
    b. Centrifugal pumps must be primed.
    c. Priming can be referred to as replacing the air in a pump with water.
    d. All of the answers are correct.

44. When flow is expected to exceed 50% of the pump's rated capacity, the transfer valve should be
    a. in pressure mode.
    b. in volume mode.
    c. open.
    d. closed.

45. The two main sides of a pump are called the _____ and _____.
    a. intake, discharge
    b. suction, draft
    c. positive, negative
    d. priming, drafting

46. The rate at which the pump is operating is called
    a. pressure.
    b. slippage.
    c. speed.
    d. flow.

47. According to NFPA 1901, a pump must be able to deliver 50% of its rated capacity at _____ psi.
    a. 100
    b. 150
    c. 200
    d. 250

48. Pump panels can be located in which of the following locations?
    a. front of the apparatus
    b. top of the apparatus
    c. rear of the apparatus
    d. all are correct

49. Flow meter use on apparatus has increased because
    a. they are easier to calibrate than Bourdon tube gauges.
    b. they provide greater versatility of pump panel location.
    c. they require less hydraulic calculations during pump operations.
    d. all are correct.

50. Common valve types on pumping apparatus include
    a. ball, butterfly, and piston.
    b. gated, hammer, and piston.
    c. ball, piston, and quarter.
    d. gated, quarter, and full.

51. To determine the size of the pump, count the number of 2 ½-inch discharges and multiply by
    a. a factor of 2.
    b. a factor of 4.
    c. 250 gpm.
    d. 500 gpm.

52. A potential disadvantage associated with using the radiator-fill system is that it can
    a. overpressure the engine cooling system.
    b. decrease pump pressure.
    c. allow antifreeze/coolant to enter the intake side of the pump.
    d. all are correct.

53. Apparatus foam systems include all of the following *except*
    a. around-the-pump.
    b. balanced volume.
    c. in-line eductor.
    d. direct injection/compressed-air.

54. Balanced-pressure foam systems include which of the following?
    a. pressure proportioning
    b. by-pass
    c. demand
    d. all are correct

55. Priming devices on modern pumping apparatus are typically rotary vane or rotary gear positive-displacement pumps, while older apparatus may use
    a. rotary lobe or piston primers.
    b. eductor or piston primers.
    c. exhaust or vacuum primers.
    d. both a and c are correct.

56. Examples of common instrumentation found on pump panels include all of the following *except*
    a. pressure gauges.
    b. flow meters.
    c. indicators.
    d. foam system.

57. Devices found on pump panels that are used to initiate, restrict, or direct water flow are called
    a. gauges.
    b. flow meters.
    c. peripherals.
    d. control valves.

58. The _____ control valve is found on multistage pumps and redirects water from within the pump between the pressure mode and the volume mode.
    a. tank-to-pump
    b. pump-to-tank
    c. transfer
    d. discharge

59. The highest recommended operating pressure for attack lines is
    a. 100 psi.
    b. 250 psi.
    c. 275 psi.
    d. 300 psi.

60. NFPA _____ includes guidelines for the care, use, and maintenance of fire hose.
    a. 1500
    b. 1961
    c. 1962
    d. 1965

61. Common fire service hose construction for attack lines include all of the following *except*
    a. single-jacketed lined hose.
    b. rubber-covered hose.
    c. double-jacketed line fire hose.
    d. all are correct.

62. Which of the following is correct concerning the couplings in the picture?
    a. The male coupling is on the right and the female coupling is on the left.
    b. The male coupling is on the left and the female coupling is on the right.
    c. This is an example of sexless couplings.
    d. This is an example of Storz couplings.

63. All of the following statements about siamese appliances are true *except*
    a. they are used to combine two or more lines into a single line.
    b. they have two or more female inlet connections.
    c. they have only one male discharge connection.
    d. they are used to divide one line into two or more lines.

64. Which of the following is an example of a four-way hydrant valve?

    a.

    b.

    c.

    d.

65. Smooth-bore nozzles on a master stream device have an operating pressure of
    a. 50 psi.
    b. 80 psi.
    c. 100 psi.
    d. 125 psi.

66. The combination nozzle that provides a specific flow at 100 psi operating pressure regardless of the stream pattern is called a(n) _____ nozzle.
    a. fixed-flow or constant-flow
    b. selectable-flow
    c. automatic
    d. smooth-bore

67. Which of the following is *not* considered one of the basic parts of a hose?
    a. outer protective shell
    b. reinforced inner liner
    c. couplings
    d. safety relief ring

68. A(n) _____ is used to combine two or more lines into one line.
    a. wye
    b. siamese
    c. adapter
    d. increaser

69. Eductors utilize the _____ principle to draw foam into a water stream.
    a. VENTURI
    b. centrifugal
    c. cavitation
    d. hydrostatic

70. The designed operating pressure for a particular nozzle is called nozzle
    a. pressure.
    b. reaction.
    c. flow.
    d. reach.

71. Tankers or mobile water supply apparatus have a minimum tank size of _____ gallons, according to NFPA 1901.
    a. 200
    b. 300
    c. 1,000
    d. 2,000

72. Which type of hydrant has a stem nut located on its bonnet and a drain hole below ground level?
    a. wet-barrel hydrant
    b. dry-barrel hydrant
    c. static hydrant
    d. dead-end hydrant

73. Static water supply reliability can be affected by
    a. freezing.
    b. excessive silt and debris.
    c. flooding.
    d. all are correct.

74. A hydrant capable of flowing 1,000 gpm would have a(n) _____ bonnet and discharge caps, according to NFPA 291.
    a. red
    b. orange
    c. green
    d. light blue

75. Which of the following best describes a static hydrant?
    a. a wet- or dry-barrel hydrant when not in use
    b. a pre-piped line that extends into a static water source
    c. a hydrant that is rarely used and/or on a dead-end main
    d. no such hydrant is used by the fire service

76. The type of relay in which all the water from one pump is not always delivered into the next pump is called
    a. a closed-relay system.
    b. a constant-pressure relay system.
    c. a constant-flow relay system.
    d. an open-relay system.

77. Tanker shuttle equipment includes all the following *except*
    a. set siphons.
    b. portable dump tanks.
    c. nurse tanker.
    d. CAFS.

78. The estimated flow of water needed for a specific incident is called
    a. critical flow.
    b. incident flow.
    c. available flow.
    d. required flow.

79. Shuttle flow capacity is limited by the volume of water being delivered and the
    a. size of the pump.
    b. available water supply.
    c. number tankers.
    d. time it takes to complete a shuttle cycle.

80. Which of the following is *not* a good rule to help safely operate pumping apparatus?
    a. Never operate the pump without water.
    b. Always maintain awareness of instrumentation during pumping operations.
    c. Never leave the pump unattended.
    d. Always open, close, and turn controls swiftly.

81. According to NFPA 1901, a dry pump must be able to achieve a prime in 30 seconds for pumps rated at less than
    a. 500 gpm.
    b. 1,000 gpm.
    c. 1,250 gpm.
    d. 1,500 gpm.

82. The annual pump test is a 40-minute test consisting of which of the following?
    a. 20 minutes at 100% capacity
       20 minutes at 50% capacity
    b. 20 minutes at 100% capacity
       10 minutes at 70% capacity
       10 minutes at 50% capacity
    c. 20 minutes at 100% capacity
       10 minutes at 75% capacity
       10 minutes at 50% capacity
    d. 20 minutes at 100% capacity
       10 minute at 165 psi (overload test)
       5 minutes at 70% capacity
       5 minutes at 50% capacity

83. Most main pumps on modern pumping apparatus receive power
    a. via a PTO.
    b. from the drive engine.
    c. from a separate engine.
    d. all are correct.

84. The phenomenon of a pump running away from the water supply is known as
    a. water hammer.
    b. cavitation.
    c. vapor pressure.
    d. water slippage.

85. The _____ sprinkler system requires the activation of a detection system and the fusing of at least one sprinkler head before water is discharged.
    a. dry-pipe
    b. pre-action
    c. deluge
    d. wet-pipe

86. The NFPA standard that focuses on fire department operations within properties protected by sprinkler and standpipe systems is
    a. NFPA 13.
    b. NFPA 13D.
    c. NFPA 13E.
    d. NFPA 15.

87. When supporting sprinkler or standpipe systems, all of the following are correct *except*
    a. check for debris in the fire department connection before connecting hose lines.
    b. start pumping immediately if smoke or fire is evident.
    c. place the transfer valve into the volume position.
    d. pump the fire department connection at a minimum of 150 psi.

88. The causes of water hammer can include each of the following *except*
    a. opening and closing a hydrant too quickly.
    b. opening and closing the priming valve too quickly.
    c. abruptly opening and closing a nozzle.
    d. opening and closing discharge control valves too quickly.

89. A standpipe that provides a 1 ½-inch hose station and is intended primarily for trained personnel during initial attack efforts is a
    a. Class 1 standpipe.
    b. Class 2 standpipe.
    c. Class 3 standpipe.
    d. Class 4 standpipe.

90. Water freezes at _____ and boils at _____ .
    a. −32°F, 212°F
    b. 0°F, 212°F
    c. 0°F, 100°F
    d. 32°F, 212°F

91. The temperature at which the vapor pressure of a liquid equals the surrounding pressure is called the
    a. boiling point.
    b. vapor pressure.
    c. evaporation.
    d. latent heat of vaporization.

92. The amount of space occupied by an object is known as
    a. density.
    b. weight.
    c. pressure.
    d. volume.

93. In the formula $F = P \times A$, the unit P is
   a. pressure.
   b. pounds.
   c. positive displacement.
   d. both a and c are correct.

94. Absolute pressure is typically expressed as
   a. in. Hg.
   b. psia.
   c. psi.
   d. psig.

95. The pressure exerted by the vertical height of a column of liquid expressed in feet is called
   a. vacuum.
   b. gauge pressure.
   c. absolute pressure.
   d. head pressure.

96. The pressure remaining in the system after water has been flowing is referred to as
   a. static pressure.
   b. residual pressure.
   c. pressure drop.
   d. normal pressure.

97. When using the hand method to calculate friction loss in 100-foot sections of 2 ½-inch hose, the thumb has a value of _____ at the time and _____ at the base.
   a. 1 (100 gpm), 2 (200 gpm)
   b. 2 (200 gpm), 4 (400 gpm)
   c. 4 (400 gpm), 8 (800 gpm)
   d. 5 (500 gpm), 10 (1,000 gpm)

98. The density of water is _____, whereas the weight of 1 gallon of water is _____.
   a. 8.34 lb/ft^3, 62.4 lb/gal
   b. 8.34 lb/gal, 62.4 lb/ft^3
   c. 1 lb, 8.34 lb/gal
   d. 62.4 lb/ft^3, 8.34 lb/gal

99. The friction loss formula $FL = 2q^2 + q$ was developed by the National Board of Fire Underwriters and was in use for many years. Which of the following is true concerning this formula?
   a. q = flow in hundreds of gallons $\frac{gpm}{100}$ per minute
   b. It was designed for use with cotton-jacketed hose.
   c. It was designed for flows greater than 100 gpm.
   d. All are correct.

100. Which of the following is *not* correct concerning the friction loss formula cq^2L?
   a. c = constant or coefficient;
      1 for 1 ½-inch hose
      2 for 2 ½-inch hose
      3 for 3-inch hose
   b. q = gm ÷ 100 (flow in hundreds of gpm)
   c. L = length of hose in hundreds of feet
   d. It was derived from a combination of Bernoulli's equation, the Darcy-Weibach equation, and the Continuity equation.

Phase I, Exam III: Answers to Questions

1. T	26. C	51. C	76. D
2. T	27. A	52. A	77. D
3. F	28. A	53. B	78. D
4. F	29. D	54. D	79. D
5. F	30. D	55. C	80. D
6. F	31. D	56. D	81. D
7. T	32. D	57. D	82. B
8. F	33. B	58. C	83. B
9. F	34. D	59. C	84. B
10. F	35. D	60. B	85. B
11. T	36. C	61. D	86. C
12. T	37. B	62. B	87. C
13. F	38. D	63. D	88. B
14. T	39. C	64. A	89. B
15. T	40. D	65. B	90. D
16. T	41. A	66. A	91. A
17. T	42. B	67. D	92. D
18. F	43. D	68. B	93. A
19. F	44. B	69. A	94. B
20. C	45. A	70. A	95. D
21. D	46. C	71. C	96. B
22. A	47. D	72. B	97. A
23. A	48. D	73. D	98. D
24. C	49. D	74. C	99. D
25. B	50. A	75. B	100. A

Phase I, Exam III:
Rationale & References for Questions

Question #1
This is the correct title for NFPA 1002. NFPA 1002 1.1. *IFPO, 2E:* Chapter 1, page 15.

Question #2
NFPA 1002 requires that pump operators be subject to periodic medical evaluations as required by NFPA 1500. NFPA 1002 1.4.2. *IFPO, 2E:* Chapter 1, pages 15 – 17.

Question #3
Engine oil can be checked if the engine has been running and is hot. However, the pump operator should wait a few minutes to allow the oil to drain from the engine into the pan to ensure an accurate reading. NFPA 1002 4.2.1. *IFPO, 2E:* Chapter 2, page 32.

Question #4
It is not advisable to turn the steering wheel so that the wheels turn. Rather, the steering wheel should be turned until just before the wheels turn. The distance should not exceed 10 degrees in either direction. For a 20-inch steering wheel, that would be approximately 2 inches of movement. NFPA 1002 4.2.1. *IFPO, 2E:* Chapter 2, pages 35 – 36.

Question #5
Although emergency vehicles were involved in over 11,000 accidents during this period, the number includes responding to incidents as opposed to just returning from incidents. NFPA 1002 4.3.1. *IFPO, 2E:* Chapter 3, pages 42 – 43.

Question #6
Pumps receiving power from a split shaft can only operate while stationary. NFPA 1002 5.2.1, 5.2.2. *IFPO, 2E:* Chapter 4, pages 92 – 93.

Question #7
This is the basic operating characteristic of positive-displacement pumps. NPFA 1002 5.2.1, 5.2.2. *IFPO, 2E:* Chapter 4, page 77.

Question #8
Discharge gauges on pump panels serve one function: to measure the positive pressure from a discharge outlet. Because the gauge reads pressure from the discharge side of the pump, only a positive pressure reading is required. NFPA 1002 5.2.1, 5.2.2. *IFPO, 2E:* Chapter 5, page 112.

Question #9
Pressure-relief and pressure governors are the two most common type of pressure-regulating devices. NFPA 1002 5.2.1, 5.2.2, 5.2.4. *IFPO, 2E:* Chapter 5, pages 122 – 125.

Question #10
Pressure and flow will not increase. Pressure and flow will actually drop based on adjacent hydrant use on the dead-end main. NFPA 1002 5.2.1, 5.2.2. *IFPO, 2E:* Chapter 7, pages 172 – 173.

Question #11
The red button on the throttle should only be used for emergency situations to decrease pump pressure. Using the button to increase pressure may cause surges serious enough to cause damage or injury. NFPA 1002 5.2.1, 5.2.2. *IFPO, 2E:* Chapter 8, pages 215 – 216.

Question #12

A flow meter or a pitot tube can be used during this test. NFPA 1002 5.1.1. *IFPO, 2E:* Chapter 8, pages 230 – 231.

Question #13

The auxiliary cooling system helps keep the engine temperature within operating limits and will not help keep the water in the pump from heating. Maintaining water flow within the pump is the best way to prevent water temperatures from significantly increasing. This can be accomplished by either opening the tank-to-pump and pump-to-tank valves or opening an unused discharge line. NFPA 1002 5.2.1, 5.2.2. *IFPO, 2E:* Chapter 9, page 243.

Question #14

A standpipe can be supplied in a manner similar to automatic sprinkler systems and through a direct connection to a fire department connection. NFPA 1002 5.2.4. *IFPO, 2E:* Chapter 9, page 253.

Question #15

Both the purity of water and atmospheric conditions can affect the properties of water. Because of this, properties of domestically prepared water are commonly used in fire protection hydraulic calculations. NPFA 1002 5.2.1, 5.2.2. *IFPO, 2E:* Chapter 10, pages 262 – 263.

Question #16

This concept is one of five pressure principles related to the manner in which liquids behave while under pressure. The higher the level of water, the greater the pressure at the bottom. NFPA 1002 5.2.1, 5.2.2. *IFPO, 2E:* Chapter 10, page 271.

Question #17

This is one of four fundamental friction-loss principles. In other words, the speed (velocity) of water traveling through the hose governs friction loss rather than the quantity of water. NFPA 1002 5.2.1, 5.2.2. *IFPO, 2E:* Chapter 10, pages 280 – 282.

Question #18

Appliance friction loss is the reduction in pressure resulting from increased turbulence caused by the appliance, and it should be considered when calculating pump discharge pressures on the fireground. NFPA 1002 5.2.1, 5.2.2. *IFPO, 2E:* Chapter 11, pages 309 – 310.

Question #19

- *Hydraulics* is that branch of science dealing with the principles of fluid at rest or in motion.
- *Hydrodynamics* is that branch of hydraulics that deals with the principles and laws of fluids in motion.
- *Hydrostatics* is that branch of hydraulics that deals with the principles and laws of fluids at rest and the pressures they exert or transmit.

NFPA 2001 5.2. *IFPO, 2E:* Chapter 10, page 261.

Question #20

NFPA 1002 does not set an age limit; it requires that pump operators meet the requirements of NFPA 1001, Fire Fighter I, not II. NFPA 1002 1.4.1. *IFPO, 2E:* Chapter 1, pages 11 – 15.

Question #21

Over the years, several names or titles have been used for the position responsible for driving and operating pumping apparatus, including:

- Wagon driver or tender
- Chauffeur
- Apparatus operator
- Motorized pump operator
- Driver/pump operator
- Apparatus engineer
- Engineer
- Pump operator
- Driver

NFPA 1002 1.4. *IFPO, 2E:* Chapter 1, pages 5 – 6.

Question #22

Laws are rules that are legally binding and enforceable. Standards are guidelines that are not legally binding or enforceable. NFPA 1002 4.2, 4.3, 5.1. *IFPO, 2E:* Chapter 1, page 15.

Question #23

The more frequently a vehicle is used, the more frequently it should be inspected. NFPA 1500 suggests that vehicles used on a daily basis be checked daily. NFPA 1002 4.2.1, 5.1.1. *IFPO, 2E:* Chapter 2, page 25.

Question #24

The three basic steps in the inspection process are:

1. Pre-inspection (reviewing previous inspection reports)
2. Conducting the actual inspection
3. Document and reporting the results

The activity of topping fluid levels is considered a service function. NFPA 1002 4.2.1, 5.1.1. *IFPO, 2E:* Chapter 2, pages 27 – 28.

Question #25

Belts should be checked for proper alignment and tension only when the engine is shut off. Inspecting belts with the engine running creates both an unsafe act and condition. NFPA 1002 4.2.1. *IFPO, 2E:* Chapter 2, page 32.

Question #26

Oil, fuel, and coolant levels should not be checked while the engine is running for safety and accuracy reasons. Some manufacturers suggest that transmission level should checked while the engine is running. NFPA 1002 4.2.1. *IFPO, 2E:* Chapter 2, page 34.

Question #27

A penny can be used to help determine tire tread wear. If the tread is at or beyond the top of Lincoln's head, the tread depth is at or about $\frac{2}{32}$ of an inch. DOT requires that all major treads have a depth of $\frac{4}{32}$ inches. NFPA 1002 4.2.1. *IFPO, 2E:* Chapter 2, page 30.

Question #28

Inspections are conducted to verify the status of a component, as in verifying water, oil, and fuel levels. Servicing activities are conducted to help maintain vehicles in peak operating condition, such as when cleaning, lubricating, and topping off fluids. Tests are conducted to determine the performance of components, as in annual pump service tests. NFPA 1002 4.2.1, 5.1.1. *IFPO, 2E:* Chapter 2, page 23.

Question #29

NFPA 1500 suggests that the following safety-related components be inspected on a routine basis:

- Tires
- Brakes
- Warning systems
- Windshield wipers
- Headlights and clearance lights
- Mirrors

Typically, manufacturers recommend the following components be inspected:

- Engine oil
- Coolant level
- Transmission oil
- Brake system
- Belts

NFPA 1002 4.2.1, 5.1.1. *IFPO, 2E:* Chapter 2, pages 28 – 29.

Question #30

While charging vehicle batteries, hydrogen gas can be generated. Ventilation is important to reduce the accumulation of this combustible gas. NFPA 1002 4.2.1. *IFPO, 2E:* Chapter 2, page 33.

Question #31

It is important that manufacturer's recommendations be followed when conducting vehicle preventive maintenance inspections. NFPA 1002 4.2.1, 5.1.1. *IFPO, 2E:* Chapter 2, pages 28 – 29.

Question #32

Newer batteries are sealed and do not require internal inspection. NFPA 1002, 4.2.1. *IFPO, 2E:* Chapter 2, page 32.

Question #33

Brake pedals should be tested during preventive maintenance inspections. Brake pedals should be depressed, and should stop prior to reaching the floorboard. NFPA 1002 4.2.1. *IFPO, 2E:* Chapter 2, page 35.

Question #34

The faster the apparatus travels, the greater the distance it takes to stop. If the speed is doubled, the stopping distance will increase approximately four times. NFPA 1002 4.3.1. *IFPO, 2E:* Chapter 3, page 55.

Question #35

Most state laws identify an authorized emergency vehicle, such as a fire department vehicle, as one with appropriate identification and warning devices. NFPA 1002 4.3. *IFPO, 2E:* Chapter 3, pages 44 – 45.

Question #36

Centrifugal force is the tendency of an object to move outward from the center. Loss of traction occurs when friction is lost between the tires and the road surface. NFPA 1002 4.3. *IFPO, 2E:* Chapter 3, pages 56 – 57.

Question #37

When a vehicle loses the ability of its tires to hold traction on the road, steering is virtually impossible. Excessive speed, poor road conditions, and tire tread can all affect traction.

NFPA 1002 4.3. *IFPO, 2E:* Chapter 3, page 56.

Question #38

Examples of environmental conditions that affect vision and traction include the following:

- Night driving
- Fog or rain
- Snow and ice
- Glare
- Wind
- Sand

NFPA 1002 4.3. *IFPO, 2E:* Chapter 3.

Question #39

NFPA 1500 requires that emergency-vehicle drivers come to a complete stop when any intersection hazard is present. Specifically, the vehicle must come to a complete stop when any of the following exists:

- As directed by a law enforcement officer
- At traffic red lights or stop signs
- At negative right-of-way and blind intersections
- When all lanes of traffic in an intersection cannot be accounted for
- When a stopped school bus with flashing warning lights is encountered
- At unguarded railroad guard crossings (also for nonemergency)
- When other intersection hazards are present

NFPA 1002 4.3. *IFPO, 2E:* Chapter 3, page 48.

Question #40

One method to help ensure the safe control of a stationary apparatus is to properly align the front wheels as follows:

- When parked next to curb – rotate the front wheels so that they point toward the curb.
- When no curb is present – the front wheels should be positioned to roll the apparatus away from the road.

NFPA 1002 4.3. *IFPO, 2E:* Chapter 3, pages 58 – 59.

Question #41

Centrifugal pumps are classified as dynamic pumps, which tend to produce lower pressures and higher flows based on their operating characteristics. NFPA 1002 5.2.1, 5.2.2. *IFPO, 2E:* Chapter 4, page 76.

Question #42

The volute, impeller, and shroud are all parts of a centrifugal pump. NFPA 1002 5.2.1, 5.2.2. *IFPO, 2E:* Chapter 4, pages 83 – 87.

Question #43

Priming is the process of removing air and getting water into the pump. Because of their close fitting parts, positive-displacement pumps can pump both air and water, which means they can prime themselves. Centrifugal pumps have loose-fitting parts and cannot pump air, which means they must be primed before they can pump. NFPA 1002, 5.2.1, 5.2.2. *IFPO, 2E:* Chapter 4, page 75.

Question #44

Transfer valves are not closed or open. Rather, they are used to redirect water to either a series (pressure) or parallel (volume) mode. Operating at flows great than 50% of the pump's rated capacity indicates larger water flows. In such cases, the transfer valve should be in volume mode. NFPA 1002 5.2.1, 5.2.2. *IFPO, 2E:* Chapter 4, pages 89 – 90.

Question #45

The main sides of a pump are called the intake and discharge. The intake side of the pump is the point at which water enters the pump; it is also referred to as the *supply side* and the *suction side* of the pump. The discharge side of the pump is the location where water leaves the pump. NFPA 1002 5.2. *IFPO, 2E:* Chapter 4, page 75.

Question #46

Basic pump terms include the following:

- *Flow* refers to the rate and quantity of water delivered by the pump, and is expressed in gallons per minute (gpm).

- *Pressure* refers to the amount of force generated by the pump or the resistance encountered on the discharge side of the pump; it is expressed in pounds per square inch (psi).

- *Speed* refers to the rate at which the pump is operating, and is typically expressed in revolutions per minute (rpm).

- *Slippage* is the term used to describe the leaking of water between the surfaces of the internal moving parts of a pump.

NFPA 1002 5.2. *IFPO, 2E:* Chapter 4, page 75.

Question #47

According to NFPA 1901, a pump must have a rated capacity as follows:

- 100% of its rated capacity at 150 psi

- 70% of its rated capacity at 200 psi

- 50% of its rated capacity at 250 psi

NFPA 1002 5.2.1, 5.2.2. *IFPO, 2E:* Chapter 4, page 91.

Question #48

With the new electronic and digital controls, pump panels can be located almost anywhere on the apparatus. NFPA 1002 5.2.1, 5.2.2. *IFPO, 2E:* Chapter 5, pages 105 – 110.

Question #49

All the answer selections are reasons for the increased use of flow meters. NFPA 1002 5.2.1, 5.2.2. *IFPO, 2E:* Chapter 5, pages 113 – 114.

Question #50

The most common valve types on pumping apparatus include ball, butterfly, piston, and gated valves. Hammer, quarter, and full are not types of valves used on pumping apparatus. NFPA 1002 5.2.1, 5.2.2. *IFPO, 2E:* Chapter 5, page 114.

Question #51

As a rule of thumb, a pump will have one 2 ½-inch discharge for each 250 gpm of rated capacity. A 1,500 gpm pumper will usually have six 2 ½-inch discharges: $6 \times 250 = 1,500$. NFPA 1002 5.2.1, 5.2.2. *IFPO, 2E:* Chapter 5, pages 118 – 119.

Question #52

The engine cooling system is normally under pressure. Adding additional water from the discharge side of the pump can overpressure the system. NFPA 1002 5.2.1, 5.2.2, 5.2.4. *IFPO, 2E:* Chapter 5, pages 125 – 126.

Question #53
Foam systems found on pumping apparatus include:

- Pre-mixed
- In-line eductor
- Around-the-pump proportioning
- Balanced pressure
- Direct injection/compressed-air

NFPA 1002 5.2.3. *IFPO, 2E:* Chapter 5, pages 126 – 130.

Question #54
Each of the answer selections is a type of balanced-pressure foam systems. NFPA 1002 5.2.3. *IFPO, 2E:* Chapter 5, pages 126 – 131.

Question #55
Older apparatus may use exhaust or vacuum primers to prime the main pump. NFPA 1002 5.2.1, 5.2.2, 5.2.4. *IFPO, 2E:* Chapter 5, pages 120 – 122.

Question #56
Instrumentation, such as gauges, flow meters, and indicators, is used to ensure that the pump is operating efficiently while providing appropriate pressure and flows. A foam system is not considered an instrumentation device. NFPA 1002 5.2.1, 5.2.2. *IFPO, 2E:* Chapter 5, page 99.

Question #57
Control valves are devices found on pump panels that are used to initiate, restrict, or direct water flow. NFPA 1002 5.2.1, 5.2.2. *IFPO, 2E:* Chapter 5, page 99.

Question #58
- *Tank-to-pump* control valves allow water to flow from the on-board water supply to the intake side of the pump.
- *Pump-to-tank* (tank fill) control valves allow water to flow from the discharge side of the pump to the tank.
- *Transfer control valves* are found on multistage pumps; they redirect water from the pump between the pressure mode and the volume mode.

NFPA 1002 5.2.1, 5.2.2. *IFPO, 2E:* Chapter 5, page 119.

Question #59
Attack hose should not be operated at pressures exceeding 275 psi. NFPA 1002 5.2.1, 5.2.2, 5.2.4. *IFPO, 2E:* Chapter 6, page 140.

Question #60
- *NFPA 1500* is the standard for occupational safety and health programs.
- *NFPA 1961* is the standard for the care, use, and maintenance of fire hose.
- *NFPA 1962* is the standard for the inspection, care, and use of fire hose, couplings, and nozzles and the service testing of hose.
- *NFPA 1965* is the standard for fire-hose appliances.

NFPA 1002 5.2.1, 5.2.2, 5.2.4. *IFPO, 2E:* Chapter 6, pages 138 – 141.

Question #61
Each of the answer selections is a type of hose construction used in the fire service today. NFPA 1002 5.2.1, 5.2.2, 5.2.4. *IFPO, 2E:* Chapter 6, pages 138 – 141.

Question #62

The male coupling is on the left and is a solid, nonmoving element with external threads. The female coupling is on the right with a stationary base and a movable (turns) female coupling with internal threads. NFPA 1002 5.2.1, 5.2.2, 5.2.4. *IFPO, 2E:* Chapter 6, pages 141 – 142.

Question #63

Siamese appliances are used to combine lines, not to divide them. NFPA 1002 5.2.1, 5.2.2, 5.2.4. *IFPO, 2E:* Chapter 6, pages 148 – 149.

Question #64

- The appliance in answer selection b is a *water thief.*
- The appliance in answer selection c is a large-gated *wye.*
- The appliance in answer selection d is a *reducer.*

NFPA 1002 5.2.1, 5.2.2, 5.2.4. *IFPO, 2E:* Chapter 6, page 151.

Question #65

- 50 psi is the operating pressure for smooth-bore hand-lines.
- 80 psi is the operating pressure for smooth-bore master streams.
- 100 psi is the operating pressure for combination nozzles.
- 125 psi is not a normal operating pressure for nozzles used in the fire service.

NFPA 1002 5.2.1, 5.2.2, 5.2.4. *IFPO, 2E:* Chapter 6, pages 153 – 157.

Question #66

- *Fixed-flow* or *constant-flow nozzles* provides a constant flow regardless of stream pattern.
- *Selectable-flow nozzles* allow the operator to adjust flow at the nozzle.
- *Automatic nozzles* maintain a constant nozzle pressure over a wide variety of flows.
- *Smooth-bore nozzles* are not a type of combination nozzle.

NFPA 1002 5.2.1, 5.2.2, 5.2.4. *IFPO, 2E:* Chapter 6, pages 154 – 157.

Question #67

The basic parts of a hose include:

- Reinforced inner liner
- Outer protective shell
- Couplings attached to both ends

NPFA 1002 5.2.1, 5.2.2, 5.2.4. *IFPO, 2E:* Chapter 6, pages 140 – 141.

Question #68

- A *siamese* is used to combine two or more lines into a single line.
- A *wye* is used to divide one hose line into two or more lines.

NFPA 1002 5.2.1, 5.2.2, 5.2.4. *IFPO, 2E:* Chapter 6, pages 148 – 149.

Question #69

The eductor uses the VENTURI principle to draw foam into the water stream. NFPA 1002 5.2.3. *IFPO, 2E:* Chapter 6, page 151.

Question #70

- *Nozzle pressure* is the designed operating pressure for a particular nozzle.
- *Nozzle flow* is the amount of water flowing from a nozzle.
- *Nozzle reach* is the distance water travels after leaving a nozzle.
- *Nozzle reaction* is the tendency of a nozzle to move in a direction opposite to water flow.

NFPA 1002 5.2.1. *IFPO, 2E:* Chapter 6, pages 153 – 154.

Question #71

NFPA specifies minimum tank capacity for apparatus as follows:

- 200 gallons – initial attack
- 300 gallons – pumper
- 1,000 gallons – mobile water apparatus/tanker

NFPA 1002 5.2.1. *IFPO, 2E:* Chapter 7, page 167.

Question #72

- The *dry-barrel hydrant* has a stem nut on its bonnet that is used to allow water to enter the barrel. The drain hole below ground allows water to drain from the hydrant after use.
- A *wet-barrel hydrant* has individual stem nuts to control discharge outlets.
- *Static-source hydrants* typically do not have control valves, and dead-end hydrants could be either wet- or dry-barrel.

NFPA 1002 5.2.1. *IFPO, 2E:* Chapter 7, pages 175 – 176.

Question #73

Each of the answer selections can affect the access and use of a static water supply. NFPA 1002 5.2.1, 5.2.2. *IFPO, 2E:* Chapter 7, pages 186 – 187.

Question #74

NFPA 291 suggest the following classification and color coding of hydrants:

Class A	Rated Capacity	Color of Bonnets and Caps
AA	1,500 gpm or greater	Light blue
A	1,000 to 1,499 gpm	Green
B	500 to 999 gpm	Orange
C	Less than 500 gpm	Red

NFPA 1002 5.2.1, 5.2.2. *IFPO, 2E:* Chapter 7, page 178.

Question #75

A static hydrant helps with drafting operations by improving the efficiency and reliability of the static source. Examples include eliminating the need to set up hard suction hose, improving access, removing silt and debris from drafting location, and allowing some operation during seasonal changes. NFPA 1002 5.2.1, 5.2.2, 5.2.4. *IFPO, 2E:* Chapter 7, pages 187 – 188.

Question #76

An open-relay system allows excessive pressure and flow to be diverted out of the relay. This reduces the effects of pressure changes as well as the number of required changes. NFPA 1002 5.2.2. *IFPO, 2E:* Chapter 7, pages 189 – 194.

Question #77

Each of the answer selections is a piece of equipment that is used in tanker shuttles except CAFS, or compressed air foam system. NFPA 1002 5.2.1. *IFPO, 2E:* Chapter 7, pages 195 – 201.

Question #78

Required flow is the estimated flow of water needed for a specific incident. NFPA 1002 5.2.1. *IFPO, 2E:* Chapter 7, page 164.

Question #79

The flow capacity of a shuttle is limited by the volume of water being delivered and the time it takes to complete a shuttle cycle. NFPA 1002 5.2.1. *IFPO, 2E:* Chapter 7, pages 200 – 201.

Question #80

Controls should never be opened, closed, or turned swiftly or abruptly. In doing so, mechanical damage may occur to the device and/or it may cause a water hammer that results in additional damage and/or injury. NFPA 1002 5.2.1, 5.2.2. *IFPO, 2E:* Chapter 8, page 208.

Question #81
NFPA 1901 requires that pumps rated less than 1,500 achieve a prime within 30 seconds.
NFPA 1002 5.1.1. *IFPO, 2E:* Chapter 8, page 214.

Question #82
The 40-minute annual pump test consist of:

- 20 minutes at 100% capacity at 150 psi

- 10 minutes at 70% capacity at 200 psi

- 10 minutes at 50% capacity at 250 psi

Pumps with a rated capacity of 750 gpm or greater must also undergo a 5-minute overload test at 165 psi. NFPA 1002 5.1.1. *IFPO, 2E:* Chapter 8, pages 228 – 229.

Question #83
Although all methods can provide power to a pump, most main pumps are powered by the drive engine through either a split shaft or directly from the crankshaft. NFPA 1002 5.2.1, 5.2.2, 5.2.4. *IFPO, 2E:* Chapter 8, pages 210 – 211.

Question #84
Cavitation is caused by insufficient intake flow to match the discharge flow and is often characterized as a pump running away from the water supply. NFPA 1002 5.1.1. *IFPO, 2E:* Chapter 9, pages 245 – 246.

Question #85
- *Pre-action systems* are similar to dry-pipe systems in that air or compressed gas is maintained in the system. However, an automatic detection system (smoke, heat, flame, etc.) or manual system (pull box) must operate to allow water to enter the system. At this point, it is similar to a wet-pipe system; when a sprinkler head fuses, water discharges.

- *Dry-pipe systems* maintain air or compressed gas under pressure within the system. When a head fuses, water enters the system and discharges through any fused heads.

- A *deluge system* maintains all sprinkler heads open. When a detection system operates, water enters the system and is discharged through all the open heads.

- *Wet-pipe systems* maintain water in the system at all times. When a sprinkler head is fused, water immediately discharges.

NFPA 1002 5.2.4. *IFPO, 2E:* Chapter 9, pages 148 – 253.

Question #86
- *NFPA 13* deal with installation of sprinkler systems.

- *NFPA 13D* focuses on the installation of sprinkler systems in one- and two-family dwellings.

- *NFPA 14* deals with standpipe and hose systems.

NFPA 1002 5.2.4. *IFPO, 2E:* Chapter 9, pages 248 – 253.

Question #87
In most cases, when supporting sprinkler and standpipe systems, volume is required over pressure, and the transfer valve should be placed in the volume mode. NFPA 1002 5.2.4. *IFPO, 2E:* Chapter 9, pages 251 – 252.

Question #88
Each of the answer selections can cause a water hammer except operating the priming device, which should not generate any significant increase or decrease in pressure. NFPA 1002 5.1.1. *IFPO, 2E:* Chapter 9, pages 244 – 245.

Question #89

Class 1 standpipes provide 2 ½-inch hose connections and are intended for use by firefighters or fire brigade members.

Class 2 standpipes provide a 1 ½-inch hose station and are intended primarily for trained personnel during initial attack efforts.

Class 3 standpipes provide 1 ½-inch and 2 ½-inch hose connections for use by firefighters or fire brigade members.

NFPA 1002 5.2.4. *IFPO, 2E:* Chapter 9, page 253.

Question #90

The basic characteristics and properties of water are listed below. These are considered approximate in that water purity, atmospheric conditions, and rounding can affect the values.

- Virtually incompressible

- Freezes at 32°F

- Expands when frozen

- Boils at 212°F (considered water boiling point)

- 1 gallon weighs 8.34 lbs

- Density is 62.4 lb/ft^3

- Number of gallons in 1 cubic foot is 7.38 gallons/ft^3

NFPA 1002 5.2.1, 5.2.2. *IFPO, 2E:* Chapter 10, pages 262 – 263.

Question #91

- *Evaporation* is the physical change of state from a liquid to a vapor.

- *Vapor pressure* (VP) is the pressure exerted on the atmosphere by molecules as they evaporate from the surface of the liquid.

- *Boiling point* (BP) is the temperature at which the vapor pressure of a liquid equals the surrounding pressure.

NFPA 1002 5.2.1, 5.2.2. *IFPO, 2E:* Chapter 10, pages 262 – 263.

Question #92

- *Density* is the weight of a substance expressed in units of mass per volume.

- *Weight* is the downward force exerted on an object by the Earth's gravity, typically expressed in pounds.

- *Volume* is the amount of space occupied by an object.

- *Pressure* is the force exerted by a substance in units of weight per area, typically expressed in pounds per square inch (psi).

NFPA 1002 5.2.1, 5.2.2. *IFPO, 2E:* Chapter 10, pages 264 – 269.

Question #93

Pressure is the force exerted by a substance in *units of weight per area*, typically expressed in *pounds per square inch* (psi or lb/in^2). *Force* is the total weight of the substance. So, the pressure is multiplied by the area to calculate the force or weight.

F = Force (weight in pounds)

P = Pressure

A = Area (square inches)

NFPA 1002 5.2.1, 5.2.2. *IFPO, 2E:* Chapter 10, pages 269 – 270.

Question #94

- *Gauge pressure* is the measurement of pressure that does not include atmospheric pressure, typically expressed as psig.

- *Absolute pressure* is the measurement of pressure that includes atmospheric pressure, typically expressed as psia.

- *Vacuum* is the measurement of pressure that is less than atmospheric pressure, typically expressed in inches of mercury (in. Hg).

- *Head pressure* is the pressure exerted by the vertical height of a column of liquid expressed in feet.

NFPA 1002 5.2.1, 5.2.2. *IFPO, 2E:* Chapter 10, pages 275 – 277.

Question #95

- *Head pressure* is the pressure exerted by the vertical height of a column of liquid expressed in feet.

- *Atmospheric pressure* is the pressure exerted by the atmosphere on the Earth.

- *Gauge pressure* is the measurement of pressure that does not include atmospheric pressure, typically expressed as psig.

- *Absolute pressure* is the measurement of pressure that includes atmospheric pressure and is typically expressed as psia.

- *Vacuum* is the measurement of pressure that is less than atmospheric pressure, typically expressed in inches of mercury (in. Hg).

NFPA 1002 5.2.1, 5.2.2. *IFPO, 2E:* Chapter 10, pages 275 – 277.

Question #96

- *Residual pressure* is the pressure remaining in the system after water has been flowing.

- *Static pressure* is the pressure in a system when no water is flowing.

- *Pressure drop* is the difference between the static pressure and the residual pressure when measured at the same location.

- *Normal pressure* is the water flow pressure found in a municipal water supply during normal consumption demands.

NFPA 1002 5.2.1, 5.2.2. *IFPO, 2E:* Chapter 10, pages 276 – 277.

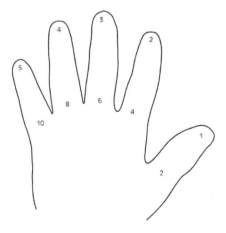

Question #97

The hand method is a fireground method used to estimate friction loss in 100-foot sections of 2 ½-inch hose. Simply multiply the two figures on a finger for the approximate friction loss pressure for each 100-section of 2 ½-inch hose. NFPA 1002 5.2.1, 5.2.2. *IFPO, 2E:* Chapter 10, pages 302 – 303.

Question #98

One gallon of water weighs 8.34 pounds and is expressed 8.34 lb/gal. The density of water is 62.4 pounds and is expressed 62.34 lb/ft^3 (cubic feet). NFPA 1002 5.2. *IFPO, 2E:* Chapter 10, pages 264 – 266.

Question #99

Improvements in hose construction have reduced friction loss within the hose. This requires the use of a more accurate formula and $FL = 2q^2 + q$. NFPA 1002 5.2.1, 5.2.2. *IFPO, 2E:* Chapter 11, pages 304 – 305.

Question #100

This friction loss formula was derived from a combination of Bernoulli's equation, the Darcy-Weibach equation, and the Continuity equation. It is both simple and reasonably accurate for varying hose sizes, and it can be use for both preplanning and fireground calculations.

$$FL = c \times q^2 \times L$$

where FL = friction loss

           c = constant for a specific hose diameter

           q = gm ÷ 100 (flow in hundreds of gpm)

           L = length of hose in hundreds of feet

NFPA 1002 5.2.1, 5.2.2. *IFPO, 2E:* Chapter 11, pages 306 – 309.

APPLICATION & ANALYSIS

Section two is evaluating for a higher level of learning. Within this section, we are testing to determine an understanding of comparing material, describing processes, explaining procedures, and interpreting results. A test-taker mastering this section should have a better grasp of the material and a greater depth of understanding. Referring to Table I-1 (Bloom's Taxonomy, Cognitive Domain), we are addressing the following levels:

- application
- analysis

Phase Two, Exam One

1. Fire pump operations are more complex and unpredictable than are other fireground operations.
   a. True
   b. False

2. All of the activities related to fire pump operations occur in the first few minutes after arrival on scene.
   a. True
   b. False

3. In general, the fire department is ultimately responsible for the establishment, implementation, and monitoring of a preventive maintenance program.
   a. True
   b. False

4. Replacing the battery on a fire department vehicle is usually performed by the pump operator.
   a. True
   b. False

5. Safety-related components and manufacturers' recommendations comprise the majority of inspection items included in a preventive maintenance program.
   a. True
   b. False

6. Apparatus fuel tanks should be refilled per department policy.
   a. True
   b. False

7. NFPA 1901 requires the following warning on pump panels:

   "**WARNING:** Death or serious injury might occur if proper operating procedures are not followed. The pump operator as well as individuals connecting supply or discharge hoses to the apparatus must be familiar with water hydraulics hazards and component limitations."
   a. True
   b. False

8. One unique feature of a pump connected to the front crankshaft is that the engine speed does not need to be at idle to be engaged.
   a. True
   b. False

9. The pressure at any point in a liquid at rest is equal in every direction.
   a. True
   b. False

10. The pressure of a liquid acting on a surface is proportional to that surface.
    a. True
    b. False

11. On modern apparatus, it is no longer required to inspect the radiator and coolant levels.
    a. True
    b. False

12. NFPA 1500 includes driving requirements that are only applicable during emergency response.
    a. True
    b. False

13. Which of the following is correct concerning NFPA 1002 requirements for fire department drivers/operators?
    a. license for vehicles they operate; medically fit; meet the requirements of Firefighter I per NFPA 1001; at least 21 years old
    b. meet the requirements of Firefighter I per NFPA 1001; at least 21 years old; have a high school diploma or GED
    c. license for vehicles they operate; medically fit; meet the requirements of Firefighter I per NFPA 1001
    d. license for vehicles they operate; meet the requirements of Firefighter II per NFPA 1001; earned a high school diploma or GED

14. The three primary duties of a pump operator include
    a. preventive maintenance, transporting firefighters, and driving the apparatus.
    b. preventive maintenance, driving the apparatus, and operating the pump.
    c. securing a water supply, discharge maintenance, and pump operations.
    d. driving the apparatus, operating the pump, and securing a water supply.

15. _____ are _____ that are legally binding and enforceable.
    a. Laws, rules
    b. Laws, guidelines
    c. Standards, guidelines
    d. Standards, ordinances

16. Inspections are conducted to verify the _____ of components, whereas servicing activities are conducted to help maintain the vehicle in peak operating _____ .
    a. condition, performance
    b. status, performance
    c. performance, condition
    d. status, condition

17. Documenting problems found during preventive maintenance inspections is important for each of the following reasons *except*
    a. it may help to determine maintenance trends.
    b. it is used to document inspection hours.
    c. it may be required for warranty claims.
    d. it is required by NFPA 1500.

18. When inspecting the steering on an apparatus, the steering wheel
    a. free play should be no more than 1 inch.
    b. free play should not exceed 10 degrees.
    c. should be fully turned in both directions.
    d. all of the answers are correct.

19. Safety should be considered when conducting preventive maintenance inspections and testing. Which of the following could be considered a common safety consideration for preventive maintenance inspections and tests?
    a. Ensure work area is free from hazards.
    b. Check for loose equipment before raising a tilt cab.
    c. Always be careful when opening the radiator cap.
    d. All of these are important safety considerations.

20. To regain control of a vehicle during a skid, you should
    a. disengage the clutch.
    b. turn the front wheels in the opposite direction of the skid.
    c. take your foot off the accelerator.
    d. increase speed slightly.

21. Fire department apparatus accidents occur for a variety of reasons. Several common factors associated with apparatus accidents include all the following *except*
    a. not following laws and standards related to emergency response.
    b. not being certified to NFPA 1002.
    c. not being fully aware of both driver and apparatus limitations.
    d. a lack of appreciation for driving conditions such as weather and traffic.

22. The distance an apparatus travels from the time the brain sends the message to depress the brakes until the brakes are depressed is called
    a. perception distance.
    b. reaction distance.
    c. braking distance.
    d. total stopping distance.

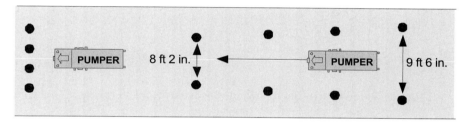

23. The driving exercise noted above is called
    a. diminishing clearance.
    b. alley dock.
    c. confined-space turnaround.
    d. serpentine.

24. Fill in the missing rated capacity information:
    100% at _____ psi
    _____% at 200 psi
    50% at _____ psi
    a. 100, 200, 50
    b. 100, 70, 250
    c. 150, 75, 250
    d. 150, 70, 250

25. A two-stage centrifugal pump is discharging 1,000 gpm at 200 psi in volume mode. Each impeller will deliver
    a. 1,000 gpm at 200 psi.
    b. 500 gpm at 200 psi.
    c. 1,000 gpm at 100 psi.
    d. 500 gpm at 100 psi.

26. A centrifugal pump mounted at the rear of an apparatus would most likely be powered
    a. through a PTO.
    b. directly from the crankshaft.
    c. through a split-shaft transmission.
    d. by any of these methods.

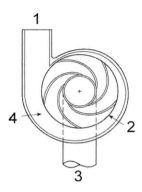

27. Identify the impeller, and indicate its direction of rotation.
    a. 4 – clockwise
    b. 2 – clockwise
    c. 2 – counterclockwise
    d. not enough information provided

28. NFPA _____ suggests a standard color code to match discharges with their respective gauges.
    a. 1500
    b. 1901
    c. 1911
    d. 1961

29. Which of the following is *not* considered an indicator on the pump panel?
    a. water level instrument
    b. tachometer
    c. flow meter
    d. pump engagement light

30. While priming a pump, several indicators that the pump is primed include each of the following *except*
    a. a positive reading on the pressure gauge.
    b. air/oil discharge under the vehicle.
    c. main pump sounds as if it is under load.
    d. priming motor sounds as if it is slowing.

31. This control device operates the
    a. transfer valve.
    b. pressure governor.
    c. pressure relief device.
    d. primer.

32. A 1,000 gpm pumper will usually have how many 2 ½-inch discharges?
    a. four
    b. five
    c. six
    d. seven

33. Each of the following can be used for intake hose *except*
    a. LDH.
    b. soft-sleeve hose.
    c. hard suction hose.
    d. unlined cotton-jacketed hose.

34. Components used to support varying hose configurations are called
    a. hose tools.
    b. pump devices.
    c. appliances.
    d. adapters.

35. The basic components of an eductor include
    a. a metering valve and pickup hose.
    b. a metering valve, pickup hose, and foam concentrator.
    c. a foam concentrator and pickup hose.
    d. a pickup hose, metering valve, and foam gauge.

36. When a nozzle is not provided sufficient nozzle pressure
    a. less flow will be delivered.
    b. reach will be reduced.
    c. a poor pattern will develop.
    d. each of these answers are correct.

37. A pumper with a 1,000-gallon tank is flowing 125 gpm through a 1 ¾-inch hand-line. How long will the on-board water supply last?
    a. 8 minutes
    b. 10 minutes
    c. 20 minutes
    d. no correct answer is provided

38. When conducting hydrant flow testing, all of the following equipment may be used *except*
    a. hydrant diffusers.
    b. hydrant wrenches.
    c. a fire department pumper.
    d. a pitot gauge for each hydrant.

39. One formula used to determine the distance between two or more pumpers in a relay operation is
    a. (PDP – 20) × 100/FL.
    b. (PDP – 20) × L.
    c. total length ÷ number of pumpers.
    d. number of pumpers ÷ average length of hose between pump.

40. In a tanker shuttle, the shuttle flow capacity is the volume of water that can be pumped
    a. without running out of water.
    b. by all the available tankers.
    c. by the average of available tankers.
    d. within 15 minutes of operation.

41. The weight of water in a 1,000-gallon pumper is about
    a. 835 lbs.
    b. 8,350 lbs.
    c. 50,000 lbs.
    d. 83,500 lbs.

42. Although a variety of pump sizes and configurations exists, the same basic steps must be taken to move water from the supply to the discharge point. What critical pump operation step is out of order?

    Step 1   Position apparatus, set parking brake, and let engine return to idle
    Step 2   Provide water to intake side of pump (on-board, hydrant, draft)
    Step 3   Engage the pump
    Step 4   Set transfer valve (if so equipped)
    Step 5   Open discharge lines
    Step 6   Throttle to desired pressure
    Step 7   Set the pressure-regulating device
    Step 8   Maintain appropriate flows and pressures

    a. Change step 4 with 6
    b. Change step 6 with 7
    c. Change step 2 with 3
    d. Change step 7 with 8

43. Which of the following is incorrect concerning pump operations?
    a. Always maintain constant vigilance to safety.
    b. Open, close, and turn controls abruptly.
    c. Never operate the pump without water.
    d. Always keep water moving when operating the pump at high speeds.

44. As a general rule of thumb, the transfer valve should be in volume when
    a. flows are greater than 50% of a pump's rated capacity.
    b. pressures are less than 150 psi.
    c. pressures are greater 150 psi.
    d. both a and b.
    e. both a and c.

45. When supplying one discharge line using the on-board water tank, the transfer valve should be
    a. turned off.
    b. turned on.
    c. in volume mode.
    d. in pressure mode.

46. The power for pumps can be transferred from the drive engine by each of the following methods *except*
    a. front crankshaft.
    b. split-shaft.
    c. power take-off (PTO).
    d. midship transfer (MST).

47. A pump connected to the front crankshaft of the drive engine is usually mounted
    a. on top of the drive engine.
    b. toward the back of the apparatus.
    c. in the middle of the apparatus.
    d. on the front of the apparatus.

48. Each of the flowing is a sign that cavitation may be occurring *except*
    a. excessive pump vibrations.
    b. no corresponding increase in pressure when the engine speed is increased.
    c. priming device slows down or seems to be operating under a load.
    d. excessive pump vibrations or rattling sounds such as sand or gravel going through the pump.

49. Cavitation can be caused by all of the following *except*
    a. insufficient flow from a hydrant.
    b. excessive lift while drafting.
    c. intake lines are too small.
    d. not enough discharge lines operating.

50. When connecting to a fire department connection, the pump operator should do all of the following *except*
    a. connect at least one hose line.
    b. pump at 150 psi unless otherwise directed.
    c. begin pumping immediately when fire and smoke are visible.
    d. all of these actions are correct.

51. According to NFPA 14, Standard for the Installation of Standpipe and Hose Systems, there are three classes of standpipe system. Which of the following is correct?
    a. They are classified based on intended use.
    b. They all must have a minimum operating pressure of 150 psi.
    c. They all must have at least one 1 ½-inch connection.
    d. They all must be clearly marked "Fire Department Use Only."

52. Increased pressure will _____ the ability of a liquid to vaporize, while decreased pressure will _____ the ability of a liquid to vaporize.
    a. increase, reduce
    b. reduce, increase
    c. initiate, stop
    d. stop, initiate

53. The _____ sprinkler system maintains compressed air or inert gas within the pipes. When a pull box is operated, water enters the system but does not discharge through sprinkler heads.
    a. dry-pipe
    b. pre-action
    c. deluge
    d. wet-pipe

54. The _____ sprinkler system maintains compressed air or inert gas within the pipes. When a sprinkler head is fused, the air/gas escapes, lowering the pressure and allowing water to enter the system and discharge through fused heads.
    a. dry-pipe
    b. pre-action
    c. deluge
    d. wet-pipe

55. Which branch of science focuses on principles and laws of fluids in motion?
    a. hydraulics
    b. hydrodynamics
    c. hydrostatics
    d. hydrolastics

56. Which of the following physical characteristics and properties of water are incorrect?
    a. It boils at approximately 212°F.
    b. One gallon weighs 8.34 lbs.
    c. Its density is 7.38 gallons/ft^3.
    d. It freezes at approximately 32°F.

57. The amount of heat absorbed when changing from a liquid to a vapor state best describes
     a. latent heat of fusion.
     b. specific heat.
     c. latent heat of vaporization.
     d. conduction.

58. The formula $P = \dfrac{F}{A}$ is used to calculate
     a. pressure.
     b. pounds or weight.
     c. force.
     d. both a and b are correct.

59. The formula used to calculate the weight of water of a given volume is
     a. $W = D \times V$.
     b. $W = D \div V$.
     c. $W = \dfrac{D}{V}$.
     d. $W = \dfrac{V}{D}$.

60. How many gallons of water are in 1 cubic foot of water?
     a. 8.35
     b. 62.4
     c. 7.48
     d. not enough information provided

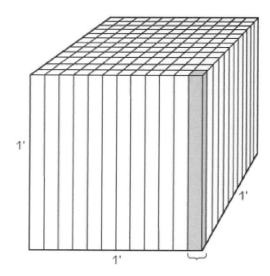

61. What is the pressure at the bottom of this shaded 1-square-inch column of water?
     a. .433 lb
     b. 1 lb
     c. 5 lbs
     d. 8.34 lbs

62. The formula used to determine flow (gpm) when smooth-bore nozzles are used is $Q = 29.7 \times d^2 \times \sqrt{NP}$. An acceptable value for $\sqrt{NP}$ to use for calculating flow on the fireground for hand lines is
     a. 5.
     b. 7.
     c. 9.
     d. 30.

63. The Iowa State formula for calculating needed flow is expressed as

    a. $NF = \dfrac{V}{100}$.

    b. $NF = \dfrac{V}{100} \times 3$.

    c. $NF = \dfrac{A}{3}$.

    d. $NF = \dfrac{A}{3} \div 100$.

64. The National Fire Academy formula for calculating needed flow is expressed as

    a. $NF = \dfrac{V}{100}$.

    b. $NF = \dfrac{V}{100} \times 3$.

    c. $NF = \dfrac{A}{3}$.

    d. $NF = \dfrac{A}{3} \div 100$.

65. A gauge reading of 100 psia at sea level is equivalent to a gauge reading of
    a. 85.3 psig.
    b. 114.7 psig.
    c. 100 psig.
    d. none of the answers are correct.

66. Which of the following is the correct formula for calculating smooth-bore nozzle reaction?
    a. $NR = 1.57 \times d \times NP$
    b. $NR = gpm \times \sqrt{NP} \times .0505$
    c. $NR = gpm \times \sqrt{NP \times .0505}$
    d. $NR = 1.57 \times d^2 \times NP$

67. Using the drop-ten method, calculate friction loss in 100-foot sections of 2 ½-inch hose flowing 200 gpm.
    a. 100 psi
    b. 20 psi
    c. 10 psi
    d. 5 psi

68. Using the hand method, calculate friction loss in a 100-foot section of 2 ½-inch hose flowing 200 gpm.
    a. 2 psi
    b. 4 psi
    c. 6 psi
    d. 8 psi

69. Calculate friction loss in a 100-foot section of 2 ½-inch hose flowing 400 gpm using the hand method.
    a. 4 psi
    b. 8 psi
    c. 16 psi
    d. 32 psi

70. During pump operations, the drop in hydrant pressure from static to residual can be used to estimate the additional flow the hydrant is capable of providing. A 10% or less drop means the hydrant may be able to deliver as much as
    a. three times the original flow.
    b. two times the original flow.
    c. one time the original flow.
    d. half the original flow.

71. What is the nozzle reaction for a ½-inch smooth-bore nozzle discharging water with a nozzle pressure of 50 psi using the formula $NR = 1.57 \times d^2 \times NP$?
    a. 20 lbs
    b. 25 lbs
    c. 50 lbs
    d. 70 lbs

72. The nozzle reaction for a combination nozzle discharging 95 gpm with a nozzle pressure of 100 psi is _____ when using the formula $NR = gpm \times \sqrt{NP} \times 0.0505$.
    a. 48 lbs
    b. 78 lbs
    c. 85 lbs
    d. 126 lbs

73. Calculate friction loss in 100-foot sections of 2 ½-inch hose flowing 250 gpm using the drop-ten method.
    a. 100 psi
    b. 20 psi
    c. 15 psi
    d. 5 psi

74. Two methods are used on the fireground to calculate pressure changes as the result of changes in elevation. The elevation formula using feet is expressed as
    a. $EL = 5 \times H$, where H = height in number of floor levels above or below the pump.
    b. $EL = 5 \times H$, where H = head pressure.
    c. $EL = 0.5 \times H$, where H = distance in feet above or below the pump.
    d. $EL = 0.5 \times H$, where H = head pressure.

75. An adaptation of the formula $FL = 2q^2 + q$ specifically for use with 3-inch hose is
    a. $FL = 2q^2$.
    b. $FL = q^2$.
    c. $FL = q^2 + q$.
    d. $FL = q^2 + \frac{1}{2}q$.

76. Needed flow, using the Iowa State formula, can be used
    a. for an entire structure.
    b. for a section of a structure.
    c. based on the square footage of the involved area.
    d. both a and b are correct.

77. Using the formula cq^2L, calculate the friction loss for a 100-foot section of 1 ½-inch flowing 100 gpm, where c = 24.
    a. 24 psi
    b. 35 psi
    c. 48 psi
    d. not enough information was provided

78. Calculate the friction loss for a 100-foot section of 2 ½-inch hose flowing 250 gpm using the formula cq^2L, where c = 2.
    a. 13 psi
    b. 23 psi
    c. 38 psi
    d. not enough information provided

79. Using the formula $2q^2 + q$, calculate the friction loss for a 100-foot section of 2 ½-inch hose flowing 100 gpm.
    a. 3 psi
    b. 9 psi
    c. 10 psi
    d. 15 psi

80. Calculate the friction loss for a 100-foot section of 2 ½-inch flowing 250 gpm using the formula $2q^2 + q$.
    a. 5 psi
    b. 15 psi
    c. 20 psi
    d. 35 psi

81. Calculate the friction in 100 feet of 4-inch hose flowing 500 gpm using the formula cq^2L, where c = 0.2.
    a. 5 psi
    b. 10 psi
    c. 15 psi
    d. 20 psi

82. Using the *condensed q* formula, calculate the friction loss for 300 gpm flowing through a 100-foot section of 3-inch hose.
    a. 4 psi
    b. 9 psi
    c. 15 psi
    d. 19 psi

83. A hose line is raised to the sixth floor. What is the pressure gain/loss?
    a. 25 psi
    b. 30 psi
    c. 50 psi
    d. 55 psi

84. What is the pressure gain/loss for a line raised 50 feet above the apparatus?
    a. 25 psi
    b. 50 psi
    c. 100 psi
    d. not enough information provided

85. What is the pressure gain/loss for a hose line 75 feet below the apparatus?
    a. 37.5 psi
    b. 50 psi
    c. −37.5 psi
    d. −50 psi

86. A hose line is taken down to the third-floor basement level. What is the pressure gain/loss?
    a. 5 psi
    b. 10 psi
    c. −5 psi
    d. −10 psi

87. Calculate the friction loss for 250 gpm flowing through a 100-foot section of 3-inch hose using the condensed q formula:
    a. 4 psi
    b. 6 psi
    c. 15 psi
    d. 19 psi

88. Using the formula $Q = 30 \times d^2 \times \sqrt{NP}$, calculate the flow through a smooth-bore nozzle with a ¾-inch tip at 50 psi nozzle pressure.
    a. 75 gpm
    b. 118 gpm
    c. 125 gpm
    d. 135 gpm

89. Calculate the flow through a smooth-bore nozzle on a master stream device operating on the fireground with a ¾-inch tip using the formula $Q = 30 \times d^2 \times \sqrt{NP}$.
    a. 150 gpm
    b. 160 gpm
    c. 170 gpm
    d. not enough information provided

90. In the formula $PDP = NP + FL + AFL \pm EL$, NP refers to
    a. nozzle pressure.
    b. nominal pressure.
    c. normal pressure.
    d. natural pressure.

91. What is the PDP for a hand-line with the following elements?

    1. NP: ¾-inch smooth-bore nozzle
    2. FL: 150 feet of 1 ¾-inch hose line flowing 118 gpm with friction of 32 psi
    3. EL: elevation gain of about 5 psi

    a. 32 psi
    b. 37 psi
    c. 87 psi
    d. 137 psi

92. What is the PDP for a 300-foot line of 1 ¾-inch hose flowing 100 gpm through a combination nozzle when the friction loss is 47 psi for the line?
    a. 32 psi
    b. 37 psi
    c. 87 psi
    d. 147 psi

93. When traction is lost in a curve, _____ force will move the apparatus in an outward direction.
    a. surge
    b. gravitational
    c. cavitational
    d. centrifugal

94. The _____ distance is the distance the apparatus travels from the time the hazard is seen until the brain recognizes it as a hazard.
    a. perception
    b. reaction
    c. braking
    d. total stopping

95. The "W" in the SAE oil classification refers to
    a. oil rated for flow at 0°F.
    b. oil rated for flow at 2°F.
    c. oil rated for flow at 4°F.
    d. oil rated for flow at 6°F.

96. The use of flow meters on apparatus has increased because
    a. they eliminate the need for fireground hydraulics to determine pump discharge pressure.
    b. they provide increased versatility on locating pump control panels.
    c. they are easier to calibrate than Bourdon tube gauges.
    d. all are correct.

97. A fire department connection is usually a(n)
    a. wye with two inlets.
    b. OS&Y or PIV valve.
    c. gated siamese with two male couplings.
    d. clappered siamese with female inlets.

98. When two or more lines are to be combined into one line, the appliance to use is a
    a. gated wye.
    b. siamese.
    c. hydrant thief.
    d. distribution manifold.

99. Increased friction from turbulent flows in a hose can be attributed to all of the following *except*
    a. laminar flow.
    b. kinks and bends.
    c. couplings and adapters.
    d. rough interior.

100. Feathering or gating is the process used to adjust discharge settings by
    a. slightly increasing/decreasing pump speed.
    b. slowly opening/closing nozzles.
    c. using discharge control valves.
    d. none of the answers is correct.

Phase II, Exam I: Answers to Questions

1.	F	26.	C	51.	A	76.	D
2.	F	27.	B	52.	B	77.	A
3.	T	28.	B	53.	B	78.	A
4.	F	29.	C	54.	A	79.	A
5.	T	30.	B	55.	B	80.	B
6.	T	31.	C	56.	C	81.	A
7.	T	32.	A	57.	C	82.	B
8.	F	33.	D	58.	A	83.	A
9.	T	34.	C	59.	A	84.	A
10.	F	35.	A	60.	C	85.	C
11.	F	36.	D	61.	A	86.	D
12.	F	37.	A	62.	B	87.	B
13.	C	38.	C	63.	A	88.	B
14.	B	39.	A	64.	C	89.	A
15.	A	40.	A	65.	A	90.	A
16.	D	41.	B	66.	D	91.	C
17.	B	42.	C	67.	C	92.	D
18.	B	43.	B	68.	D	93.	D
19.	D	44.	D	69.	D	94.	A
20.	C	45.	D	70.	A	95.	A
21.	B	46.	D	71.	A	96.	D
22.	B	47.	D	72.	A	97.	D
23.	A	48.	C	73.	C	98.	B
24.	D	49.	D	74.	C	99.	A
25.	B	50.	D	75.	B	100.	C

Phase II, Exam I:
Rationale & References for Questions

Question #1

This is a misperception about fire pump operations. In reality, scientific theory and principles prevail in pump operations. When understood, pump operations can be both predictable and controllable. NFPA 1002 5.1. *IFPO, 2E:* Chapter 1, page 4.

Question #2

It is a misperception about fire pump operations that most of the activities related to fire pump operations occur in the first few minutes upon scene arrival. After initial operations are set up, pump operators must:

- Continually observe instrumentation.

- Adjust flows and pressure as appropriate for safety of personnel and equipment.

- Be prepared to readily adapt to changing fireground situations.

- Monitor and plan for water supply needs and long-term operations.

- Maintain a constant vigilance to safety.

NFPA 1002 5.1. *IFPO, 2E:* Chapter 1, page 4.

Question #3

The fire department, through the fire chief, has the ultimate responsibility for the establishment, implementation, and monitoring of a preventive maintenance program. NFPA 1002 4.2.1, 5.1.1. *IFPO, 2E:* Chapter 2, page 22.

Question #4

The specific preventive maintenance activities conducted by pump operators and mechanics depend on the level of training and the type of preventive maintenance activity being conducted. In general, certified mechanics conduct those activities that require apparatus to be taken out of service, require several hours to complete, or are detailed and complicated repairs. Checking and adding engine oil or battery fluid may be conducted by the pump operator, whereas changing the engine oil or replacing the battery is most likely performed by a mechanic. NFPA 1002 4.2.1, 5.1.1. *IFPO, 2E:* Chapter 2, page 22.

Question #5

The two main criteria for determining what components to include in a preventive maintenance inspection are 1) safety-related components and 2) manufacturers' inspection recommendations. NFPA 1002 4.2.1, 5.1.1. *IFPO, 2E:* Chapter 2, pages 28 – 29.

Question #6

Apparatus fuel tanks should be refilled according to department policy. When no refill policy exists, apparatus fuel tanks should be refilled when the fuel level reaches the three-quarter mark, that is, when the fuel gauge drops from full to the three-quarter full mark. NFPA 1002 4.2.1. *IFPO, 2E:* Chapter 2, page 35.

Question #7

Because of the potential danger, the statement is now required on all new pump panels. NFPA 1002 5.2.1, 5.2.2. *IFPO, 2E:* Chapter 4, page 102.

Question #8

In order to engage the pump, the engine speed must be reduced to normal idle speed. NFPA 1002 5.2.1, 5.2.2, 5.2.4. *IFPO, 2E:* Chapter 8, page 211.

Question #9

Basic pressure principles include:

- Pressure at any point in a liquid at rest is equal in every direction. In other words, water pressure is exerted in every direction – downward and outward as well as upward.
- Pressure of a liquid acting on a surface is perpendicular to that surface.
- External pressure applied to a confined liquid (fluid) is transmitted equally throughout the liquid.
- Pressure at any point beneath the surface of a liquid in an open container is directly proportional to its depth.
- Pressure exerted at the bottom of a container is independent of the shape or volume of the container.

NFPA 1002 5.2. *IFPO, 2E:* Chapter 10, pages 271 – 274.

Question #10

Basic pressure principles include:

- Pressure at any point in a liquid at rest is equal in every direction.
- Pressure of a liquid acting on a surface is *perpendicular* to that surface.
- External pressure applied to a confined liquid (fluid) is transmitted equally throughout the liquid.
- Pressure at any point beneath the surface of a liquid in an open container is directly proportional to its depth.
- Pressure exerted at the bottom of a container is independent of the shape or volume of the container.

NFPA 1002 5.2. *IFPO, 2E:* Chapter 10, pages 271 – 274.

Question #11

Pump operators should continue to inspect the radiator and coolant levels as indicated by department policy and manufacturer's recommendations. NFPA 1002 4.2.1, 4.2.2, 5.1.1. Chapter 2, page 33.

Question #12

NFPA 1500 includes requirements for both emergency and nonemergency use of apparatus. NFPA 1002 4.3.1, 4.3.6. *IFPO, 2E:* Chapter 3, pages 46 – 49.

Question #13

NFPA 1002 requires the following:

- Must be licensed to drive all vehicles they are expected to operate
- Subject to medical evaluation of operator
- Must meet the requirements of Firefighter I per NFPA 1001

Typically, local jurisdictions set specific requirements for age and education levels. NFPA 1002 1.4.1, 1.4.2, 5.1. *IFPO, 2E:* Chapter 1, pages 11 – 16.

Question #14

The three main or primary duties of the pump operator as outlined in NFPA 1002 include:

1. Preventative Maintenance (NFPA 1002 4.2).

2. Driving (NFPA 1002 4.3).

3. Pump Operations (NFPA 1002 5.1 and 5.2).

NFPA 1002 4.2, 4.3, 5.1, 5.2. *IFPO, 2E:* Chapter 1, pages 7 – 11.

Question #15

Laws are rules that are legally binding and enforceable. Standards are guidelines that are not legally binding or enforceable. NFPA 1002 4.2, 4.3, 5.1. *IFPO, 2E:* Chapter 1, page 15.

Question #16

Inspections are conducted to verify the status of a component; for example, to verify water, oil, and fuel levels.

Servicing activities are conducted to help maintain vehicles in peak operating condition; for example, cleaning, lubricating, and topping off fluids.

Tests are conducted to determine the performance of components, as in annual pump service tests. NFPA 1002 4.2.1, 5.1.1. *IFPO, 2E:* Chapter 2, page 23.

Question #17

Documenting preventive maintenance activities is important because the practice:

- Helps keep track of needed maintenance and repairs.

- May help to determine trends.

- Is required by NFPA 1500.

- May be required for warranty claims.

NFPA 1002 4.2.2. *IFPO, 2E:* Chapter 2, page 25.

Question #18

It is not advisable to turn the steering wheel so that the wheels turn. Rather, the steering wheel should be turned until just before the wheels turn. The distance should not exceed 10 degrees in either direction. For a 20-inch steering wheel, that would be approximately 2 inches of movement. NFPA 1002 4.2.1. *IFPO, 2E:* Chapter 2, pages 35 – 36.

Question #19

Safety should be considered when conducting preventive maintenance inspections and tests. Common safety considerations for preventive maintenance inspections and tests include:

- General safety considerations:
  - Not hurrying or rushing through inspections
  - Ensuring work area is free from hazards
  - Always keeping safety in mind
- Specific safety considerations:
  - Checking for loose equipment before raising a tilt cab
  - Not smoking around engine compartment and fuels
  - Wearing appropriate clothing (no loose jewelry, safety glasses, gloves)
  - Considering vapor and electrical hazards
  - Always being careful when opening the radiator cap
  - Using proper tools
  - Securing all equipment and closing all doors prior to moving the apparatus

NFPA 1002 4.2.1. *IFPO, 2E:* Chapter 2, page 37.

Question #20

The only action that will most likely help regain traction and control is removing the foot from the accelerator. This allows the engine to slow the vehicle. NFPA 1002 4.3.1. *IFPO, 2E:* Chapter 3, page 56.

Question #21

Although fire department vehicle accidents occur for a variety of reasons, several common factors that appear in the vast majority of accidents include:

- Not following laws and standards related to emergency response

- Not being fully aware of both driver and apparatus limitations

- Lack of appreciation for driving conditions such as weather and traffic

Not being certified to NFPA 1002 has not been linked as a common factor associated with emergency-vehicle accidents. NFPA 1002 4.3. *IFPO, 2E:* Chapter 3, pages 43 – 44.

Question #22

Total stopping distance is measured from the time a hazard is detected until the vehicle comes to a complete stop. Total stopping distance consists of:

- *Perception distance* (distance apparatus travels from the time the hazard is seen until the brain recognizes it as a hazard)

- *Reaction distance* (distance apparatus travels from the time the brain sends the message to depress the brakes until the brakes are depressed)

- *Braking distance* (distance of travel from the time the brake is depressed until the vehicle comes to a complete stop)

NFPA 1002 4.3. *IFPO, 2E:* Chapter 3, pages 55 – 65.

Question #23

Several driving exercises are typically used to assess the ability to safely operate and control the vehicle. These include the following:

- *Alley dock.* Assesses the ability to back the vehicle into a restricted area, such as a fire station or down an alley.

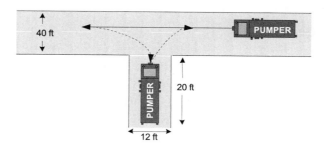

- *Serpentine.* Assesses the ability to drive around obstacles, such as parked cars and tight corners.

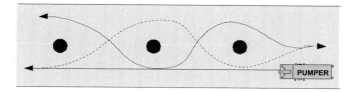

- *Confined-space turnaround.* Assesses the ability to turn the vehicle around within a confined space, such as a narrow street or driveway.

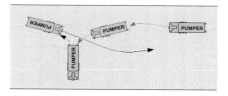

- *Diminishing clearance.* Assesses the ability to drive the vehicle in a straight line, such as on a narrow street or road.

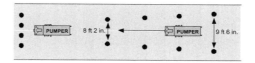

NFPA 1002 4.3.2, 4.3.3, 4.3.4, 4.3.5. *IFPO, 2E:* Chapter 3, pages 65 – 66.

Question #24

According to NFPA 1901, a pump must have a rated capacity as follows:

- 100% of its rated capacity at 150 psi
- 70% of its rated capacity at 200 psi
- 50% of its rated capacity at 250 psi

NFPA 1002 5.2.1, 5.2.2. *IFPO, 2E:* Chapter 4, page 91.

Question #25

- In *volume mode*, each individual impeller will add the flow it generates to the total discharge, with the pressure remaining constant among the impellers.
- In *pressure mode*, each subsequent impeller pumps the same flow from the previous impeller while adding the pressure it generates.

NFPA 1002 5.2.1, 5.2.2. *IFPO, 2E:* Chapter 4, pages 89 – 90.

Question #26

- Pumps connected to a split-shaft transmission are usually located midship or aft.
- Pumps connected directly to the crankshaft are usually located at the front of the engine.
- Pumps connected to a PTO are usually mounted at the front or midship. Pumps are usually smaller when powered by a PTO.

NFPA 1002 5.2.1, 5.2.1. *IFPO, 2E:* Chapter 4, pages 92 – 93.

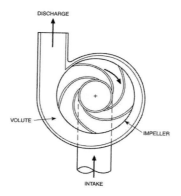

Question #27

The impeller (#2) is traveling in a clockwise rotation. NFPA 1002 5.2.1, 5.2.2. *IFPO, 2E:* Chapter 4, pages 86 – 89.

Question #28

NFPA 1901 sets minimum gauge sizes, requires intake relief valves, and suggests a standard color code to match discharges with their gauges. NFPA 1002 5.2.1, 5.2.2. *IFPO, 2E:* Chapter 5, page 102.

Question #29

Indicators are loosely categorized as all instrumentation on the pump panel other than pressure gauges or flow meters. NFPA 1002 5.2.1, 5.2.2. *IFPO, 2E:* Chapter 5, page 114.

Question #30

Several ways to determine when a pump is primed include the following indicators:

- A positive reading on the pressure gauge
- Priming motor sounds as if it is slowing
- Main pump sounds as if it is under load
- Oil/water discharge under the vehicle (not air/oil)

NFPA 1002 5.2.1, 5.2.2. *IFPO, 2E:* Chapter 5, page 121.

Question #31

This control device operates and set the pressure relief device. NFPA 1002 5.2.1, 5.2.2. *IFPO, 2E:* Chapter 5, pages 122 – 123.

Question #32

As a rule of thumb, a pump will have one 2 ½-inch discharge for each 250 gpm of rated capacity. A 1,000 gpm pumper will usually have four 2 ½-inch discharges, 4 × 250 = 1,000. NFPA 1002 5.2.1, 5.2.2. *IFPO, 2E:* Chapter 5, pages 118 – 119.

Question #33

Intake (supply) hose must have a minimal rating of 185 psi. Each of the hoses listed meets this requirement except unlined cotton-jacketed hose. NFPA 1002 5.2.1, 5.2.2, 5.2.4. *IFPO, 2E:* Chapter 6, pages 139 – 140.

Question #34

Appliances are accessories and components used to support varying hose configurations and include wyes, siamese, adapters, and double males/females. NFPA 1002 5.2.1, 5.2.2, 5.2.4. *IFPO, 2E:* Chapter 6, pages 147 – 148.

Question #35

The two basic parts of an eductor are the *metering valve* (controls the percentage of foam drawn into the eductor) and *pickup hose* (noncollapsable tube used to move the foam to the eductor). NFPA 1002 5.2.3. *IFPO, 2E:* Chapter 6, pages 151 – 152.

Question #36

When correct nozzle pressures are not provided, the following can occur:

- Insufficient nozzle pressure:
    - Less flow
    - Reduced reach
    - Poor pattern development
- Excessive nozzle pressure:
    - Poor pattern development
    - Excessive nozzle reaction

NFPA 1002 5.2.1. *IFPO, 2E:* Chapter 6, page 156.

Question #37

A pumper with a 1,000-gallon tank flowing 125 gpm will last about 8 minutes.

$$125\text{gallons per minute} \overline{)\ 1000\text{gallons}} = 8 \text{ minutes}$$

NFPA 1002 5.2.1. *IFPO, 2E:* Chapter 7, page 167.

Question #38

Hydrant flow testing equipment includes:

- Pressure gauge mounted on an outlet cap (calibrated within the past 12 months)
- Pitot gauge for each hydrant
- Hydrant diffuser
- Hydrant wrenches
- Portable radios

A fire department pumper is not required when conducting hydrant flow testing. NFPA 1002 5.2.1. *IFPO, 2E:* Chapter 7, pages 180 – 181.

Question #39

One formula to determine the distance between pumpers in a relay operation is:

(PDP – 20) × 100/FL

PDP = pump discharge pressure

20 = reserved intake pressure at the next pump

100 = length of one section of hose

FL = friction loss per 100-foot section of hose

NFPA 1002 5.2.1, 5.2.2. *IFPO, 2E:* Chapter 7, page 194.

Question #40

Shuttle flow capacity is the volume of water that can be pumped without running out of water.
NFPA 1002 5.2.1, 5.2.2. *IFPO, 2E:* Chapter 7, pages 200 – 201.

Question #41

One gallon of water weighs approximately 8.35 lbs. 1,000 gallons × 8.35 lbs = 8,350 lbs.
NFPA 1002 5.2.1. *IFPO, 2E:* Chapter 7, page 195.

Question #42

The pump is usually engaged before securing a water supply.

Step 1   Position apparatus, set parking brake, and let engine return to idle

Step 2   Engage the pump

Step 3   Provide water to intake side of pump (on-board, hydrant, draft)

Step 4   Set transfer valve (if so equipped)

Step 5   Open discharge lines

Step 6   Throttle to desired pressure

Step 7   Set the pressure-regulating device

Step 8   Maintain appropriate flows and pressures

NFPA 1002 5.2.1, 5.2.2, 5.2.4. *IFPO, 2E:* Chapter 8, page 207.

Question #43

Controls should always be opened, closed, and turned *slowly.*

Several caveats that come close to being universal among departments and manufacturers for the operation of centrifugal pumps include:

- Never operate the pump without water.

- Always keep water moving when operating the pump at high speeds.

- Never open, close, or turn controls abruptly.

- Always maintain awareness of instrumentation during pumping operations.

- Never leave the pump unattended.

- Always maintain constant vigilance to safety.

NFPA 1002 5.2.1, 5.2.2, 5.2.4. *IFPO, 2E:* Chapter 8, page 208.

Question #44

Transfer valve operation rule of thumb:

- Use *volume mode* when flows are greater than 50% of a pump's rated capacity and pressures are less than 150 psi.

- Use *pressure mode* when flows are less than 50% of a pump's rated capacity and pressures are greater than 150 psi.

NFPA 1002 5.2.1, 5.2.2, 5.2.4. *IFPO, 2E:* Chapter 8, pages 214 – 215.

Question #45

Because of the limited supply of water in the on-board tank, the transfer valve should be in pressure (series) mode. NFPA 1002 5.2.1, 5.2.2, 5.2.4. *IFPO, 2E:* Chapter 8, page 220.

Question #46
The three methods used to transfer power to the pump are:

- PTO

- Front crankshaft

- Split-shaft

NFPA 1002 5.2.1, 5.2.2, 5.2.4. *IFPO, 2E:* Chapter 8, pages 210 – 212.

Question #47
Pumps connected to the front crankshaft are typically mounted on the front of the apparatus. NFPA 1002 5.2.1, 5.2.2, 5.2.4. *IFPO, 2E:* Chapter 8, page 211.

Question #48
All the answer selections may be signs that a pump is cavitating, except the priming device. The priming device would not be operating when conditions are conducive to cavitation. NFPA 1002 5.1.1. *IFPO, 2E:* Chapter 9, pages 245 – 247.

Question #49
Cavitation is caused by insufficient intake flow to match the discharge flow. Not operating enough discharge lines would not cause cavitation, in that the pump has excess capacity to pump and adequate intake flow. NFPA 1002 5.1.1. *IFPO, 2E:* Chapter 9, pages 245 – 246.

Question #50
Supporting a sprinkler system may require the following:

- Connecting at least one line to the FD connection

- Pumping at 150 psi unless otherwise directed

- Pumping immediately when fire and smoke are visible

- Not shutting down the sprinkler system for improved visibility

- Placing transfer valve into the volume mode

- Ensuring that water supply to the pump does not reduce the sprinkler system water supply

NFPA 1002 5.2.4. *IFPO, 2E:* Chapter 9, pages 252 – 253.

Question #51
Standpipe systems are classified based on intended use as follows:

- *Class 1* standpipes provide 2 ½-inch connections for trained firefighters/fire brigades and have an initial flow rate of 500 gpm.

- *Class 2* standpipes provide 1 ½-inch connections for initial attack; they have a minimum flow rate of 100 gpm.

- *Class 3* standpipes provide 1 ½-inch and 2 ½-inch connections for trained firefighters/fire brigades; they have an initial flow rate of 500 gpm.

NFPA 1002 5.2.4. *IFPO, 2E:* Chapter 9, page 253.

Question #52
Increased pressure reduces the ability of a liquid to evaporate (increase the boiling point). Decreased pressure increases the ability of a liquid to evaporate (lowers the boiling point). NFPA 1002 5.2.1. *IFPO, 2E:* Chapter 9, page 245.

Question #53

- *Dry-pipe systems* maintain air or compressed gas under pressure within the system. When a head fuses, water enters the system and discharges through any fused heads.

- *Pre-action systems* are similar to dry-pipe systems in that air or compressed gas is maintained in the system. However, an automatic detection system (smoke, heat, flame, etc.) or manual system (pull box) must operate to allow water to enter the system. At this point, it is similar to a wet-pipe system; when a sprinkler head fuses, water discharges.

- *Deluge systems* maintain all sprinkler heads in an open position. When a detection system operates, water enters the system and is discharged through all the open heads.

- *Wet-pipe systems* maintain water in the system at all times. When a sprinkler head is fused, water immediately discharges through the fused heads.

NFPA 1002 5.2.4. *IFPO, 2E:* Chapter 9, pages 148 – 253.

Question #54

- *Dry-pipe systems* maintain air or compressed gas under pressure within the system. When a head fuses, water enters the system and discharges through any fused heads.

- *Pre-action systems* are similar to dry-pipe systems in that air or compressed gas is maintained in the system. However, an automatic detection system (smoke, heat, flame, etc.) or manual system (pull box) must operate to allow water to enter the system. At this point, it is similar to a wet-pipe system; when a sprinkler head fuses, water discharges.

- *Deluge systems* maintain all sprinkler heads in an open position. When a detection system operates, water enters the system and is discharged through all the open heads.

- *Wet-pipe systems* maintain water in the system at all times. When a sprinkler head is fused, water immediately discharges through the fused heads.

NFPA 1002 5.2.4. *IFPO, 2E:* Chapter 9, pages 148 – 253.

Question #55

- *Hydraulics* is that branch of science dealing with the principles of fluid at rest or in motion.

- *Hydrodynamics* is that branch of hydraulics that deals with the principles and laws of fluids in motion.

- *Hydrostatics* is that branch of hydraulics that deals with the principles and laws of fluids at rest and the pressures they exert or transmit.

NFPA 1002 5.2. *IFPO, 2E:* Chapter 10, page 261.

Question #56

The basic characteristics and properties of water are listed below. These are considered approximate in that water purity, atmospheric conditions, and rounding can affect the values.

- Virtually incompressible

- Freezes at 32°F

- Expands when frozen

- Boils at 212°F

- 1 gallon weighs 8.34 lbs

- Density is 62.4 lb/ft^3

- Number of gallons in 1 cubic foot is 7.38 gallons/ft^3

NFPA 1002 5.2.1, 5.2.2. *IFPO, 2E:* Chapter 10, pages 262 – 263.

Question #57

- *Latent heat of fusion* is the amount of heat that is absorbed by a substance when changing from a solid to a liquid state.
- *Specific heat* is the amount of heat required to raise the temperature of a substance by 1°F. The specific heat of water is 1 Btu/lb.
- *Latent heat of vaporization* is the amount of heat absorbed when changing from a liquid to a vapor state.
- *Conduction* is the transfer of heat through a medium by direct contact.

NFPA 1002 5.2.1, 5.2.2. *IFPO, 2E:* Chapter 10, pages 263 – 264.

Question #58

Pressure is the force exerted by a substance in *units of weight per area*, typically expressed in *pounds per square inch* (psi or lb/in^2).

P = Pressure (psi)
F = Force (weight in pounds)
A = Area (square inches)

NFPA 1002 5.2.1, 5.2.2. *IFPO, 2E:* Chapter 10, pages 269 – 270.

Question #59

To calculate weight, the formula $W = D \times V$ is used. NFPA 1002 5.2. *IFPO, 2E:* Chapter 10, page 265.

Question #60

There are 7.48 gallons of water in 1 cubic foot. NFPA 1002 5.2. *IFPO, 2E:* Chapter 10, page 267.

Question #61

One cubic foot of water weights 62.4 lbs and contains 144 cubic-inch columns.
62.4 pounds per cubic foot divided by 144 cubic inch columns of water = .433 lbs per cubic inch column of water. NFPA 1002 5.2. *IFPO, 2E:* Chapter 10, page 273.

Question #62

The operating pressure for smooth-bore nozzles on hand-lines is 50 psi. The square root of 50 is 7.071. An acceptable value for fireground calculations is 7. NFPA 1002 5.2.4. *IFPO, 2E:* Chapter 10, pages 296 – 298.

Question #63

Iowa State formula: $NF = \dfrac{V}{100}$

where   NF = needed flow in gpm
        V = volume of the area in cubic feet
        100 = is a constant in ft^3/gpm

NFA formula: $NF = \dfrac{A}{3}$

where   NF = needed flow in gpm
        A = area of the structure in square feet
        3 = constant in ft^2/gpm

NFPA 1002 5.2.4. *IFPO, 2E:* Chapter 10, pages 292 – 293.

Question #64

NFA formula: $NF = \dfrac{A}{3}$

where   NF = needed flow in gpm
A = area of the structure in square feet
3 = constant in ft^2/gpm

Iowa State formula: $NF = \dfrac{V}{100}$

where   NF = needed flow in gpm
V = volume of the area in cubic feet
100 = is a constant in ft^3/gpm

NFPA 1002 5.2.4. *IFPO, 2E:* Chapter 10, pages 292 – 293.

Question #65

A gauge that measures psia at sea level would have a reading of 14.7 psia because psia (absolute pressure) includes atmospheric pressure.

A gauge reading of 100 psi (psig) is actually 114.7 psia because psig (gauge pressure) will read 0 psi at sea level. NFPA 1002 5.2. *IFPO, 2E:* Chapter 10, page 275.

Question #66

$NR = 1.57 \times d^2 \times NP$

where   NR = nozzle reaction
1.57 = constant
d = diameter of nozzle orifice in inches
NP = operating nozzle pressure in psi

NFPA 1002 5.2. *IFPO, 2E:* Chapter 10, pages 282 – 283.

Question #67

The drop-ten method simply subtracts 10 from the first two numbers of gpm flow. It is not as accurate as other methods, but provide a simple rule of thumb for fireground use.

200 gpm = 20 – 10 = 10 psi friction loss per 100-foot sections of 2 ½-inch hose

NFPA 1002 5.2.1, 5.2.2. *IFPO, 2E:* Chapter 10, page 303.

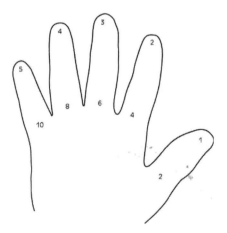

Question #68

The hand method is a fireground method used to estimate friction loss in 100-foot sections of 2 ½-inch hose. Simply select the fingertip representing the gpm flow and then multiply the two figures on the finger for the approximate friction loss pressure for each 100-foot section of 2 ½-inch hose.

200 gpm = 2 × 4 = 8 psi per 100-foot section of 2 ½-inch hose

NFPA 1002 5.2.1, 5.2.2. *IFPO, 2E:* Chapter 10, pages 302 – 303.

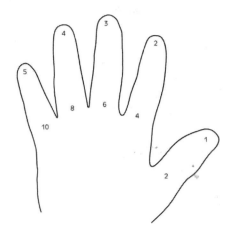

Question #69
The hand method is a fireground method used to estimate friction loss in 100-foot sections of 2 ½-inch hose. Simply select the fingertip representing the gpm flow and then multiply the two figures on the finger for the approximate friction loss pressure for each 100-foot section of 2 ½-inch hose.

400 gpm = 4 × 8 = 32 psi per 100-foot section of 2 ½-inch hose

NFPA 1002 5.2.1, 5.2.2. *IFPO, 2E:* Chapter 10, pages 302 – 303.

Question #70
Based on the percent drop in pressure, additional flows may be available from a hydrant as follows:

- 0% – 10% drop three times the original flow
- 11% – 15% drop two times the original flow
- 16% – 25% drop one time the original flow

NFPA 1002 5.2. *IFPO, 2E:* Chapter 10, page 277.

Question #71
$$NR = 1.57 \times d^2 \times NP$$

$$NR = 1.57 \times .5^2 \times 50$$

$$NR = 1.57 \times 0.25 \times 50$$

$$NR = 19.625 \text{ or } 20 \text{ lb}$$

NFPA 1002 5.2. *IFPO, 2E:* Chapter 10, pages 282 – 283.

Question #72
$$NR = gpm \times \sqrt{NP} \times 0.0505$$

$$NR = 95 \times \sqrt{100} \times 0.0505$$

$$NR = 95 \times 10 \times 0.0505$$

$$NR = 47.97 \text{ or } 48 \text{ lb}$$

NFPA 1002 5.2. *IFPO, 2E:* Chapter 10, pages 282 – 283.

Question #73
The drop-ten method simply subtracts 10 from the first two numbers of gpm flow. It is not as accurate as other methods, but provides a simple rule of thumb for fireground use.

250 gpm = 25 – 10 = 15 psi friction loss per 100-foot sections of 2 ½-inch hose

NFPA 1002 5.2.1, 5.2.2. *IFPO, 2E:* Chapter 10, page 303.

Question #74

Elevation formula by floor level:

$EL = 5 \times H$

where   EL = the gain or loss of elevation in psi

5 = gain or loss in pressure for each floor level

H = height in number of floor levels above or below the pump

Elevation formula in feet:

$EL = 0.5 \times H$

where   EL = the gain or loss of elevation in psi

0.5 = pressure exerted at base of 1-cubic-inch column of water 1 foot high

H = height in feet above or below the pump

NFPA 1002 5.2.4. *IFPO, 2E:* Chapter 11, page 311.

Question #75

Improvements in hose construction and larger diameter hose effectively reduced friction loss within the hose, creating the need for more accurate friction loss formulas. The more accurate formula for 3-inch hose is $FL = q^2$. NFPA 1002 5.2.1, 5.2.2. *IFPO, 2E:* Chapter 11, pages 304 – 305.

Question #76

The Iowa State formula can be used for an entire structure or a section of the structure, and is based on volume. NFPA 1002 5.2. *IFPO, 2E:* Chapter 11, pages 292 – 293.

Question #77

cq^2L

$24 \times \left(\dfrac{100}{100}\right)^2 \times \left(\dfrac{100}{100}\right)$

$24 \times (1)^2 \times (1)$

$24 \times 1 \times 1$

24 psi friction loss

NFPA 1002 5.2. *IFPO, 2E:* Chapter 11, pages 306 – 307.

Question #78

cq^2L

$2 \times \left(\dfrac{250}{100}\right)^2 \times \left(\dfrac{100}{100}\right)$

$2 \times (2.5)^2 \times (1)$

$2 \times 6.25 \times 1$

12.5 psi friction loss (rounded to 13 psi)

NFPA 1002 5.2. *IFPO, 2E:* Chapter 11, pages 306 – 307.

Question #79

$$2q^2 + q$$

$$2 \times \left(\frac{100}{100}\right)^2 + \left(\frac{100}{100}\right)$$

$$2 \times (1)^2 + (1)$$

$$2 \times 1 + 1$$

3 psi friction loss

NFPA 1002 5.2. *IFPO, 2E:* Chapter 11, pages 304 – 305.

Question #80

$$2q^2 + q$$

$$2 \times \left(\frac{250}{100}\right)^2 + \left(\frac{250}{100}\right)$$

$$2 \times (2.5)^2 + (2.5)$$

$$2 \times 6.25 + 2.5$$

15 psi friction loss

NFPA 1002 5.2. *IFPO, 2E:* Chapter 11, pages 304 – 305.

Question #81

$$cq^2 L$$

$$.2 \times \left(\frac{500}{100}\right)^2 \times \left(\frac{100}{100}\right)$$

$$.2 \times (5)^2 \times (1)$$

$$.2 \times 25 \times 1$$

5 psi friction loss

NFPA 1002 5.2. *IFPO, 2E:* Chapter 11, pages 306 – 307.

Question #82

$$q^2$$

$$\left(\frac{300}{100}\right)^2$$

$$(3)^2$$

9 psi friction loss

NFPA 1002 5.2. *IFPO, 2E:* Chapter 11, pages 306 – 307.

Question #83

Elevation formula by floor level:

$EL = 5 \times H$

where EL = the gain or loss of elevation in psi

5 = gain or loss in pressure for each floor level

H = height in number of floor levels above or below the pump

$EL = 5 \times 5$

$EL = 25$ psi

NFPA 1002 5.2. *IFPO, 2E:* Chapter 11, page 311.

Question #84

Elevation formula in feet:

$EL = 0.5 \times H$

where EL = the gain or loss of elevation in psi

0.5 = pressure exerted at base of 1-cubic-inch column of water 1 foot high

H = height in feet above or below the pump

$EL = 0.5 \times 50$

$EL = 25$ psi

NFPA 1002 5.2. *IFPO, 2E:* Chapter 11, page 311.

Question #85

Elevation formula in feet:

$EL = 0.5 \times H$

where EL = the gain or loss of elevation in psi

0.5 = pressure exerted at base of 1-cubic-inch column of water 1 foot high

H = height in feet above or below the pump

$EL = 0.5 \times -75$

$EL = -37.5$ psi

NFPA 1002 5.2. *IFPO, 2E:* Chapter 11, page 311.

Question #86

Elevation formula by floor level:

$EL = 5 \times H$

where EL = the gain or loss of elevation in psi

5 = gain or loss in pressure for each floor level

H = height in number of floor levels above or below the pump

$EL = 5 \times -2$

$EL = -10$ psi

NFPA 1002 5.2. *IFPO, 2E:* Chapter 11, page 311.

Question #87

q^2

$\left(\dfrac{250}{100}\right)^2$

$(2.5)^2$

6.25 psi friction loss

NFPA 1002 5.2. *IFPO, 2E:* Chapter 11, pages 306 – 307.

Question #88

$Q = 30 \times d^2 \times \sqrt{NP}$

$Q = 30 \times 0.75^2 \times \sqrt{50}$

$Q = 30 \times 0.56 \times 7$

$Q = 117.6$ or 118 gpm

NFPA 1002 5.2. *IFPO, 2E:* Chapter 11, pages 296 – 298.

Question #89

$Q = 30 \times d^2 \times \sqrt{NP}$

$Q = 30 \times 0.75^2 \times \sqrt{80}$

$Q = 30 \times 0.56 \times 8.94$

$Q = 150$ gpm

NFPA 1002 5.2. *IFPO, 2E:* Chapter 11, pages 296 – 298.

Question #90

$PDP = NP + FL + AFL \pm EL$

where PDP = pump discharge pressure
       NP = nozzle pressure
       FL = friction loss in hose
       AFL = appliance friction loss
       EL = elevation gain or loss

NFPA 1002 5.2. *IFPO, 2E:* Chapter 12, pages 317 – 319.

Question #91

$PDP = NP + FL + AFL \pm EL$

where PDP = pump discharge pressure
       NP = nozzle pressure
       FL = friction loss in hose
       AFL = appliance friction loss
       EL = elevation gain or loss

$PDP = 50$ psi $+ 32$ psi $+ 5$ psi
$PDP = 87$ psi

NFPA 1002 5.2. *IFPO, 2E:* Chapter 12, pages 317 – 319.

Question #92

$PDP = NP + FL + AFL \pm EL$

where PDP = pump discharge pressure
       NP = nozzle pressure
       FL = friction loss in hose
       AFL = appliance friction loss
       EL = elevation gain or loss

$PDP = 100$ psi $+ 47$ psi
PDP $= 147$ psi

NFPA 1002 5.2. *IFPO, 2E:* Chapter 12, pages 317 – 319.

Question #93

Centrifugal force is the tendency of an object, when moving in a circular pattern, to move outward from the center. NFPA 1002 4.3.1, 4.3.6. *IFPO, 2E:* Chapter 3, pages 56 – 57.

Question #94

Total stopping distance is measured from the time a hazard is detected until the vehicle comes to a complete stop. Total stopping distance consists of:

- *Perception distance* (distance apparatus travels from the time the hazard is seen until the brain recognizes it as a hazard)
- *Reaction distance* (distance apparatus travels from the time the brain sends the message to depress the brakes until the brakes are depressed)
- *Braking distance* (distance of travel from the time the brake is depressed until the vehicle comes to a complete stop)

NFPA 1002 4.3. *IFPO, 2E:* Chapter 3, pages 55 – 65.

Question #95

The "W" after the number in the SAE oil classification means the oil is rated for flow at 0°F. NFPA 1002 4.2.1, 5.1.1. *IFPO, 2E:* Chapter 2, pages 32 – 33.

Question #96

Flow meters are increasingly finding their way onto pump panels because they reduce fireground hydraulic calculations and, with the use of "smart" electronics, they are easy to calibrate and increase the versatility of pump-panel location. NFPA 1002 5.2. *IFPO, 2E:* Chapter 5, page 113.

Question #97

Fire department connections are usually a siamese appliance with two 2 ½-inch clapper inlets. NFPA 1002 5.2.4. *IFPO, 2E:* Chapter 9, page 252.

Question #98

A siamese is used to combine two or more lines into one line. NFPA 1002 5.2. *IFPO, 2E:* Chapter 6, page 149.

Question #99

Laminar flow usually causes minimal friction loss, whereas each of the other items causes turbulent flow and increases friction loss. NFPA 1002 5.2. *IFPO, 2E:* Chapter 10, pages 179 – 282.

Question #100

Feathering or gating is the process of slowly opening/closing discharge control valves to adjust discharge settings. NFPA 1002 5.2. *IFPO, 2E:* Chapter 9, pages 237 – 238.

Phase Two, Exam Two

1. Compared to suppression activities, fire pump operations are more complex and unpredictable.
   a. True
   b. False

2. Preventive maintenance refers to the duty of ensuring that the apparatus, pump, and related components are in a ready state and peak operating efficiency.
   a. True
   b. False

3. In general, the fire department mechanic is ultimately responsible for the establishment, implementation, and monitoring of a preventive maintenance program.
   a. True
   b. False

4. The pump operator will replace a vehicle battery in most departments.
   a. True
   b. False

5. Changing the oil and replacing the oil filter is usually performed by the fire department mechanic.
   a. True
   b. False

6. Most state laws exempt emergency-vehicle drivers from obeying the same laws as other vehicle operators.
   a. True
   b. False

7. NFPA 1002 requires the following warning on pump panels:

   "**WARNING:** Death or serious injury might occur if proper operating procedures are not followed. The pump operator as well as individuals connecting supply or discharge hoses to the apparatus must be familiar with water hydraulics hazards and component limitations."
   a. True
   b. False

8. External pressure applied to a confined liquid (fluid) is transmitted equally throughout the liquid.
   a. True
   b. False

9. Pressure exerted at the bottom of a container is dependent on the shape or volume of the container.
   a. True
   b. False

10. Apparatus coolant levels should never be checked when the apparatus is hot.
    a. True
    b. False

11. In general, a larger-diameter supply hose will provide more friction loss over longer distances than will a smaller supply hose.
    a. True
    b. False

12. According to NFPA 1002, driver operators must meet which of the following requirements?
    a. Firefighter II per NFPA 1001; medically fit; at least 21 years old
    b. Firefighter I per NFPA 1001; licensed for vehicles they operate; medically fit
    c. license for vehicles they operate; at least 21 years of age; Firefighter II per NFPA 1001
    d. Firefighter I per NFPA 1001; licensed for vehicles they operate; earned a high school diploma or GED

13. _____ are _____ that are not legally binding and enforceable.
    a. Laws, rules
    b. Laws, guidelines
    c. Standards, guidelines
    d. Standards, ordinances

14. Tests are conducted to determine the _____ of components, while servicing activities are conducted to help maintain the vehicle in peak operating _____.
    a. performance, condition
    b. status, performance
    c. performance, status
    d. status, condition

15. All of the following are legitimate reasons for documenting problems found during preventive maintenance inspections *except*:
    a. It is required by NFPA 1500.
    b. It may help to determine trends.
    c. It helps keep track of needed maintenance and repairs.
    d. It helps ensure apparatus are kept clean.

16. NFPA 1901 requires that fuel tanks on apparatus must be sufficient in size to drive the pump for at least _____ hours at its rated capacity when pumping at draft.
    a. 1 hour
    b. 2 hours
    c. 2 ½ hours
    d. 4 hours

17. Safety should be considered when conducting preventive maintenance inspections and testing. Each of the following could be considered a common safety consideration for preventive maintenance inspections and tests *except*:
    a. Do not wear loose clothing or jewelry.
    b. Check for loose equipment before raising a tilt cab.
    c. Ensure work area is free from hazards and always keep safety in mind.
    d. Take vapor readings in the engine compartment before testing spark plugs.

18. Which of the following NFPA standards requires that pump operators be able to conduct and document routine tests, inspections, and servicing functions to ensure the apparatus is in a ready state?
    a. NFPA 1002
    b. NFPA 1500
    c. NFPA 1901
    d. NFPA 1911

19. Regaining the control of a vehicle during a skid can be accomplished by
    a. disengaging the clutch.
    b. turning the front wheels in the opposite direction of the skid.
    c. taking the foot off the accelerator.
    d. increasing speed slightly.

20. Several common factors associated with the vast majority of apparatus accidents include all the following *except*
    a. not being certified to NFPA 1002.
    b. not following laws and standards related to emergency response.
    c. not being fully aware of both driver and apparatus limitations.
    d. lack of appreciation for driving conditions such as weather and traffic.

21. The distance of travel from the time the brake is depressed until the vehicle comes to a complete stop is called
    a. perception distance.
    b. reaction distance.
    c. braking distance.
    d. total stopping distance.

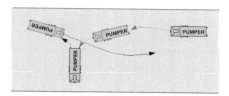

22. The driving exercise shown below is called
    a. diminishing clearance.
    b. alley dock.
    c. confined-space turnaround.
    d. serpentine.

23. Fill in the missing rated capacity information:
    _____% at 150 psi
    70% at _____ psi
    _____% at 250 psi
    a. 100%, 220 psi, 50%
    b. 100%, 200 psi, 50 %
    c. 100%, 175 psi, 50%
    d. 50%, 200 psi, 100%

24. A two-stage centrifugal pump is discharging 500 gpm at 250 psi in pressure mode. The first impeller will discharge 500 gpm at 125 psi and the second impeller will discharge
    a. 500 gpm at 125 psi.
    b. 500 gpm at 250 psi.
    c. 250 gpm at 125 psi.
    d. 250 gpm at 250 psi.

25. A 1,250 gpm midship-mounted centrifugal pump would most likely be powered
    a. through a PTO.
    b. directly from the crankshaft.
    c. through a split shaft transmission.
    d. by any of these methods.

26. A 1,250 gpm pumper will usually have how many 2 ½-inch discharges?
    a. four
    b. five
    c. six
    d. seven

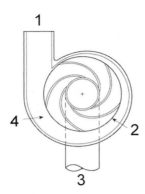

27. Identify the intake and discharge for this line diagram of a centrifugal pump.
    a. 1 – Discharge, 3 – Intake
    b. 1 – Intake, 2 – Discharge
    c. 2 – Intake, 4 – Discharge
    d. not enough information provided

28. Instrumentation on pump panels can be grouped into three categories. Which of the following is not one of the categories?
    a. pressure gauges
    b. indicators
    c. emergency warning lights
    d. flow meters

29. Each of the following is an example of a control valve commonly found on pumping apparatus *except* a
    a. ball.
    b. butterfly.
    c. ram.
    d. gated.

30. Several indicators that the pump is primed include each of the following *except*
    a. a positive reading on the pressure gauge.
    b. water/oil discharge under the vehicle.
    c. main pump sounds as if it is under load.
    d. priming motor sounds as if it is increasing.

31. Correctly label items 1 and 3.
    a. 1 – main discharge gauge
        3 – individual discharge gauge
    b. 1 – main intake gauge
        3 – individual discharge gauge
    c. 1 – main intake gauge
        3 – individual flow meter
    d. 1 – main discharge gauge
        3 – individual flow meter

32. Which of the following can be used as an attack hose, according to NFPA 1961?
    a. LDH
    b. 3 ½-inch hose
    c. 1 ¾-inch hose
    d. each of these can be use for an intake hose

33. Which NFPA standard requires fire departments to establish and maintain an accurate record of each hose section?
    a. NFPA 1961
    b. NFPA 1962
    c. NFPA 1963
    d. NFPA 1964

34. When excessive pressure is provided to a nozzle,
    a. nozzle reaction will remain constant.
    b. nozzle reaction will increase.
    c. a poor pattern will develop.
    d. only b and c are correct
    e. all are correct

35. A pumper with a 1,000-gallon tank is flowing 250 gpm through a 2 ½-inch hose line. How long will the on-board water supply last?
    a. 4 minutes
    b. 5 minutes
    c. 10 minutes
    d. no correct answer is provided

36. The equipment required to conduct hydrant flow testing includes all of the following except a
    a. fire department pumper.
    b. hydrant diffuser.
    c. pressure gauge mounted on an outlet cap (calibrated within the past 12 months).
    d. pitot gauge for each hydrant.

37. Each of the following factors contribute to the availability of water supply *except*
    a. flow.
    b. pressure.
    c. quantity.
    d. pump capacity.

38. A static hydrant is best described as a
    a. hydrant that is identified as being out of service.
    b. a wet- or dry-barrel hydrant in which no water is flowing.
    c. prepiped line that extends into a water source.
    d. no such hydrant exists

39. In a relay operation, the second pumper, all in-line pumpers, and the attack pumper should maintain at least _____ psi intake pressure.
    a. 10
    b. 20
    c. 30
    d. 50

40. The time it takes to fill a tanker, drive to the dump site, and empty its water is known as
    a. total shuttle time.
    b. shuttle flow capacity.
    c. shuttle cycle time.
    d. maximum tanker shuttle time.

41. The same basic steps must be taken to move water from the supply to the discharge point. What critical pump operation step is out of order?

    Step 1   Position apparatus, set parking brake, and let engine return to idle
    Step 2   Engage the pump
    Step 3   Provide water to intake side of pump (on-board, hydrant, draft)
    Step 4   Throttle to desired pressure
    Step 5   Open discharge lines
    Step 6   Set transfer valve (if so equipped)
    Step 7   Set the pressure regulating device
    Step 8   Maintain appropriate flows and pressures

    a. change step 4 with 6
    b. change step 6 with 7
    c. change step 2 with 3
    d. change step 7 with 8

42. Which of the following is an incorrect statement concerning pump operations?
    a. Always maintain constant vigilance to safety.
    b. Always maintain awareness of instrumentation during pumping operations.
    c. Always keep water moving when operating the pump at high speeds.
    d. Always operate the pump without water.

43. As a general rule of thumb, the transfer valve should be in pressure mode when
    a. flows are expected to be less than 50% of a pump's rated capacity.
    b. pressures are less than 150 psi.
    c. pressures are greater 150 psi.
    d. both a and b
    e. both a and c

44. The transfer valve should be _____ when supplying one discharge line using the on-board water tank.
    a. turned off
    b. turned on
    c. in volume mode
    d. in pressure mode

45. Pumps can receive power from the drive engine through each of the following *except* the
   a. transfer valve.
   b. split-shaft.
   c. power take-off (PTO).
   d. front crankshaft.

46. A pump connected to a split-shaft transmission is usually mounted
   a. on top of the drive engine.
   b. toward the back of the apparatus.
   c. in the middle of the apparatus.
   d. either in the middle or toward the back of the apparatus.

47. All of the following can cause cavitation *except*
   a. poor hydrant supply.
   b. excessive lift while drafting.
   c. intake lines that are too small.
   d. too few discharge lines operating.

48. Signs that might suggest cavitation may be occurring include all the following *except*
   a. engine speed automatically increases during a pumping operation.
   b. priming device slows down or seems to be operating under a load.
   c. no corresponding increase in pressure when the engine speed is increased.
   d. excessive pump vibrations or rattling sounds like sand or gravel going through the pump.

49. When connecting to a fire department connection, the pump operator should do all of the following *except*
   a. pump at 150 psi unless otherwise directed.
   b. connect at least one hose line.
   c. wait to start pumping until interior crews verify sprinkler activation, even when fire and smoke is visible.
   d. place transfer valve into the volume mode.

50. According to NFPA 14, Standard for the Installation of Standpipe and Hose Systems, there are three classes of standpipe systems. Which classification(s) is/are for use only by trained firefighters or fire brigades?
   a. all of them
   b. only Class 3 standpipes
   c. both Class 1 and Class 3
   d. only Class 2

51. Increased pressure will _____ the boiling point of water while decreased pressure will _____ the boiling point.
   a. increase, reduce
   b. reduce, increase
   c. initiate, stop
   d. stop, initiate

52. With OS&Y valves, if a large section of the stem is showing the valve is
   a. closed or mostly closed.
   b. open or mostly open.
   c. broken.
   d. No such valve exists.

53. _____ is the branch of hydraulics that deals with the principles and laws of fluids in motion while _____ deals with the principles and laws of fluids at rest and the pressures they exert or transmit.
    a. Hydraulics, hydrodynamics
    b. Hydrodynamics, hydrostatics
    c. Hydrostatics, hydrodynamics
    d. Hydraulics, hydrolastic

54. Which of the following physical characteristics and properties of water are incorrect?
    a. boils at approximately 212°F
    b. 1 gallon weighs 8.34 lbs
    c. density is 62.4 lb/ft^3
    d. number of gallons in 1 cubic foot is 8.34 gallons/ft^3

55. The amount of heat required to raise the temperature of a substance by 1°F refers to
    a. latent heat of fusion.
    b. specific heat.
    c. latent heat of vaporization.
    d. radiation.

56. The formula $F = P \times A$ is used to calculate
    a. pressure
    b. density.
    c. force.
    d. both a and c are correct

57. The weight of water for a given volume can be calculated using which of the following formulas?
    a. $W = \dfrac{V}{D}$
    b. $W = D \div V$
    c. $W = \dfrac{D}{V}$
    d. $W = D \times V$

58. The formula used to determine flow (gpm) when smooth-bore nozzles are used is $Q = 29.7 \times d^2 \times \sqrt{NP}$. Acceptable values for using the formula on the fireground are
    a. 30 for 29.7
       7 for $\sqrt{NP(handlines)}$
       9 for $\sqrt{NP(handlines)}$
    b. 30 for 29.7
       8 for $\sqrt{NP(handlines)}$
       10 for $\sqrt{NP(handlines)}$
    c. 30 for 29.7
       9 for $\sqrt{NP(handlines)}$
       7 for $\sqrt{NP(handlines)}$
    d. 2 for d^2
       7 for $\sqrt{NP(handlines)}$
       9 for $\sqrt{NP(handlines)}$

59. The $NF = \frac{A}{3}$ formula is called the
    a. National Fire Academy formula used to calculate needed flow.
    b. Iowa State formula used to calculate needed flow.
    c. needed foam formula.
    d. nozzle force formula.

60. Which of the following is the correct formula for calculating nozzle reaction for a combination nozzle?
    a. $NR = 1.57 \times d \times NP$
    b. $NR = gpm \times \sqrt{NP} \times .0505$
    c. $NR = gpm \times \sqrt{NP \times .0505}$
    d. $NR = 1.57 \times d^2 \times NP$

61. Two methods are used on the fireground to calculate pressure changes as the result of changes in elevation. The elevation formula using floor levels is expressed as
    a. $EL = 5 \times H$, where H = height in number of floor levels above or below the pump.
    b. $EL = 5 \times H$, where H = head pressure.
    c. $EL = 0.5 \times H$, where H = distance in feet above or below the pump.
    d. $EL = 0.5 \times H$, where H = head pressure.

62. In the formula $PDP = NP + FL + AFL \pm EL$, which of the following is not correct?
    a. AFL = appliance friction loss
    b. EL = elevation gain or loss
    c. FL = friction loss and can be calculated using a number of estimation methods and fireground formulas
    d. NP = either 50 psi for smooth-bore nozzles or 100 psi combination nozzles

63. During pump operations, when discharge lines are closed or flows reduced, the water temperature within the pump can quickly rise. To compensate for this, the pump operator should
    a. open the tank-to-pump and the pump-to tank valves to recirculate water from the discharge side to the tank and then back to the intake side of the pump.
    b. flow water from an unused discharge.
    c. use the auxiliary cooling system.
    d. both a and b are correct

64. The Iowa State formula for calculating needed flow is expressed as
    a. $NF = \frac{V}{100}.$

    b. $NF = \frac{V}{100} \times 3.$

    c. $NF = \frac{A}{3}.$

    d. $NF = \frac{A}{3} \div 100.$

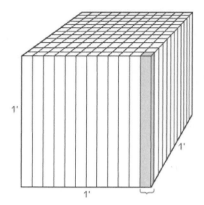

65. How many 1-square-inch columns are in 1 cubic foot?
    a. 12
    b. 120
    c. 144
    d. 62.4

66. A gauge reading of 100 psi is equivalent to a gauge reading of _____ at sea level.
    a. 85.3 psia
    b. 114.7 psia
    c. 100 psia
    d. none of the answers are correct

67. During pump operations, the drop in hydrant pressure from static to residual can be used to estimate the additional flow the hydrant is capable of providing. A 20% or less drop means the hydrant may be able to deliver as much as
    a. three times the original flow.
    b. two times the original flow.
    c. one time the original flow.
    d. half the original flow.

68. The Iowa State formula estimates needed flow
    a. for an entire structure.
    b. for a section of a structure.
    c. based on the square footage of the involved area.
    d. both a and b are correct

69. Using the hand method, calculate friction loss in 100-foot sections of 2 ½-inch hose flowing 500 gpm.
    a. 10 psi
    b. 15 psi
    c. 25 psi
    d. 50 psi

70. Using the drop-ten method, calculate friction loss in 100-foot sections of 2 ½-inch hose flowing 350 gpm.
    a. 250 psi
    b. 35 psi
    c. 25 psi
    d. 15 psi

71. Calculate the friction loss for a 100-foot section of 2 ½-inch hose flowing 300 gpm, using the formula cq^2L (c = 2).
    a. 2 psi
    b. 3 psi
    c. 6 psi
    d. 18 psi

72. Nozzle reaction for smooth-bore nozzles can be calculated using which of the following?
    a. $NR = 1.57 \times d \times NP$
    b. $NR = gpm \times \sqrt{NP} \times .0505$
    c. $NR = gpm \times \sqrt{NP \times .0505}$
    d. $NR = 1.57 \times d^2 \times NP$

73. Calculate the friction loss for a 100-foot section of 2 ½-inch hose flowing 300 gpm using the formula $2q^2 + q$.
    a. 5 psi
    b. 16 psi
    c. 21 psi
    d. 36 psi

74. Using the formula cq^2L, calculate the friction loss for a 100-foot section of 1 ½-inch hose flowing 100 gpm, where c = 24.
    a. 24 psi
    b. 35 psi
    c. 48 psi
    d. not enough information was provided

75. Calculate friction loss in 100-foot sections of 2 ½-inch hose flowing 250 gpm using the hand method.
    a. 4 psi
    b. 8.5 psi
    c. 12.5 psi
    d. 32 psi

76. Calculate friction loss in 100-foot sections of 2 ½-inch hose flowing 500 gpm, using the drop-ten method.
    a. 40 psi
    b. 45 psi
    c. 50 psi
    d. 55 psi

77. Using the formula $2q^2 + q$, calculate the friction loss for a 100-foot section of 2 ½-inch hose flowing 450 gpm.
    a. 35 psi
    b. 41.5 psi
    c. 45 psi
    d. 55.5 psi

78. Calculate the friction in a 100-foot section of 4-inch hose flowing 475 gpm, using the formula cq^2L, where c is equal to 0.2.
    a. 4.5 psi
    b. 10.5 psi
    c. 15.5 psi
    d. 20.5 psi

79. Using the condensed q formula, calculate the friction loss for 200 gpm flowing through 100 feet of 3-inch hose.
    a. 4 psi
    b. 9 psi
    c. 15 psi
    d. 19 psi

80. Calculate the friction loss for 225 gpm flowing through a 100-foot section of 3-inch hose using the condensed q formula.
    a. 5 psi
    b. 8 psi
    c. 15 psi
    d. 20 psi

81. What is the pressure gain/loss for a line raised 25 feet above the apparatus?
    a. 12.5 psi
    b. 25 psi
    c. 50 psi
    d. not enough information provided

82. What is the pressure gain/loss for a line 45 feet below the apparatus?
    a. 20 psi
    b. 45 psi
    c. –22.5 psi
    d. –45 psi

83. A hose line is raised to the second floor. What is the pressure gain/loss?
    a. 5 psi
    b. 10 psi
    c. 15 psi
    d. 20 psi

84. A hose line is taken down to the fourth-floor basement level. What is the pressure gain/loss?
    a. 5 psi
    b. 10 psi
    c. –5 psi
    d. –15 psi

85. What is the PDP for a hand-line with the following elements:
    1. NP: 1 ¼-inch smooth-bore nozzle
    2. FL: 500 feet of 3-inch hose line flowing 320 gpm with friction of 41 psi
    a. 41 psi
    b. 61 psi
    c. 91 psi
    d. 101 psi

86. The PDP for a 500-foot line of 3-inch hose flowing 250 gpm through a combination nozzle when the friction loss is 25 psi for the hose line is _____.
    a. 15 psi
    b. 25 psi
    c. 75 psi
    d. 125 psi

87. When a _____ sprinkler system operates, water discharges through all the sprinkler heads.
    a. dry-pipe
    b. pre-action
    c. deluge
    d. wet-pipe

88. In a _____ system, water is in the system at all times. When a sprinkler head is fused, water immediately discharges.
    a. dry-pipe
    b. pre-action
    c. deluge
    d. wet-pipe

89. Using the formula $Q = 30 \times d^2 \times \sqrt{NP}$, calculate the flow through a smooth-bore nozzle with a 1-inch tip on a master-stream device operating on the fireground.
    a. 150 gpm
    b. 175 gpm
    c. 200 gpm
    d. 270 gpm

90. Calculate the flow through a smooth-bore nozzle with a ¾-inch tip and 90 psi nozzle pressure using the formula $Q = 30 \times d^2 \times \sqrt{NP}$.
    a. 150 gpm
    b. 160 gpm
    c. 170 gpm
    d. not enough information provided

91. What is the nozzle reaction for a 1-inch smooth-bore nozzle discharging water with a nozzle pressure of 50 psi using the formula $NR = 1.57 \times d^2 \times NP$?
    a. 50 lb
    b. 79 lb
    c. 85 lb
    d. 100 lb

92. The nozzle reaction for a combination nozzle discharging 250 gpm with a nozzle pressure of 100 psi is _____ when using the fireground nozzle reaction formula $NR = gpm \times 0.5$.
    a. 50 lb
    b. 79 lb
    c. 85 lb
    d. 125 lb

93. A vehicle that is navigating a curve loses traction. _____ force will move the apparatus in an outward direction.
    a. Centrifugal
    b. Gravitational
    c. Cavitational
    d. Momentum

94. The _____ distance is measured from the time a hazard is detected until the vehicle comes to a complete stop.
    a. perception
    b. reaction
    c. braking
    d. total stopping

95. When operating in temperatures around 0°F, an engine oil rating of _____ would most likely be used.
    a. SAE 5
    b. CF
    c. SAE 30
    d. SAE 20W50

96. According to NFPA 1901, new apparatus must have quick-buildup air tanks that can reach operating pressure within _____ seconds.
    a. 15
    b. 45
    c. 60
    d. 90

97. The use of flow meters on apparatus has increased because of each of the following *except:*
    a. They eliminate the need for fireground hydraulics to determine pump discharge pressure.
    b. They provide increased versatility on locating pump control panels.
    c. They are easier to calibrate than Bourdon tube gauges.
    d. They eliminate the need for pressure gauges.

98. Within a municipal water distribution system, secondary feeder mains are typically
    a. up to 4 inches.
    b. 6 to 8 inches.
    c. 12 to 14 inches.
    d. 16 inches or larger.

99. When combining two or more lines a _____ is most often used.
    a. gated wye
    b. siamese
    c. hydrant thief
    d. distribution manifold

100. The process of slowly opening/closing discharge control valves to adjust discharge settings is called
    a. drafting.
    b. cavitation.
    c. pressure regulation.
    d. feathering or gating.

Phase II, Exam II: Answers to Questions

1.	F	26.	B	51.	A	76.	A
2.	T	27.	A	52.	B	77.	C
3.	F	28.	C	53.	B	78.	A
4.	F	29.	C	54.	D	79.	A
5.	T	30.	D	55.	B	80.	A
6.	F	31.	A	56.	C	81.	A
7.	F	32.	C	57.	D	82.	C
8.	T	33.	B	58.	A	83.	A
9.	F	34.	D	59.	A	84.	D
10.	T	35.	A	60.	B	85.	C
11.	F	36.	A	61.	A	86.	D
12.	B	37.	D	62.	D	87.	C
13.	C	38.	C	63.	D	88.	D
14.	A	39.	B	64.	A	89.	D
15.	D	40.	C	65.	C	90.	B
16.	C	41.	A	66.	B	91.	B
17.	D	42.	D	67.	C	92.	D
18.	A	43.	E	68.	D	93.	A
19.	C	44.	D	69.	D	94.	D
20.	A	45.	A	70.	C	95.	D
21.	C	46.	D	71.	D	96.	C
22.	C	47.	D	72.	D	97.	D
23.	B	48.	B	73.	C	98.	C
24.	B	49.	C	74.	A	99.	B
25.	C	50.	C	75.	C	100.	D

Phase II, Exam II:
Rationale & References for Questions

Question #1
This is a misperception about fire pump operations. In reality, scientific theory and principles prevail in pump operations. When understood, pump operations can be both predictable and controllable. NFPA 1002 5.1. *IFPO, 2E:* Chapter 1, page 4.

Question #2
Preventive maintenance is an often overlooked and underemphasized duty. The goal of preventive maintenance is to ensure the apparatus is in a ready state at all times. NFPA 1002 4.2. *IFPO, 2E:* Chapter 1, page 7.

Question #3
The fire department, through the fire chief, has the ultimate responsibility for the establishment, implementation, and monitoring of a preventive maintenance program. NFPA 1002 4.2.1, 5.1.1. *IFPO, 2E:* Chapter 2, page 22.

Question #4
The specific preventive maintenance activities conducted by pump operators and mechanics depend on the level of training and the type of preventive maintenance activity being conducted. In general, certified mechanics conduct those activities that require apparatus to be taken out of service, require several hours to complete, or are detailed and complicated repairs. Checking and adding engine oil or battery fluid may be conducted by the pump operator, whereas changing the engine oil or replacing the battery is most likely performed by a mechanic. NFPA 1002 4.2.1, 5.1.1. *IFPO, 2E:* Chapter 2, page 22.

Question #5
The specific preventive maintenance activities conducted by pump operators and mechanics depend on the level of training and the type of preventive maintenance activity being conducted. In general, certified mechanics conduct those activities that require apparatus to be taken out of service, require several hours to complete, or are detailed and complicated repairs. Checking and adding engine oil or battery fluid may be conducted by the pump operator, whereas changing the engine oil or replacing the battery is most likely performed by a mechanic. NFPA 1002 4.2.1, 5.1.1. *IFPO, 2E:* Chapter 2, page 22.

Question #6
Most state laws require that emergency-vehicle drivers obey the same laws as other vehicle operators unless specifically exempt from doing so. State laws typically define several conditions that must exist for exemptions to be extended and include:

- Only authorized emergency vehicles are covered
- The exemptions are only provided when responding to an emergency
- Audible and visual warning devices must be operating when taking advantage of the exemption.

NFPA 1002 4.3. *IFPO, 2E:* Chapter 3, pages 44 – 45.

Question #7
The requirement to place the above warning on pump panels is actually contained in NFPA 1901, not NFPA 1002. NFPA 1002 5.2.1, 5.2.2. *IFPO, 2E:* Chapter 4, page 102.

Question #8

Basic pressure principles include:

- Pressure at any point in a liquid at rest is equal in every direction.

- Pressure of a liquid acting on a surface is perpendicular to that surface.

- External pressure applied to a confined liquid (fluid) is transmitted equally throughout the liquid.

  This principle provides the basis for understanding the transmission of pressure through a network of fire hoses.

- Pressure at any point beneath the surface of a liquid in an open container is directly proportional to its depth.

- Pressure exerted at the bottom of a container is independent of the shape or volume of the container.

  NFPA 1002 5.2. *IFPO, 2E:* Chapter 10, pages 271 – 274.

Question #9

Basic pressure principles include:

- Pressure at any point in a liquid at rest is equal in every direction.

- Pressure of a liquid acting on a surface is perpendicular to that surface.

- External pressure applied to a confined liquid (fluid) is transmitted equally throughout the liquid.

- Pressure at any point beneath the surface of a liquid in an open container is directly proportional to its depth.

- Pressure exerted at the bottom of a container is independent of the shape or volume of the container.

  NFPA 1002 5.2. *IFPO, 2E:* Chapter 10, pages 271 – 274.

Question #10

The engine coolant level should only be checked when the engine is cool. NFPA 1002 4.2.1, 4.2.2, 5.1.1. *IFPO, 2E:* Chapter 2, page 33.

Question #11

Larger-diameter hoses experience less friction loss and allow more water to flow over greater distances. NFPA 1002 5.2. *IFPO, 2E:* Chapter 10, pages 179 – 282.

Question #12

NFPA 1002 requires the following of driver operators:

- Be licensed to drive all vehicles they are expected to operate

- Subject to medical evaluation

- Meet the requirements of Firefighter I per NFPA 1001

Typically, local jurisdictions set specific requirements for age and education levels. NFPA 1002 1.4.1, 1.4.2, 5.1. *IFPO, 2E:* Chapter 1, pages 11 – 16.

Question #13

Laws are rules that are legally binding and enforceable. Standards are guidelines that are not legally binding or enforceable. NFPA 1002 4.2, 4.3, 5.1. *IFPO, 2E:* Chapter 1, page 15.

Question #14

Inspections are conducted to verify the status of a component, for example, verifying water, oil, and fuel levels.

Servicing activities are conducted to help maintain vehicles in peak operating condition, such as cleaning, lubricating, and topping off fluids.

Tests are conducted to determine the performance of components, such as annual pump service tests. NFPA 1002 4.2.1, 5.1.1. *IFPO, 2E:* Chapter 2, page 23.

Question #15
Documenting preventive maintenance activities is important because the practice:

- Helps keep track of needed maintenance and repairs.

- May help to determine trends.

- Is required by NFPA 1500.

- May be required for warranty claims.

NFPA 1002 4.2.2. *IFPO, 2E:* Chapter 2, page 25.

Question #16
According to NFPA 1901, the fuel tank must be sufficient in size to drive the pump for at least 2 ½ hour at its rated capacity when pumping at draft. NFPA 1002 4.2.1. *IFPO, 2E:* Chapter 2, page 35.

Question #17
Safety should be considered when conducting preventive maintenance inspections and tests. Common safety considerations for preventive maintenance inspections and test include:

General safety considerations:

- Not hurrying or rushing through inspections

- Ensuring work area is free from hazards

- Always keeping safety in mind

Specific safety considerations:

- Checking for loose equipment before raising a tilt cab

- Not smoking around engine compartment and fuels

- Wearing appropriate clothing (no loose jewelry, wear safety glasses, gloves)

- Considering vapor and electrical hazards

- Always being careful when opening the radiator cap

- Using proper tools

- Securing all equipment and closing all doors prior to moving the apparatus

NFPA 1002 4.2.1. *IFPO, 2E:* Chapter 2, page 37.

Question #18
NFPA 1002 requires that pump operators be able to conduct and document routine tests, inspections, and servicing functions to ensure the apparatus is in a ready state. NFPA 1002 4.2.1, 5.1.1. *IFPO, 2E:* Chapter 2, pages 23 to 25.

Question #19
The only action that will most likely help regain traction and control is removing the foot from the accelerator, which allows the engine to slow the vehicle. NFPA 1002 4.3.1. *IFPO, 2E:* Chapter 3, page 56.

Question #20
Although fire department vehicle accidents occur for a variety of reasons, several common factors that appear in the vast majority of accidents include:

- Not following laws and standards related to emergency response

- Not being fully aware of both driver and apparatus limitations

- Lack of appreciation for driving conditions such as weather and traffic

Lack of NFPA 1002 certification has not been linked as a common factor associated with emergency-vehicle accidents. NFPA 1002 4.3. *IFPO, 2E:* Chapter 3, pages 43 – 44.

Question #21

Total stopping distance is measured from the time a hazard is detected until the vehicle comes to a complete stop. Total stopping distance consists of:

- *Perception distance* (distance apparatus travels from the time the hazard is seen until the brain recognizes it as a hazard)

- *Reaction distance* (distance apparatus travels from the time the brain sends the message to depress the brakes until the brakes are depressed)

- *Braking distance* (distance of travel from the time the brake is depressed until the vehicle comes to a complete stop)

NFPA 1002 4.3. *IFPO, 2E:* Chapter 3, page 55 – 65.

Question #22

Several driving exercises are typically used to assess the driver's ability to safely operate and control the vehicle. These include the following:

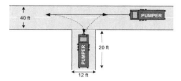

- *Alley Dock.* Assesses the ability to back the vehicle into a restricted area, such as a fire station or down an alley.

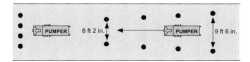

- *Serpentine.* Assesses the ability to drive around obstacles, such as parked cars and tight corners.

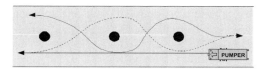

- *Confined-Space Turnaround.* Assesses the ability to turn the vehicle around within a confined space, such as a narrow street or driveway.

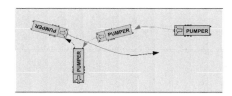

- *Diminishing Clearance.* Assesses the ability to drive the vehicle in a straight line, such as on a narrow street or road.

NFPA 1002 4.3.2, 4.3.3, 4.3.4, 4.3.5. *IFPO, 2E:* Chapter 3, pages 65 – 66.

Question #23
According to NFPA 1901, a pump must have a rated capacity as follows:

- 100% of its rated capacity at 150 psi

- 70% of its rated capacity at 200 psi

- 50% of its rated capacity at 250 psi

NFPA 1002 5.2.1, 5.2.2. *IFPO, 2E:* Chapter 4, page 91.

Question #24

- In *volume mode*, each individual impeller will add the flow it generates to the total discharge, with the pressure remaining constant among the impellers.

- In *pressure mode,* each subsequent impeller pumps the same flow from the previous impeller while adding the pressure it generates.

NFPA 1002 5.2.1, 5.2.2. *IFPO, 2E:* Chapter 4, pages 89 – 90.

Question #25

- Pumps connected to a split-shaft transmission are usually located midship or aft.

- Pumps connected directly to the crankshaft are usually located at the front of the engine.

- Pumps connected to a PTO are usually mounted at the front or midship. Pumps are usually smaller when powered by a PTO.

NFPA 1002 5.2.1, 5.2.1. *IFPO, 2E:* Chapter 4, pages 92 – 93.

Question #26
As a rule of thumb, a pump will have one 2 ½-inch discharge for each 250 gpm of rated capacity. A 1,250 gpm pumper will usually have five 2 ½-inch discharges, 5 × 250 = 1,250. NFPA 1002 5.2.1, 5.2.2. *IFPO, 2E:* Chapter 5, pages 118 – 119.

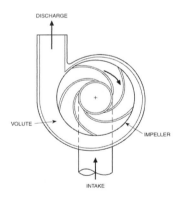

Question #27
The discharge is #1 and the intake is #3 in the diagram. NFPA 1002 5.2.1, 5.2.2. *IFPO, 2E:* Chapter 4, page 86 – 89.

Question #28
Instrumentation on pump panels can be grouped into three categories as follows:

- Pressure gauges

- Flow meters

- Indicators

NFPA 1002 5.2.1, 5.2.2. *IFPO, 2E:* Chapter 5, pages 99 – 100.

Question #29
Ball, butterfly, and gated valves are commonly found on pumping apparatus. NFPA 1002 5.2.1, 5.2.2. *IFPO, 2E:* Chapter 5, page 114.

Question #30
Several way to determine when a pump is primed include the following indicators:

- A positive reading on the pressure gauge

- Priming motor sounds as if it is slowing (not increasing)

- Main pump will sound as if it is under load

- Oil/water discharge under the vehicle

NFPA 1002 5.2.1, 5.2.2. *IFPO, 2E:* Chapter 5, page 121.

Question #31
The two largest gauges on a pump panel are usually the master/main intake (*left*) and discharge (*right*) gauges. The smaller gauges are usually individual discharge gauges. NFPA 1002 5.2.1, 5.2.2. *IFPO, 2E:* Chapter 5, page 101.

Question #32
The only attack hose listed is the 1 ¾-inch hose. NFPA 1002 5.2.1, 5.2.2, 5.2.4. *IFPO, 2E:* Chapter 6, pages 139 – 140.

Question #33
NFPA 1962 Standard for the Inspection, Care, and Use of Fire Hose, Couplings, and Nozzles and the Service Testing of Hose. NFPA 1002 5.1.1. *IFPO, 2E:* Chapter 6, page 146.

Question #34
When correct nozzle pressures are not provided, the following can occur.

Insufficient nozzle pressure:

- Less flow

- Reduced reach

- Poor pattern development

Excessive nozzle pressure:

- Poor pattern development

- Excessive nozzle reaction

NFPA 1002 5.2.1. *IFPO, 2E:* Chapter 6, page 156.

Question #35
A pumper with 1,000 gallon tank flowing 250 gpm will last about 4 minutes.

250gallons per minute ⟌ 1000gallons = 4 minutes. NFPA 1002 5.2.1. *IFPO, 2E:* Chapter 7, page 167.

Question #36

Hydrant flow testing equipment include:

- Pressure gauge mounted on an outlet cap (calibrated within the past 12 months)
- Pitot gauge for each hydrant
- Hydrant diffuser
- Hydrant wrenches
- Portable radios

A fire department pumper is not required when conducting hydrant flow testing. NFPA 1002 5.2.1. *IFPO, 2E:* Chapter 7, pages 180 – 181.

Question #37

Factors that contribute to the availability of water include flow, pressure, quantity, and location or accessibility of the water supply. Pump capacity is not a factor of water supply availability. NFPA 1002 5.2.1. *IFPO, 2E:* Chapter 7, page 165.

Question #38

Static hydrants are prepiped lines that extend into a static water supply source. NFPA 1002 5.2.1. *IFPO, 2E:* Chapter 7, pages 186 – 187.

Question #39

A general rule of thumb is to maintain 20 psi intake pressure on pumpers within a relay. NFPA 1002 5.2.3. *IFPO, 2E:* Chapter 7, page 192.

Question #40

Shuttle cycle time is the total time it takes to dump water and return with another load. NFPA 1002 5.2.1, 5.2.2. *IFPO, 2E:* Chapter 7, pages 200 – 201.

Question #41

Usually, the transfer valve is set before discharge pressure is increased.

Step 1   Position apparatus, set parking brake, and let engine return to idle
Step 2   Engage the pump
Step 3   Provide water to intake side of pump (on-board, hydrant, draft)
Step 4   Set transfer valve (if so equipped)
Step 5   Open discharge lines
Step 6   Throttle to desired pressure
Step 7   Set the pressure regulating device
Step 8   Maintain appropriate flows and pressures

NFPA 1002 5.2.1, 5.2.2, 5.2.4. *IFPO, 2E:* Chapter 8, page 207.

Question #42

Centrifugal pumps should never operate without water.

Several caveats that come close to being universal among departments and manufactures for the operation of centrifugal pumps include:

- Never operate the pump without water.
- Always keep water moving when operating the pump at high speeds.
- Never open, close, or turn controls abruptly.
- Always maintain awareness of instrumentation during pumping operations.
- Never leave the pump unattended.
- Always maintain constant vigilance to safety.

NFPA 1002 5.2.1, 5.2.2, 5.2.4. *IFPO, 2E:* Chapter 8, page 208.

Question #43

A transfer-valve operation rule of thumb is:

- Use *volume mode* when flows are greater than 50% of a pump's rated capacity, and pressures are less than 150 psi.

- Use *pressure mode* when flows are less than 50% of a pump's rated capacity, and pressures are greater than 150 psi.

NFPA 1002 5.2.1, 5.2.2, 5.2.4. *IFPO, 2E:* Chapter 8, pages 214 – 215.

Question #44

Because of the limited supply of water in the on-board tank, the transfer valve should be in pressure (series) mode. NFPA 1002 5.2.1, 5.2.2, 5.2.4. *IFPO, 2E:* Chapter 8, page 220.

Question #45

The three methods used to transfer power to the pump are:

- PTO

- Front crankshaft

- Split-shaft

NFPA 1002 5.2.1, 5.2.2, 5.2.4. *IFPO, 2E:* Chapter 8, pages 210 – 212.

Question #46

Pumps connected to a split-shaft transmission are typically mounted in the middle or toward the back of the apparatus. NFPA 1002 5.2.1, 5.2.2, 5.2.4. *IFPO, 2E:* Chapter 8, page 211.

Question #47

Cavitation is caused by insufficient intake flow to match the discharge flow. Not operating enough discharge lines would not cause cavitation in that the pump has excess capacity to pump and adequate intake flow. NFPA 1002 5.1.1. *IFPO, 2E:* Chapter 9, pages 245 – 246.

Question #48

All the answer selections may be signs that a pump is cavitating, except the priming device. The priming device would not be operating when conditions are conducive to cavitation. NFPA 1002 5.1.1. *IFPO, 2E:* Chapter 9, pages 245 – 247.

Question #49

When fire and smoke are showing, the pump operator should immediately starting pumping to the sprinkler system.

Supporting a sprinkler system may require the following:

- Connecting at least one line to the FD connection

- Pumping at 150 psi unless otherwise directed

- Pumping immediately when fire and smoke are visible

- Not shutting down sprinkler system for improved visibility

- Placing transfer valve into the volume mode

- Assuring that water supply to the pump does not reduce the sprinkler system water supply

NFPA 1002 5.2.4. *IFPO, 2E:* Chapter 9, pages 252 – 253.

Question #50

Standpipe systems are classified based on intended use as follows:

- *Class 1* standpipes provide 2 ½-inch connections for trained firefighters/fire brigades, at an initial flow rate of 500 gpm.
- *Class 2* standpipes provide 1 ½-inch connections for initial attack, at a minimum flow rate of 100 gpm.
- *Class 3* standpipes provide 1 ½-inch and 2 ½-inch connections for trained firefighters/fire brigades, at an initial flow rate of 500 gpm

NFPA 1002 5.2.4. *IFPO, 2E:* Chapter 9, page 253.

Question #51

Increased pressure reduces the ability of a liquid to evaporate (increase the boiling point). Decreased pressure increases the ability of a liquid to evaporate (lowers the boiling point). NFPA 1002 5.2.1. *IFPO, 2E:* Chapter 9, page 245.

Question #52

The outside stem and yoke (OS&Y) valve allows for quick determination of valve position. If the stem is out and exposed, the valve is open. If the stem is in or only protruding a short distance, the valve is closed. NFPA 1002 5.1.1. *IFPO, 2E:* Chapter 9, page 249.

Question #53

- *Hydraulics* is that branch of science dealing with the principles of fluid at rest or in motion.
- *Hydrodynamics* is that branch of hydraulics that deals with the principles and laws of fluids in motion.
- *Hydrostatics* is that branch of hydraulics that deals with the principles and laws of fluids at rest and the pressures they exert or transmit.

NFPA 1002 5.2. *IFPO, 2E:* Chapter 10, page 261.

Question #54

The basic characteristics and properties of water are listed below. These are considered approximate in that water purity, atmospheric conditions, and rounding can effect the values.

- Virtually incompressible
- Freezes at 32°F
- Expands when frozen
- Boils at 212°F
- 1 gallon weighs 8.34 lbs
- Density is 62.4 lb/ft^3
- Number of gallons in 1 cubic foot is 7.38 gallons/ft^3

NFPA 1002 5.2.1, 5.2.2. *IFPO, 2E:* Chapter 10, pages 262 – 263.

Question #55

- *Latent heat of fusion* is the amount of heat that is absorbed by a substance when changing from a solid to a liquid state.
- *Specific heat* is the amount of heat required to raise the temperature of a substance by 1°F. The specific heat of water is 1 Btu/lb.
- *Latent heat of vaporization* is the amount of heat absorbed when changing from a liquid to a vapor state.
- *Radiation* is heat transfer through electromagnetic waves.

NFPA 1002 5.2.1, 5.2.2. *IFPO, 2E:* Chapter 10, pages 263 – 264.

Question #56

Pressure is the force exerted by a substance in *units of weight per area,* typically expressed in *pounds per square inch* (psi or lb/in^2). Force is the total weight of the substance. So, the pressure is multiplied by the area to calculate the force or weight.

F = Force (weight in pounds)

P = Pressure (psi)

A = Area (square inches)

NFPA 1002 5.2.1, 5.2.2. *IFPO, 2E:* Chapter 10, pages 269 – 270.

Question #57

To calculate weight, the formula $W = D \times V$ is used. NFPA 1002 5.2. *IFPO, 2E:* Chapter 10, page 265.

Question #58

The operating pressure for smooth-bore nozzles on hand-lines is 50 psi. The square root of 50 is 7.071. An acceptable value for fireground calculations is 7.

The operating pressure for smooth-bore nozzles on master-stream devices is 80 psi. The square root of 80 is 8.944. An acceptable value for fireground calculations is 9. NFPA 1002 5.2. *IFPO, 2E:* Chapter 10, pages 296 – 298.

Question #59

NFA formula: $NF = \dfrac{A}{3}$

where NF = needed flow in gpm

            A = area of the structure in square feet

            3 = constant in ft^2/gpm

Iowa State formula: $NF = \dfrac{V}{100}$

where NF = needed flow in gpm

            V = volume of the area in cubic feet

            100 = is a constant in ft^3/gpm NFPA 1002 5.2. *IFPO, 2E:* Chapter 10, pages 292 – 293.

Question #60

Nozzle reaction calculation

Smooth-bore nozzles:

$NR = 1.57 \times d^2 \times NP$

where NR = nozzle reaction

            1.57 = constant

            d = diameter of nozzle orifice in inches

            NP = operating nozzle pressure in psi

Combination nozzles:

$NR = gpm \times \sqrt{NP} \times 0.0505$

where NR = nozzle reaction

            0.0505 = constant

            gpm = flow in gallons per minute

            NP = operating nozzle pressure in psi

NFPA 1002 5.2. *IFPO, 2E:* Chapter 10, pages 282 – 283.

Question #61

Elevation formula by floor level:

$EL = 5 \times H$

where EL = the gain or loss of elevation in psi

          5 = gain or loss in pressure for each floor level

          H = height in number of floor levels above or below the pump

Elevation formula in feet:

$EL = 0.5 \times H$

where EL = the gain or loss of elevation in psi

          0.5 = pressure exerted at base of 1-cubic-inch column of water 1 foot high

          H = height in number of floor levels above or below the pump

NFPA 1002 5.2. *IFPO, 2E:* Chapter 11, page 311.

Question #62

The operating pressure for smooth-bore nozzles on hand-lines is 50 psi, while operating pressure for smooth-bore nozzles on master-stream devices is 80 psi. Also, the low-pressure combination nozzle has an operating pressure of 75 psi.

$PDP = NP + FL + AFL \pm EL$

where PDP = pump discharge pressure

          NP = nozzle pressure

          FL = friction loss in hose

          AFL = appliance friction loss

          EL = elevation gain or loss

NFPA 1002 5.2. *IFPO, 2E:* Chapter 12, pages 317 – 319.

Question #63

The auxiliary cooling system helps keep the engine temperature within operating limits and will not help keep the water in the pump from heating. Maintaining water flow within the pump is the best way to help keep water temperatures from rising significantly. NFPA 1002 5.2.1, 5.2.2. *IFPO, 2E:* Chapter 9, page 243.

Question #64

Iowa State formula: $NF = \dfrac{V}{100}$.

where NF = needed flow in gpm

          V = volume of the area in cubic feet

          100 = is a constant in ft³/gpm

NFA formula: $NF = \dfrac{A}{3}$.

where NF = needed flow in gpm

          A = area of the structure in square feet

          3 = constant in ft²/gpm

NFPA 1002 5.2. *IFPO, 2E:* Chapter 10, pages 292 – 293.

Question #65

There are 144 cubic-inch columns of water in 1 cubic foot. (12 inches × 12 inches = 144 cubic inches or in² per cubic foot). NFPA 1002 5.2. *IFPO, 2E:* Chapter 10, page 273.

Question #66

A gauge that measures psia at sea level would have a reading of 14.7 psia because psia (absolute pressure) includes atmospheric pressure.

A gauge reading of 100 psi (psig) is actually 114.7 psia because psig (gauge pressure) will read 0 psi at sea level. NFPA 1002 5.2. *IFPO, 2E:* Chapter 10, page 275.

Question #67

Based on the percent drop in pressure, additional flows may be available from a hydrant as follows:

- 0 – 10% drop   three times the original flow

- 11 – 15% drop   two times the original flow

- 16 – 25% drop   one time the original flow

NFPA 1002 5.2. *IFPO, 2E:* Chapter 10, page 277.

Question #68

The Iowa State formula can be used for an entire structure or a section of the structure and is based on volume. NFPA 1002 5.2. *IFPO, 2E:* Chapter 11, pages 292 – 293.

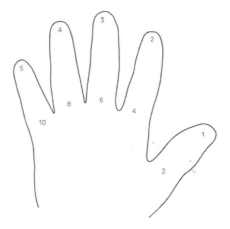

Question #69

The hand method is a fireground method used to estimate friction loss in 100-foot sections of 2 ½-inch hose. Simply select the fingertip representing the gpm flow and then multiply the two figures on the finger for the approximate friction loss pressure for each 100-foot section of 2 ½-inch hose.

500 gpm = 5 × 10 = 50 psi per 100-foot section of 2 ½-inch hose NFPA 1002 5.2.1, 5.2.2. *IFPO, 2E:* Chapter 10, pages 302 – 303.

Question #70

The drop-ten method simply subtracts 10 from the first two numbers of gpm flow. It is not as accurate as other methods, but provide a simple rule of thumb for fireground use.

350 gpm = 35 – 10 = 25 psi friction loss per 100-foot sections of 2 ½-inch hose NFPA 1002 5.2.1, 5.2.2. *IFPO, 2E:* Chapter 10, page 303.

Question #71

cq^2L

$2 \times \left(\dfrac{300}{100}\right)^2 \times \left(\dfrac{100}{100}\right)$

$2 \times (3)^2 \times (1)$

$2 \times 9 \times 1$

18 psi friction loss NFPA 1002 5.2. *IFPO, 2E:* Chapter 11, pages 206 – 307.

Question #72

Nozzle reaction calculation

Smooth-bore nozzles:

$NR = 1.57 \times d^2 \times NP$

where NR = nozzle reaction

          1.57 = constant

          d = diameter of nozzle orifice in inches

          NP = operating nozzle pressure in psi

Combination nozzles:

$NR = gpm \times \sqrt{NP} \times .0505$

where NR = nozzle reaction

          0.0505 = constant

          gpm = flow in gallons per minute

          NP = operating nozzle pressure in psi

NFPA 1002 5.2. *IFPO, 2E:* Chapter 10, pages 282 – 283.

Question #73

$2q^2 + q$

$2 \times \left(\dfrac{300}{100}\right)^2 + \left(\dfrac{300}{100}\right)$

$2 \times (3)^2 + (3)$

$2 \times 9 + 3$

21 psi friction loss NFPA 1002 5.2. *IFPO, 2E:* Chapter 11, pages 304 – 305.

Question #74

cq^2L

$24 \times \left(\dfrac{100}{100}\right)^2 \times \left(\dfrac{100}{100}\right)$

$24 \times (1)^2 \times (1)$

$24 \times 1 \times 1$

24 psi friction loss NFPA 1002 5.2. *IFPO, 2E:* Chapter 11, pages 306 – 307.

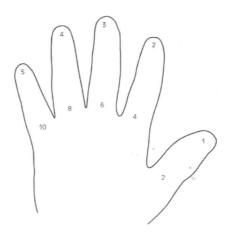

Question #75

The hand method is a fireground method used to estimate friction loss in 100-foot sections of 2 ½-inch hose. Simply select the fingertip representing the gpm flow and then multiply the two figures on the finger for the approximate friction loss pressure for each 100-foot section of 2 ½-inch hose.

250 gpm = 2.5 × 5 = 12.5 psi per 100-foot sections of 2 ½-inch hose NFPA 1002 5.2.1, 5.2.2. *IFPO, 2E:* Chapter 10, pages 302 – 303.

Question #76

The drop-ten method simply subtracts 10 from the first two numbers of gpm flow. It is not as accurate as other methods, but provides a simple rule of thumb for fireground use.

500 gpm = 50 – 10 = 40 psi friction loss per 100-foot sections of 2 ½-inch hose. NFPA 1002 5.2.1, 5.2.2. *IFPO, 2E:* Chapter 10, page 303.

Question #77

$$2q^2 + q$$

$$2 \times \left(\frac{450}{100}\right)^2 + \left(\frac{450}{100}\right)$$

$$2 \times (4.5)^2 + (4.5)$$

$$2 \times 20.25 + 4.5$$

45 psi friction loss NFPA 1002 5.2. *IFPO, 2E:* Chapter 11, pages 304 – 305.

Question #78

$$cq^2 L$$

$$.2 \times \left(\frac{475}{100}\right)^2 \times \left(\frac{100}{100}\right)$$

$$.2 \times (4.75)^2 \times (1)$$

$$.2 \times 22.56 \times 1$$

4.5 psi friction loss NFPA 1002 5.2. *IFPO, 2E:* Chapter 11, pages 306 – 307.

Question #79

q^2

$\left(\dfrac{200}{100}\right)^2$

$(2)^2$

4 psi friction loss NFPA 1002 5.2. *IFPO, 2E:* Chapter 11, pages 306 – 307.

Question #80

q^2

$\left(\dfrac{225}{100}\right)^2$

$(2.25)^2$

5 psi friction loss NFPA 1002 5.2. *IFPO, 2E:* Chapter 11, pages 306 – 307.

Question #81

Elevation formula in feet:

$EL = 0.5 \times H$

where EL = the gain or loss of elevation in psi

0.5 = pressure exerted at base of 1-cubic-inch column of water 1 foot high

H = height in feet above or below the pump

$EL = 0.5 \times 25$

$EL = 12.5$ psi

NFPA 1002 5.2. *IFPO, 2E:* Chapter 11, page 311.

Question #82

Elevation formula in feet:

$EL = 0.5 \times H$

where EL = the gain or loss of elevation in psi

0.5 = pressure exerted at base of 1-cubic-inch column of water 1 foot high

H = height in feet above or below the pump

$EL = 0.5 \times -45$

$EL = -22.5$ psi

NFPA 1002 5.2. *IFPO, 2E:* Chapter 11, page 311.

Question #83

Elevation formula by floor level:

$EL = 5 \times H$

where EL = the gain or loss of elevation in psi

5 = gain or loss in pressure for each floor level

H = height in number of floor levels above or below the pump

$EL = 5 \times 1$

$EL = 5$ psi

NFPA 1002 5.2. *IFPO, 2E:* Chapter 11, page 311.

Question #84

Elevation formula by floor level:

$EL = 5 \times H$

where EL = the gain or loss of elevation in psi

5 = gain or loss in pressure for each floor level

H = height in number of floor levels above or below the pump

$EL = 5 \times -3$

$EL = -15$ psi

NFPA 1002 5.2. *IFPO, 2E:* Chapter 11, page 311.

Question #85

$PDP = NP + FL + AFL \pm EL$

where PDP = pump discharge pressure

NP = nozzle pressure

FL = friction loss in hose

AFL = appliance friction loss

EL = elevation gain or loss

PDP = 50 psi + 41 psi

PDP = 91 psi

NFPA 1002 5.2. *IFPO, 2E:* Chapter 12, pages 317 – 319.

Question #86

$PDP = NP + FL + AFL \pm EL$

where PDP = pump discharge pressure

NP = nozzle pressure

FL = friction loss in hose

AFL = appliance friction loss

EL = elevation gain or loss

PDP = 100 psi + 25 psi

PDP = 125 psi

NFPA 1002 5.2. *IFPO, 2E:* Chapter 12, pages 317 – 319.

Question #87

- *Dry-pipe systems* maintain air or compressed gas under pressure within the system. When a head fuses, water enters the system and discharges through any fused heads.

- *Pre-action systems* are similar to dry-pipe systems in that air or compressed gas is maintained in the system. However, an automatic detection system (smoke, heat, flame, etc.) or manual system (pull box) must operate to allow water to enter the system. At this point, it is similar to a wet-pipe system; when a sprinkler head fuses, water discharges.

- *Deluge systems* maintain all sprinkler heads in an open position. When a detection system operates, water enters the system and is discharged through all the open heads.

- *Wet-pipe systems* maintain water in the system at all times. When a sprinkler head is fused, water immediately discharges through the fused heads.

NFPA 1002 5.2.4. *IFPO, 2E:* Chapter 9, pages 148 – 253.

Question #88

- *Dry-pipe systems* maintain air or compressed gas under pressure within the system. When a head fuses, water enters the system and discharges through any fused heads.

- *Pre-action systems* are similar to dry-pipe systems in that air or compressed gas is maintained in the system. However, an automatic detection system (smoke, heat, flame, etc.) or manual system (pull box) must operate to allow water to enter the system. At this point, it is similar to a wet-pipe system; when a sprinkler head fuses, water discharges.

- *Deluge systems* maintain all sprinkler heads in an open position. When a detection system operates, water enters the system and is discharged through all the open heads.

- *Wet-pipe systems* maintain water in the system at all times. When a sprinkler head is fused, water immediately discharges through the fused heads.

NFPA 1002 5.2.4. *IFPO, 2E:* Chapter 9, pages 148 – 253.

Question #89

$$Q = 30 \times d^2 \times \sqrt{NP}$$

$$Q = 30 \times 1^2 \times \sqrt{80}$$

$$Q = 30 \times 1 \times 9$$

$$Q = 270 \text{ gpm}$$

NFPA 1002 5.2. *IFPO, 2E:* Chapter 11, page 296 – 298.

Question #90

$$Q = 30 \times d^2 \times \sqrt{NP}$$

$$Q = 30 \times 0.75^2 \times \sqrt{90}$$

$$Q = 30 \times 0.56 \times 9.5$$

$$Q = 159.6 \text{ or } 160 \text{ gpm}$$

NFPA 1002 5.2. *IFPO, 2E:* Chapter 11, page 296 – 298.

Question #91

$$NR = 1.57 \times d^2 \times NP$$

$$NR = 1.57 \times 1^2 \times 50$$

$$NR = 78.5 \text{ lb (79 lbs when rounded)}$$

NFPA 1002 5.2. *IFPO, 2E:* Chapter 10, pages 282 – 283.

Question #92

$$NR = \text{gpm} \times 0.5$$

$$NR = 250 \times .5$$

$$NR = 125 \text{ lb}$$

NFPA 1002 5.2. *IFPO, 2E:* Chapter 10, pages 282 – 283.

Question #93

Centrifugal force is the tendency of an object, when moving in a circular pattern, to move outward from the center. NFPA 1002 4.3.1, 4.3.6. *IFPO, 2E:* Chapter 3, pages 56 – 57.

Question #94

Total stopping distance is measured from the time a hazard is detected until the vehicle comes to a complete stop. Total stopping distance consists of:

- *Perception distance* (distance apparatus travels from the time the hazard is seen until the brain recognizes it as a hazard)
- *Reaction distance* (distance apparatus travels from the time the brain sends the message to depress the brakes until the brakes are depressed)
- *Braking distance* (distance of travel from the time the brake is depressed until the vehicle comes to a complete stop)

NFPA 1002 4.3. *IFPO, 2E:* Chapter 3, page 55 – 65.

Question #95

The only oil listed that is rated for 0°F is SAE 20W50. The "W" after the number in the SAE oil classification means the oil is rated for flow at 0°F. NFPA 1002 4.2.1, 5.1.1. *IFPO, 2E:* Chapter 2, pages 32 – 33.

Question #96

NFPA 1901 requires new apparatus to have quick buildup times that can reach operating pressure within 60 seconds. NFPA 1002 4.2.1, 5.1.1. *IFPO, 2E:* Chapter 2, page 35.

Question #97

Flow meters are increasingly finding their way onto pump panels because they reduce fireground hydraulic calculations and, with the use of "smart" electronics, they are easy to calibrate and increase the versatility on pump panel location.

Pressure gauges are still needed to ensure safety from excessive pressures. NFPA 1002 5.2. *IFPO, 2E:* Chapter 5, page 113.

Question #98

Municipal water distribution systems use main sizes as follows:

- *Primary* feeder mains: 16 inches or larger
- *Secondary* feeder mains: 12 to 14 inches
- *Distributors*: 6 to 8 inches

NFPA 1002 5.2.1, 5.2.2, 5.2.4. *IFPO, 2E:* Chapter 7, page 172.

Question #99

A siamese is used to combine two or more lines into one line. NFPA 1002 5.2. *IFPO, 2E:* Chapter 6, page 149.

Question #100

Feathering or gating is the process of slowly opening/closing discharge control valves to adjust discharge settings. NFPA 1002 5.2. *IFPO, 2E:* Chapter 9, pages 237 – 238.

Phase Two, Exam Three

1. Fire pump operations are no more complex and unpredictable than other fireground operations.
   a. True
   b. False

2. Within the first few minutes after arrival on scene, all of the activities related to fire pump operations occur.
   a. True
   b. False

3. Replacing the battery on a fire department vehicle is usually performed by a mechanic.
   a. True
   b. False

4. Apparatus fuel tanks should be refilled when the fuel gauge indicates one-third empty.
   a. True
   b. False

5. Most state laws require that emergency-vehicle drivers obey the same laws as other vehicle operators.
   a. True
   b. False

6. NFPA 1500 requires that emergency-vehicle drivers come to a complete stop when any intersection hazard is present.
   a. True
   b. False

7. Because of the potential danger, NFPA 1901 requires the following warning on pump panels:

   "**WARNING:** Death or serious injury might occur if proper operating procedures are not followed. The pump operator as well as individuals connecting supply or discharge hoses to the apparatus must be familiar with water hydraulics hazards and component limitations."
   a. True
   b. False

8. Drafting is the process of moving or drawing water from a static water source through a hard suction hose to the intake side of the pump using a suction process.
   a. True
   b. False

9. The drop-ten method is a fireground method to estimate friction loss for 3-inch hose only.
   a. True
   b. False

10. The effect of a water hammer is based on the principle that the pressure of a liquid acting on a surface is perpendicular to that surface.
    a. True
    b. False

11. The pressure of a liquid acting on a surface is perpendicular to that surface.
    a. True
    b. False

12. The inspection of the radiator and coolant levels on modern apparatus is no longer necessary due to technological advancements.
    a. True
    b. False

13. Driver operators must meet which of the following requirements according to NFPA 1002?
    a. Firefighter I per NFPA 1001; subject to periodic medical evaluation; be licensed to drive all vehicles they are expected to operate
    b. Firefighter II per NFPA 1001; subject to periodic medical evaluation; be licensed to drive all vehicles they are expected to operate
    c. license for vehicles they operate; have turned 21 years of age; Firefighter II per NFPA 1001
    d. Firefighter I per NFPA 1001; licensed for vehicles they operator; earned a high school diploma or GED

14. _____ refers to the duty of ensuring that the apparatus, pump, and related components are in a ready state and peak operating efficiency.
    a. Preventive maintenance
    b. Pump testing
    c. Manufacturers' recommendations
    d. Operating the pump

15. Standards are _____ that are not legally binding and enforceable.
    a. rules
    b. guidelines
    c. laws
    d. ordinances

16. The establishment, implementation, and monitoring of a preventive maintenance program is the responsibility of the
    a. pump operator.
    b. mechanic.
    c. fire department.
    d. applicable standard.

17. Tests are conducted to determine the _____ of components, while inspections are conducted to verify the _____ of components
    a. condition, performance
    b. status, performance
    c. performance, status
    d. status, condition

18. Preventive maintenance documentation is important for each of the following reasons *except*:
    a. It helps keep track of needed maintenance and repairs.
    b. It is required by NFPA 1500.
    c. It may help to determine trends.
    d. It helps ensure pump operators are spending enough time conducting the inspection.

19. Key elements for defensive driving include anticipating and planning for
    a. possible hazards while driving.
    b. the need to yield the right of way.
    c. the effects of speed, braking, and weather considerations.
    d. all of the answers are correct.

20. Steering wheels on emergency apparatus should
    a. have no more than 1 inch free play.
    b. not exceed 10 degrees of free play in either direction.
    c. be turned fully in both directions during vehicle inspections to ensure proper lubrication of steering components.
    d. all of the answers are correct.

21. NFPA _____ requires that manufacturer's certification and acceptance tests be conducted on new apparatus.
    a. 1002
    b. 1500
    c. 1901
    d. 1911

22. The distance an apparatus travels from the time the hazard is seen until the brain recognizes it as a hazard is called the
    a. perception distance.
    b. reaction distance.
    c. braking distance.
    d. total stopping distance.

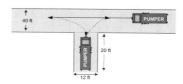

23. The driving exercise noted in the figure above is called
    a. diminishing clearance.
    b. alley dock.
    c. confined-space turnaround.
    d. serpentine.

24. One action that may help regain control of a vehicle during a skid is to
    a. slightly increase engine speed.
    b. turn front wheels into the direction of the skid.
    c. disengage the clutch and aggressively depress the brakes.
    d. turn front wheels in the opposite direction of the skid.

25. Fill in the missing rated capacity information:

    _____ % at 150 psi
    _____ % at 200 psi
    _____ % at 250 psi

    a. 100, 70, 50
    b. 100, 75, 50
    c. 50, 70, 100
    d. 50, 75, 100

26. A four-stage centrifugal pump is discharging 1,000 gpm at 200 psi in volume mode. Each impeller will deliver
    a. 500 gpm at 200 psi.
    b. 250 gpm at 200 psi.
    c. 1,000 gpm at 100 psi.
    d. 500 gpm at 100 psi.

27. A 750-gpm centrifugal pump mounted on the front of an apparatus would most likely be powered
    a. through a PTO.
    b. directly from the crankshaft.
    c. through a split-shaft transmission.
    d. by any of these methods.

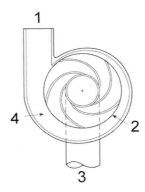

28. Identify each of the numbered items for the line diagram above of a centrifugal pump.
    a. 1 – Intake
       2 – Volute
       3 – Impeller
       4 – Discharge
    b. 1 – Discharge
       2 – Vane
       3 – Intake
       4 – Volute
    c. 1 – Discharge
       2 – Volute
       3 – Intake
       4 – Impeller
    d. 1 – Discharge
       2 – Impeller
       3 – Intake
       4 – Volute

29. The most common types of pressure gauges found on pump panels are called
    a. Bourdon tube gauges.
    b. flow meters.
    c. dual-purpose (positive/negative reading) gauges.
    d. indicating gauges.

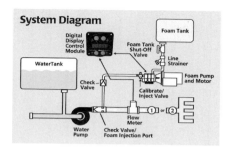

30. The foam system illustrated above is called a
    a. direct injection/compressed-air foam system (CAFS).
    b. in-line eductor system.
    c. around-the-pump proportioning system.
    d. balanced pressure system.

31. All of the following are indicators that the pump is primed *except*
    a. air/oil discharge under the vehicle.
    b. main pump sounds as if it is underload.
    c. a positive reading on the pressure gauge.
    d. priming motor sounds as if it is slowing.

32. Correctly label items 2 and 4 in the figure above.
    a. 2 – main intake gauge
       4 – main intake pressure relief device
    b. 2 – main intake gauge
       4 – main pump intake
    c. 2 – main discharge gauge
       4 – main pump intake
    d. 2 – main intake gauge
       4 – master discharge

33. A 1,500-gpm pumper will usually have how many 2 ½-inch discharges?
    a. four
    b. five
    c. six
    d. seven

34. Supply hose have a maximum operating pressure of _____ psi, while attack hose have a maximum operating pressure of _____ psi.
    a. 185, 285
    b. 185, 275
    c. 150, 250
    d. 95, 150

35. _____ requires fire departments to establish and maintain an accurate record of each hose section.
    a. NFPA 1961
    b. NFPA 1962
    c. The hose manufacturer
    d. Federal regulation

36. NFPA 1901 specifies minimum tank capacities for fire apparatus as follows:
    a. Aerial: 200 gallons
       Pumper: 300 gallons
       Tanker: 1,000 gallons
    b. Inital attack: 200 gallons
       Pumper: 300 gallons
       Tanker: 1,000 gallons
    c. Pumper: 300 gallons
       Tanker: 1,000 gallons
    d. Initial attack: 250 gallons
       Pumper: 500 gallons
       Tanker: 1,000 gallons

37. A pumper with a 500-gallon tank is pumping one preconnect at 125 gpm. How long will the on-board water supply last?
    a. 4 minutes
    b. 6 minutes
    c. 8 minutes
    d. 10 minutes

38. Each of the following items may be required when flow testing hydrants *except*
    a. pitot gauges for each hydrant.
    b. hydrant diffusers.
    c. pressure gauges mounted on an outlet cap (calibrated within the past 12 months).
    d. a fire department pumper.

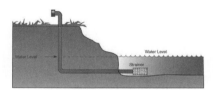

39. The picture above represents
    a. a static or dry hydrant.
    b. maximum lift.
    c. OS&Y.
    d. fire department connection.

40. In a relay operation the second pumper, all in-line pumpers, and the attack pumper should maintain at least _____ psi intake pressure.
    a. 10
    b. 20
    c. 30
    d. 50

41. The volume of water that can be pumped in a tanker shuttle without running out of water is called
    a. shuttle available flow.
    b. shuttle flow capacity.
    c. shuttle required flow.
    d. maximum available flow.

42. The same basic steps must be taken to move water from the supply to the discharge point. What critical pump operation step is out of order?

   Step 1   Position apparatus, set parking brake, and let engine return to idle
   Step 2   Engage the pump
   Step 3   Provide water to intake side of pump (on-board, hydrant, draft)
   Step 4   Set transfer valve (if so equipped)
   Step 5   Open discharge lines
   Step 6   Throttle to desired pressure
   Step 7   Maintain appropriate flows and pressures
   Step 8   Set the pressure regulating device

   a. change step 4 with 6
   b. change step 6 with 7
   c. change step 2 with 3
   d. change step 7 with 8

43. Which of the following is an incorrect statement concerning pump operations?
   a. Always maintain constant vigilance to safety.
   b. Always maintain awareness of instrumentation during pumping operations.
   c. Always keep water moving when operating the pump at high speeds.
   d. Always operate the pump without water.

44. As a general rule of thumb, the transfer valve should be in volume mode when
   a. flows are expected to be greater than 50% of a pump's rated capacity and pressures less than 150 psi.
   b. flows are expected to be less than 50% of a pump's rated capacity and pressures greater than 150 psi.
   c. flows are expected to be greater than 50% of a pump's rated capacity and pressures greater than 150 psi.
   d. flows are expected to be less than 50% of a pump's rated capacity and pressures less than 150 psi.

45. As a general rule, the transfer valve should be _____ when supplying one discharge line using the on-board water tank.
   a. in parallel
   b. in series
   c. off
   d. on

46. Each of the following items can be used to power the pump from the drive engine *except* the
   a. PTO.
   b. split-shaft.
   c. separate engine.
   d. front crankshaft.

47. Cavitation can be caused by all of the following *except*
   a. not enough discharge lines operating.
   b. excessive lift while drafting.
   c. intake lines too small.
   d. insufficient flow from a hydrant.

48. Each of the flowing are signs that cavitation may be occurring *except*
   a. engine speed automatically increases during a pumping operation.
   b. no corresponding increase in pressure occurs when the engine speed is increased.
   c. priming device slows down or seems to be operating under a load.
   d. excessive pump vibrations or rattling sounds like sand or gravel going through the pump.

49. The pump operator should do all of the following when connecting to a fire department connection, *except*
    a. place a transfer valve into the volume mode.
    b. pump at 100 psi unless otherwise directed.
    c. begin pumping immediately when fire and smoke are visible.
    d. connect at least one hose line.

50. According to NFPA 14, Standard for the Installation of Standpipe and Hose Systems, there are three classes of standpipe system. Which classification(s) is/are for initial attack?
    a. all of them
    b. only Class 3 standpipes
    c. both Class 1 and Class 3
    d. only Class 2

51. Decreased pressure will _____ the ability of a liquid to vaporize, while increased pressure will _____ the ability of a liquid to vaporize.
    a. increase, reduce
    b. reduce, increase
    c. initiate, stop
    d. stop, initiate

52. If the stem on the OS&Y is barely visible the valve is
    a. closed or mostly closed.
    b. open or mostly open.
    c. broken.
    d. no such valve exists.

53. Which branch of science focuses on principles and laws of fluids at rest?
    a. hydraulics
    b. hydrodynamics
    c. hydrostatics
    d. hydrolastic

54. Which of the following physical characteristics and properties of water are incorrect?
    a. Density of water is 62.4 lb/ft^3.
    b. 1 pound of water weighs 8.34 lbs.
    c. Water is virtually incompressible.
    d. Number of gallons in 1 cubic foot is 7.38 gallons/ft^3.

55. The amount of heat that is absorbed by a substance when changing from a solid to a liquid state is called the
    a. latent heat of fusion.
    b. specific heat.
    c. latent heat of vaporization.
    d. evaporation.

56. The formula $P = \dfrac{F}{A}$ is used to calculate
    a. pounds or weight.
    b. force.
    c. pressure.
    d. both a and b are correct.

57. The correct formula used to calculate the weight of water for a specific volume is
    a. $W = \dfrac{D}{V}$.

    b. $W = \dfrac{V}{D}$.

    c. $W = D \times V$.

    d. $W = D \div V$.

58. The formula used to determine flow (gpm) when smooth-bore nozzles are used is $Q = 29.7 \times d^2 \times \sqrt{NP}$. An acceptable value for $\sqrt{NP}$ to use for calculating flow on the fireground for master-streams is
    a. 5.
    b. 7.
    c. 9.
    d. 30.

59. In the Iowa State formula $NF = \dfrac{V}{100}$, the V is
    a. velocity.
    b. vacuum.
    c. volume.
    d. variable nozzle flows.

60. For the elevation formula $EL = 0.5 \times H$, which of the following is not correct?
    a. H = height in number of floor levels above or below the pump.
    b. 0.5 = pressure exerted at the base of 1-cubic-inch column of water 1 foot high.
    c. 0.5 is the number used for easier calculations on the fireground; the actual number is closer to .433.
    d. H = distance in feet above or below the pump.

61. In the formula $DP = NP + FL + AFL \pm EL$, all of the following are correct *except*:
    a. NP = nozzle pressure.
    b. AFL = appliance friction loss and is 10 psi for smaller appliances and 25 psi for larger appliances.
    c. FL = friction loss and can be calculated using a number of estimation methods and fireground formulas.
    d. EL = elevation gain or loss.

62. There are _____ gallons in 1 cubic foot of water.
    a. 8.35
    b. 62.4
    c. 7.48
    d. not enough information provided

63. The correct National Fire Academy formula for calculating needed flow is
    a. $NF = \dfrac{V}{100}$.

    b. $NF = \dfrac{V}{100} \times 3$.

    c. $NF = \dfrac{A}{3}$.

    d. $NF = \dfrac{A}{3} \div 100$.

64. There are _____ cubic inch columns of water in 1 cubic foot.
    a. 12
    b. 120
    c. 144
    d. 62.4

65. A gauge reading 14.7 psia at sea level is equivalent to a gauge reading of _____ psig.
    a. −14.7
    b. 0
    c. 29.4
    d. none of the answers are correct

66. During pump operations, the drop in hydrant pressure from static to residual can be used to estimate the additional flow the hydrant is capable of providing. A 12% or less drop means the hydrant may be able to deliver as much as
    a. three times the original flow.
    b. two times the original flow.
    c. one time the original flow.
    d. half the original flow.

67. The Iowa State formula estimates needed flow, in gpm,
    a. for an entire structure.
    b. for a section of a structure.
    c. based on the square footage of the involved area.
    d. both a and b are correct.

68. Using the hand method, calculate friction loss in 100-foot sections of 2 ½-inch hose flowing 350 gpm.
    a. 3.5 psi
    b. 7 psi
    c. 10.5 psi
    d. 24.5 psi

69. Calculate friction loss in 100-foot sections of 2 ½-inch hose flowing 150 gpm, using the drop-ten method.
    a. 5 psi
    b. 15 psi
    c. 25 psi
    d. 50 psi

70. Using the formula cq^2L, calculate the friction loss for a 100-foot section of 4-inch hose flowing 500 gpm.
    a. 1 psi
    b. 2 psi
    c. 5 psi
    d. 10 psi

71. Which of the following is the correct formula for calculating smooth-bore nozzle reaction?
    a. $NR = 1.57 \times d^2 \times NP$
    b. $NR = gpm \times \sqrt{NP} \times .0505$
    c. $NR = gpm \times \sqrt{NP \times .0505}$
    d. $NR = 1.57 \times d \times NP$

72. Calculate friction loss in 100-foot sections of 2 ½-inch hose flowing 450 gpm, using the hand method.
    a. 25.5 psi
    b. 40.5 psi
    c. 50.5 psi
    d. 55.5 psi

73. Using the drop-ten method, calculate friction loss in a 100-foot section of 2 ½-inch hose flowing 225 gpm.
    a. 10 psi
    b. 12 psi
    c. 22 psi
    d. 25 psi

74. Using the formula $2q^2 + q$, calculate the friction loss for 100 feet of 2 ½-inch hose flowing 225 gpm.
    a. 5 psi
    b. 7 psi
    c. 12 psi
    d. 22.5 psi

75. Calculate the friction loss for a 100-foot section of 2 ½-inch hose flowing 310 gpm using the formula $2q^2 + q$.
    a. 10 psi
    b. 21 psi
    c. 22 psi
    d. 31 psi

76. Using the formula cq^2L, calculate the friction loss for a 100-foot section of 1 ½-inch hose flowing 125 gpm.
    a. 25.5 psi
    b. 37.5 psi
    c. 48.5 psi
    d. 50.5 psi

77. Calculate the friction loss for a 100-foot of 2 ½-inch hose flowing 300 gpm using the formula cq^2L.
    a. 12 psi
    b. 15 psi
    c. 18 psi
    d. 30 psi

78. Using the condensed q formula, calculate the friction loss for 350 gpm flowing through a 100-foot section of 3-inch hose.
    a. 4 psi
    b. 9 psi
    c. 12 psi
    d. 19 psi

79. Calculate the friction loss for 400 gpm flowing through a 100-foot section of 3-inch hose using the condensed q formula.
    a. 4 psi
    b. 6 psi
    c. 16 psi
    d. 20 psi

80. What is the pressure gain/loss for a hose line raised 14 feet above the apparatus?
    a. 5 psi
    b. 7 psi
    c. 10 psi
    d. 17 psi

81. What is the pressure gain/loss for a line 10 feet below the apparatus?
    a. 5 psi
    b. 10 psi
    c. −5 psi
    d. −10 psi

82. A hose line is raised to the tenth floor. What is the pressure gain/loss?
    a. 25 psi
    b. 30 psi
    c. 45 psi
    d. 55 psi

83. A hose line is taken down one level to the basement. What is the pressure gain/loss?
    a. 5 psi
    b. 10 psi
    c. −5 psi
    d. −10 psi

84. What is the PDP for a master-stream with the following elements?

    1. NP: 1 ⅜-inch smooth-bore nozzle
    2. FL: 800 feet of 3-inch hose line flowing 510 gpm with friction of 112 psi

    a. 25.5 psi
    b. 51 psi
    c. 80 psi
    d. 192 psi

85. What is the PDP for a 1,000-foot line of 2 ½-inch hose flowing 300 gpm through a combination nozzle when the friction loss is 36 psi for the line? The hose is operating on the third floor.
    a. 100 psi
    b. 136 psi
    c. 146 psi
    d. 156 psi

86. When a detection system operates, water enters the system and is discharged through all the open heads in a _____ system.
    a. dry-pipe
    b. pre-action
    c. deluge
    d. wet-pipe

87. In a _____ system, an automatic/manual system/device must operate, and the sprinkler heads must fuse before water is discharged.
    a. dry-pipe
    b. pre-action
    c. deluge
    d. wet-pipe

88. Calculate the flow through a hand-line operating on the fireground with a smooth-bore nozzle (½-inch tip) using the formula $Q = 30 \times d^2 \times \sqrt{NP}$.
    a. 53 gpm
    b. 59 gpm
    c. 63 gpm
    d. 70 gpm

89. During a hydrant flow test, you record a pitot tube reading of 70 psi from a smooth-bore nozzle with a 1-inch tip. Use $Q = 30 \times d^2 \times \sqrt{NP}$ to determine the flow.
    a. 151 gpm
    b. 251 gpm
    c. 262 gpm
    d. 280 gpm

90. What is the nozzle reaction for a 1 ½-inch smooth-bore nozzle discharging water with a nozzle pressure of 80 psi using the formula $NR = 1.57 \times d^2 \times NP$?
    a. 200 lb
    b. 258 lb
    c. 278 lb
    d. 283 lb

91. The nozzle reaction for a combination nozzle discharging 100 gpm with a nozzle pressure of 70 psi is _____ when using the formula $NR = gpm \times \sqrt{NP} \times 0.0505$.
    a. 42 lb
    b. 50 lb
    c. 72 lb
    d. 80 lb

92. A vehicle navigating a curve loses traction and begins to move in an outward direction. This outward movement is the direct result of
    a. momentum.
    b. speed and braking.
    c. water surge.
    d. centrifugal force.

93. The distance of travel from the time the brake is depressed until the vehicle comes to a complete stop is called the
    a. perception distance.
    b. reaction distance.
    c. braking distance.
    d. total stopping distance.

94. Which of the following is an API oil classification for a diesel engine?
    a. SL
    b. CG
    c. SAE 30
    d. SAE 20W50

95. Which of the following NFPA standards specifically addresses fire apparatus preventive maintenance?
    a. NFPA 1002
    b. NFPA 1500
    c. NFPA 1901
    d. NFPA 1915

96. Flow meter use on apparatus has increased because of each of the following *except* they
    a. eliminate the need for pressure gauges.
    b. provide increased versatility on locating pump control panels.
    c. are easier to calibrate than Bourdon tube gauges.
    d. eliminate the need for fireground hydraulics to determine pump discharge pressure.

97. In most cases, a fire department connection is a
    a. gated wye with two female inlets.
    b. clappered siamese with female inlets.
    c. gated siamese with two male couplings.
    d. OS&Y or PIV valve.

98. A _____ is used to combine two or more hose lines into one hose line.
    a. gated wye
    b. siamese
    c. hydrant thief
    d. distribution manifold

99. All of the following can cause increased friction in hose *except*
    a. couplings and adapters.
    b. kinks and bends.
    c. laminar flow.
    d. rough interior.

100. Adjusting discharge pressure through the process called feathering or gating is accomplished by
    a. slightly increasing/decreasing pump speed.
    b. slowly opening/closing nozzles.
    c. using discharge control valves.
    d. none of the answers are correct.

Phase II, Exam III: Answers to Questions

1.	T	26.	B	51.	A	76.	B
2.	F	27.	B	52.	A	77.	C
3.	T	28.	D	53.	C	78.	C
4.	F	29.	A	54.	B	79.	C
5.	T	30.	A	55.	A	80.	B
6.	T	31.	A	56.	C	81.	C
7.	T	32.	B	57.	C	82.	C
8.	F	33.	C	58.	C	83.	C
9.	F	34.	B	59.	C	84.	D
10.	F	35.	B	60.	A	85.	C
11.	T	36.	B	61.	B	86.	C
12.	F	37.	A	62.	C	87.	B
13.	A	38.	D	63.	C	88.	A
14.	A	39.	A	64.	C	89.	B
15.	B	40.	B	65.	B	90.	D
16.	C	41.	B	66.	B	91.	A
17.	C	42.	D	67.	D	92.	D
18.	D	43.	D	68.	D	93.	C
19.	D	44.	A	69.	A	94.	B
20.	B	45.	B	70.	A	95.	D
21.	C	46.	C	71.	A	96.	A
22.	A	47.	A	72.	B	97.	B
23.	B	48.	C	73.	B	98.	B
24.	B	49.	B	74.	C	99.	C
25.	A	50.	D	75.	C	100.	C

Phase II, Exam III:
Rationale & References for Questions

Question #1

In reality, scientific theory and principles prevail in both pump operations and fire suppression. When understood, pump operations can actually be more predictable and controllable than a structural fire. NFPA 1002 5.1. *IFPO, 2E:* Chapter 1, page 4.

Question #2

It is a misperception about fire pump operations that most of the activities related to fire pump operations occur in the first few minutes upon scene arrival. However, after initial operations are setup, pump operators must:

- Continually observe instrumentation.
- Adjust flows and pressure as appropriate for safety of personnel and equipment.
- Be prepared to readily adapt to changing fireground situations.
- Monitor and plan for water supply needs and long-term operations.
- Maintain a constant vigilance to safety.

NFPA 1002 5.1. *IFPO, 2E:* Chapter 1, page 4.

Question #3

The specific preventive maintenance activities conducted by pump operators and mechanics depend on the level of training and the type of preventive maintenance activity being conducted. In general, certified mechanics conduct those activities that require apparatus to be taken out of service, require several hours to complete, or are detailed and complicated repairs. Checking and adding engine oil or battery fluid may be conducted by the pump operator, whereas changing the engine oil or replacing the battery is most likely performed by a mechanic. NFPA 1002 4.2.1, 5.1.1. *IFPO, 2E:* Chapter 2, page 22.

Question #4

Apparatus fuel tanks should be refilled according to department policy. When no refill policy exists, apparatus fuel tanks should be refilled when the fuel level reaches the three-quarter mark. That is, when the fuel gauge drops from full to the three-quarter full mark. NFPA 1002 4.2.1. *IFPO, 2E:* Chapter 2, page 35.

Question #5

Most state laws require that emergency-vehicle drivers obey the same laws as other vehicle operators unless specifically exempt from doing so. State laws typically define several conditions that must exist for exemptions to be extended and include:

- Only authorized emergency vehicles are covered
- The exemptions are only provided when responding to an emergency
- Audible and visual warning devices must be operating when taking advantage of the exemption

NFPA 1002 4.3. *IFPO, 2E:* Chapter 3, pages 44 – 45.

Question #6

NFPA 1500 requires that emergency-vehicle drivers come to a complete stop when any intersection hazard is present. Specifically, the vehicle must come to a complete stop when any of the following exists:

- As directed by a law enforcement officer
- At traffic red lights or stop signs
- At negative right-of-way and blind intersections
- When all lanes of traffic in an intersection cannot be accounted for
- When stopped school bus with flashing warning lights is encountered
- At unguarded railroad guard crossing (also for nonemergency)
- When other intersection hazards are present

NFPA 1002 4.3. *IFPO, 2E:* Chapter 3, page 48.

Question #7

Because of the potential danger, the statement is now required on all new pump panels. NFPA 1002 5.2.1, 5.2.2. *IFPO, 2E:* Chapter 4, page 102.

Question #8

When the priming system reduces the pressure in the pump, water is forced into the pump by atmospheric pressure. NFPA 1002 5.2.1. *IFPO, 2E:* Chapter 7, page 185.

Question #9

The drop-ten method is a fireground method to estimate friction loss in *2 ½-inch hose.* It can also be used for other size hose, although it is less accurate.

For this method, simply subtract 10 from the first 2 numbers of the gpm flow. For example, the friction loss in 100 feet of 2 ½-inch hose flowing 250 gpm is 15 psi (25 – 10 = 15). NFPA 1002 5.2.1, 5.2.2. *IFPO, 2E:* Chapter 11, pages 303 – 304.

Question #10

Basic pressure principles include:

- Pressure at any point in a liquid at rest is equal in every direction.
- Pressure of a liquid acting on a surface is perpendicular to that surface.
- External pressure applied to a confined liquid (fluid) is transmitted equally throughout the liquid.

  This principle provides the basis for understanding the transmission of pressure through a network of fire hoses.

- Pressure at any point beneath the surface of a liquid in an open container is directly proportional to its depth.
- Pressure exerted at the bottom of a container is independent of the shape or volume of the container.

NFPA 1002 5.2. *IFPO, 2E:* Chapter 10, pages 271 – 274.

Question #11

Basic pressure principles include:

- Pressure at any point in a liquid at rest is equal in every direction.

- Pressure of a liquid acting on a surface is perpendicular to that surface.

  This explains why an uncharged hose line (no pressure) is relatively flat while a charged hose line is round.

- External pressure applied to a confined liquid (fluid) is transmitted equally throughout the liquid.

- Pressure at any point beneath the surface of a liquid in an open container is directly proportional to its depth.

- Pressure exerted at the bottom of a container is independent of the shape or volume of the container.

NFPA 1002 5.2. *IFPO, 2E:* Chapter 10, pages 271 – 274.

Question #12

Pump operators should continue to inspect the radiator and coolant levels as indicated by department policy and manufacturers' recommendations. NFPA 1002 4.2.1, 4.2.2, 5.1.1. *IFPO, 2E:* Chapter 2, page 33.

Question #13

NFPA 1002 requires the following:

- Be licensed to drive all vehicles they are expected to operate

- Subject to periodic medical evaluation

- Meet the requirements of Firefighter I per NFPA 1001

Typically, local jurisdictions set specific requirements for age and education levels. NFPA 1002 1.4.1, 1.4.2, 5.1. *IFPO, 2E:* Chapter 1, pages 11 – 16.

Question #14

Preventive maintenance is an often overlooked and underemphasized duty. The goal of preventive maintenance is to ensure the apparatus is in a ready state at all times. NFPA 1002 4.2. *IFPO, 2E:* Chapter 1, page 7.

Question #15

Laws are rules that are legally binding and enforceable. Standards are guidelines that are not legally binding or enforceable. NFPA 1002 4.2, 4.3, 5.1. *IFPO, 2E:* Chapter 1, page 15.

Question #16

The fire department, through the fire chief, has the ultimate responsibility for the establishment, implementation, and monitoring of a preventive maintenance program. NFPA 1002 4.2.1, 5.1.1. *IFPO, 2E:* Chapter 2, page 22.

Question #17

Inspections are conducted to verify the *status* of a component, for example verifying water, oil, and fuel levels.

Servicing activities are conducted to help maintain vehicles in peak operating *condition*, as when cleaning, lubricating, and topping off fluids.

Tests are conducted to determine the *performance* of components, as in annual pump service tests. NFPA 1002 4.2.1, 5.1.1. *IFPO, 2E:* Chapter 2, page 23.

Question #18

Documenting preventive maintenance activities is important because:

- It helps keep track of needed maintenance and repairs.
- It may help to determine trends.
- It is required by NFPA 1500.
- It may be required for warranty claims.

NFPA 1002 4.2.2. *IFPO, 2E:* Chapter 2, page 25.

Question #19

Each of the items listed is considered important to defensive driving elements in that they will help the driver operate in a defense posture. NFPA 1002 4.3.6. *IFPO, 2E:* Chapter 3, page 54.

Question #20

It is not advisable to turn the steering wheel so that the wheels turn. Rather, the steering wheel should be turned until just before the wheels turn. The distance should not exceed 10 degrees in either direction. For a 20-inch steering wheel, that would be approximately 2 inches of movement. NFPA 1002 4.2.1. *IFPO, 2E:* Chapter 2, pages 35 – 36.

Question #21

NFPA 1901 specifies requirements that apply to all new automotive fire apparatus. NFPA 1002 4.2.1. *IFPO, 2E:* Chapter 2, pages 25.

Question #22

Total stopping distance is measured from the time a hazard is detected until the vehicle comes to a complete stop. Total stopping distance consists of:

- *Perception distance* (distance apparatus travels from the time the hazard is seen until the brain recognizes it as a hazard)
- *Reaction distance* (distance apparatus travels from the time the brain sends the message to depress the brakes until the brakes are depressed)
- *Braking distance* (distance of travel from the time the brake is depressed until the vehicle comes to a complete stop)

NFPA 1002 4.3. *IFPO, 2E:* Chapter 3, page 55 – 65.

Question #23
Several driving exercises are typically used to assess the ability of the driver to safely operate and control the vehicle. These include the following:

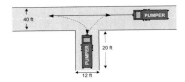

- *Alley Dock.* Assesses the ability to back the vehicle into a restricted area, such as a fire station or down an alley.

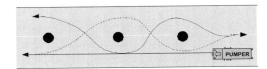

- Serpentine. Assesses the ability to drive around obstacles, such as parked cars and tight corners

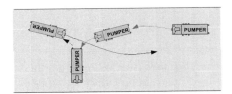

- *Confined-Space Turnaround.* Assesses the ability to turn the vehicle around within a confined space, such as a narrow street or driveway.

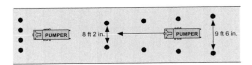

- *Diminishing Clearance.* Assesses the ability to drive the vehicle in a straight line, such as on a narrow street or road.

NFPA 1002 4.3.2, 4.3.3, 4.3.4, 4.3.5. *IFPO, 2E:* Chapter 3, pages 65 – 66.

Question #24
The only action listed that will most likely help regain traction and control is turning the front wheels into the direction of the skid. NFPA 1002 4.3.1. *IFPO, 2E:* Chapter 3, page 56.

Question #25
According to NFPA 1901, a pump must have a rated capacity as follows:

- 100% of its rated capacity at 150 psi
- 70% of its rated capacity at 200 psi
- 50% of its rated capacity at 250 psi

NFPA 1002 5.2.1, 5.2.2. *IFPO, 2E:* Chapter 4, page 91.

Question #26

- In *volume mode*, each individual impeller will add the flow it generates to the total discharge, with the pressure remaining constant among the impellers.

- In *pressure mode*, each subsequent impeller pumps the same flow from the previous impeller while adding the pressure it generates.

NFPA 1002 5.2.1, 5.2.2. *IFPO, 2E:* Chapter 4, pages 89 – 90.

Question #27

- Pumps connected to a split-shaft transmission are usually located midship or aft.

- Pumps connected directly to the crankshaft are usually located at the front of the engine.

- Pumps connected to a PTO are usually mounted at the front or midship. Pumps are usually smaller when powered by a PTO.

NFPA 1002 5.2.1, 5.2.1. *IFPO, 2E:* Chapter 4, pages 92 – 93.

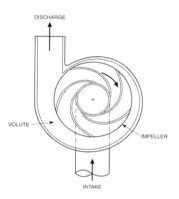

Question #28

NFPA 1002 5.2.1, 5.2.2. *IFPO, 2E:* Chapter 4, pages 86 – 89.

Question #29

Bourdon tube gauges are the most common pressure gauges found on pump panels. NFPA 1002 5.2.1, 5.2.2. *IFPO, 2E:* Chapter 5, pages 110 – 111.

Question #30

- A *pre-mixed system* consists of a tank in which foam concentrate and water are added in appropriate proportions. Often, the foam is simply added to the on-board water tank.

- An *in-line eductor system* utilizes eductors to add foam to water in appropriate proportions. This system can be either internal or external controls and requires an eductor and a foam tank (or supply). Internal systems have a foam concentrate valve and metering control. Eductors are usually installed between the pump discharge and the discharge outlet.

- *Around-the-pump proportioning systems* use an eductor (located between the intake and discharge sides of the pump) to mix foam with water. Water from the discharge side of the pump picks up foam and delivers it to the intake side of the pump, where it can then be discharged to all outlets.

- A *balanced-pressure system* mixes foam with water by means of pressure. One system uses pressure to force foam from a bladder. The second system uses a separate foam pump (two types are by-pass and demand).

- A *direct injection/compressed-air foam system* (CAFS) consists of a separate foam pump and foam tank. Direct injection moves foam from the tank; it is pumped directly into the discharge lines. CAFS adds compressed air to create a lightweight foam.

NFPA 1002 5.2.3. *IFPO, 2E:* Chapter 5, pages 126 – 130.

Question #31
Several way to determine when a pump is primed include the following indicators:

- A positive reading on the pressure gauge
- Priming motor sounds as if it is slowing
- Main pump will sound as if it is underload
- Oil/water discharge under the vehicle (not air/oil)

NFPA 1002 5.2.1, 5.2.2. *IFPO, 2E:* Chapter 5, page 121.

Question #32
The two largest gauges on a pump panel are usually the master/main intake (*left*) and discharge (*right*) gauges. The smaller gauges are usually individual discharge gauges. The large pipe on the pump panel is usually the main intake. NFPA 1002 5.2.1, 5.2.2. *IFPO, 2E:* Chapter 5, page 101.

Question #33
As a rule of thumb, a pump will have one 2 ½-inch discharge for each 250 gpm of rated capacity. A 1,500 gpm pumper will usually have six 2 ½-inch discharges, $6 \times 250 = 1,500$. NFPA 1002 5.2.1, 5.2.2. *IFPO, 2E:* Chapter 5, pages 118 – 119.

Question #34
NFPA 1962 requires that supply hoses not be operated at pressures exceeding 185 psi, and NFPA 1961 indicates that attack hose have a normal highest operating pressure of 275 psi. NFPA 1002 5.2.1, 5.2.2, 5.2.4. *IFPO, 2E:* Chapter 6, pages 139 – 140.

Question #35
NFPA 1962 Standard for the Inspection, Care, and Use of Fire Hose, Couplings, and Nozzles and the Service Testing of Hose. NFPA 1002 5.1.1. *IFPO, 2E:* Chapter 6, page 146.

Question #36
NFPA 1901 requires a minimum tank capacity for pumping apparatus as follows:

- Initial attack apparatus: 200 gallons
- Pumper fire apparatus: 300 gallons
- Mobile water apparatus/tankers: 1,000 gallons

NFPA 1002 5.2.1. *IFPO, 2E:* Chapter 7, page 167.

Question #37
A pumper with a 500-gallon tank flowing 125 gpm will last about 4 minutes.
$125 \text{gallons per minute} \overline{) 500 \text{gallons}}$ = 4 minutes. NFPA 1002 5.2.1. *IFPO, 2E:* Chapter 7, page 167.

Question #38
Hydrant flow testing equipment include:

- Pressure gauge mounted on an outlet cap (calibrated within the past 12 months)
- Pitot gauge for each hydrant
- Hydrant diffuser
- Hydrant wrenches
- Portable radios

A fire department pumper is not required when conducting hydrant flow testing. NFPA 1002 5.2.1. *IFPO, 2E:* Chapter 7, pages 180 – 181.

Question #39
The illustration is a good example of a static or dry hydrant. NFPA 1002 5.2.1. *IFPO, 2E:* Chapter 7, pages 187 – 188.

Question #40
A general rule of thumb is to maintain 20 psi intake pressure on pumpers within a relay. NFPA 1002 5.2.3. *IFPO, 2E:* Chapter 7, page 192.

Question #41
Shuttle flow capacity is the volume of water than can be pumped without running out of water. NFPA 1002 5.2.1, 5.2.2. *IFPO, 2E:* Chapter 7, pages 200 – 201.

Question #42
Usually the pressure-regulating device is set before pump discharge pressures are maintained.

Step 1	Position apparatus, set parking brake, and let engine return to idle
Step 2	Engage the pump
Step 3	Provide water to intake side of pump (on-board, hydrant, draft)
Step 4	Set transfer valve (if so equipped)
Step 5	Open discharge lines
Step 6	Throttle to desired pressure
Step 7	Set the pressure regulating device
Step 8	Maintain appropriate flows and pressures

NFPA 1002 5.2.1, 5.2.2, 5.2.4. *IFPO, 2E:* Chapter 8, page 207.

Question #43
Centrifugal pumps should never operate without water.
Several caveats that come close to being universal among departments and manufacturers for the operation of centrifugal pumps include:

- Never operate the pump without water.
- Always keep water moving when operating the pump at high speeds.
- Never open, close, or turn controls abruptly.
- Always maintain awareness of instrumentation during pumping operations.
- Never leave the pump unattended.
- Always maintain constant vigilance to safety.

NFPA 1002 5.2.1, 5.2.2, 5.2.4. *IFPO, 2E:* Chapter 8, page 208.

Question #44
Transfer valve operation rule of thumb:

- Use *volume mode* when flows are greater than 50% of a pump's rated capacity, and pressures are less than 150 psi.
- Use *pressure mode* when flows are less than 50% of a pump's rated capacity, and pressures are greater than 150 psi.

NFPA 1002 5.2.1, 5.2.2, 5.2.4. *IFPO, 2E:* Chapter 8, pages 214 – 215.

Question #45
Because of the limited supply of water in the on-board tank, the transfer valve should be in pressure (series) mode. NFPA 1002 5.2.1, 5.2.2, 5.2.4. *IFPO, 2E:* Chapter 8, page 220.

Question #46
The three methods used to transfer power to the pump are:

- PTO
- Front crankshaft
- Split-shaft

A separate engine can be used to power the pump, but it is not considered the drive engine.

NFPA 1002 5.2.1, 5.2.2, 5.2.4. *IFPO, 2E:* Chapter 8, pages 210 – 212.

Question #47

Cavitation is caused by insufficient intake flow to match the discharge flow. Not operating enough discharge lines would not cause cavitation in that the pump has excess capacity to pump and adequate intake flow. NFPA 1002 5.1.1. *IFPO, 2E:* Chapter 9, pages 245 – 246.

Question #48

All the answer selections may be signs that a pump is cavitating, except the priming device. The priming device would not be operating when conditions are conducive to cavitation. NFPA 1002 5.1.1. *IFPO, 2E:* Chapter 9, pages 245 – 247.

Question #49

Supporting a sprinkler system may require the following:

- Connecting at least one line to the FD connection
- Pumping at 150 psi unless otherwise directed
- Pumping immediately when fire and smoke are visible
- Not shutting down sprinkler system for improved visibility
- Placing transfer valve into the volume mode
- Assuring that water supply to the pump does not reduce the sprinkler system water supply

NFPA 1002 5.2.4. *IFPO, 2E:* Chapter 9, pages 252 – 253.

Question #50

Standpipe systems are classified based on intended use as follows:

- *Class 1* standpipes provide 2 ½-inch connections for trained firefighters/fire brigades, at an initial flow rate of 500 gpm.
- *Class 2* standpipes provide 1 ½-inch connections for initial attack, at a minimum flow rate of 100 gpm.
- *Class 3* standpipes provide 1 ½-inch and 2 ½-inch connections for trained firefighters/fire brigades, at an initial flow rate of 500 gpm.

NFPA 1002 5.2.4. *IFPO, 2E:* Chapter 9, page 253.

Question #51

Increased pressure reduces the ability of a liquid to evaporate (increase the boiling point). Decreased pressure increases the ability of a liquid to evaporate (lowers the boiling point). NFPA 1002 5.2.1. *IFPO, 2E:* Chapter 9, page 245.

Question #52

The outside stem and yoke (OS&Y) valve allows for quick determination of valve position. If the stem is out and exposed, the valve is open. If the stem is in or only protruding a short distance, the valve is closed. NFPA 1002 5.1.1. *IFPO, 2E:* Chapter 9, page 249.

Question #53

- *Hydraulics* is that branch of science dealing with the principles of fluid at rest or in motion.
- *Hydrodynamics* is that branch of hydraulics that deals with the principles and laws of fluids in motion.
- *Hydrostatics* is that branch of hydraulics that deals with the principles and laws of fluids at rest and the pressures they exert or transmit.

NFPA 1002 5.2. *IFPO, 2E:* Chapter 10, page 261.

Question #54

1 pound of water weighs 1 pound not 8.34 lbs!

The basic characteristics and properties of water are listed below. These are considered approximate in that water purity, atmospheric conditions, and rounding can effect the values.

- Virtually incompressible
- Freezes at 32°F
- Expands when frozen
- Boils at 212°F
- 1 gallon weighs 8.34 lbs
- Density is 62.4 lb/ft^3
- Number of gallons in 1 cubic foot is 7.38 gallons/ft^3

NFPA 1002 5.2.1, 5.2.2. *IFPO, 2E:* Chapter 10, pages 262 – 263.

Question #55

- *Latent heat of fusion* is the amount of heat that is absorbed by a substance when changing from a solid to a liquid state.
- *Specific heat* is the amount of heat required to raise the temperature of a substance by 1°F. the specific heat of water is 1 Btu/lb.
- *Latent heat of vaporization* is the amount of heat absorbed when changing from a liquid to a vapor state.
- *Evaporation* is the physical change of state from a liquid to a vapor.

NFPA 1002 5.2.1, 5.2.2. *IFPO, 2E:* Chapter 10, pages 263 – 264.

Question #56

Pressure is the force exerted by a substance in *units of weight per area*, typically expressed in *pounds per square inch* (psi or lb/in^2).

P = Pressure (psi)

F = Force (weight in pounds)

A = Area (square inches)

NFPA 1002 5.2.1, 5.2.2. *IFPO, 2E:* Chapter 10, pages 269 – 270.

Question #57

To calculate weight, the formula W = D × V is used. NFPA 1002 5.2. *IFPO, 2E:* Chapter 10, page 265.

Question #58

The operating pressure for smooth-bore nozzles on master-stream devices is 80 psi. The square root of 80 is 8.944. An acceptable value for fireground calculations is 9. NFPA 1002 5.2. *IFPO, 2E:* Chapter 10, pages 296 – 298.

Question #59

Iowa State formula: $NF = \dfrac{V}{100}$

where NF = needed flow in gpm

V = volume of the area in cubic feet

100 = is a constant in ft^3 gpm

NFPA 1002 5.2. *IFPO, 2E:* Chapter 10, pages 292 – 293.

Question #60

Elevation formula by floor level:

$$EL = 5 \times H$$

where EL = the gain or loss of elevation in psi
            5 = gain or loss in pressure for each floor level
            H = height in number of floor levels above or below the pump

Elevation formula in feet:

$$EL = 0.5 \times H$$

where EL = the gain or loss of elevation in psi
            0.5 = pressure exerted at base of 1-cubic-inch column of water 1 foot high
            H = distance in feet above or below the pump

NFPA 1002 5.2. *IFPO, 2E:* Chapter 11, page 311.

Question #61

Appliance friction varies between 5 to 15 psi for common appliances used in pump operations.

$$PDP = NP + FL + AFL \pm EL$$

where PDP = pump discharge pressure
            NP = nozzle pressure
            FL = friction loss in hose
            AFL = appliance friction loss
            EL = elevation gain or loss

NFPA 1002 5.2. *IFPO, 2E:* Chapter 12, pages 317 – 319.

Question #62

There are 7.48 gallons of water in 1 cubic foot. NFPA 1002 5.2. *IFPO, 2E:* Chapter 10, page 267.

Question #63

NFA formula: $NF = \dfrac{A}{3}$

where NF = needed flow in gpm
            A = area of the structure in square feet
            3 = constant in ft^2/gpm

Iowa State formula: $NF = \dfrac{V}{100}$

where NF = needed flow in gpm
            V = volume of the area in cubic feet
            100 = is a constant in ft^3 gpm

NFPA 1002 5.2. *IFPO, 2E:* Chapter 10, pages 292 – 293.

Question #64

There are 144 cubic-inch columns of water in 1 cubic foot. NFPA 1002 5.2. *IFPO, 2E:* Chapter 10, page 273.

Question #65

A gauge that measures psia at sea level would have a reading of 14.7 psia, because psia (absolute pressure) includes atmospheric pressure.

A gauge reading of 100 psi (psig) is actually 114.7 psia, because psig (gauge pressure) will read 0 psi at sea level. NFPA 1002 5.2. *IFPO, 2E:* Chapter 10, page 275.

Question #66

Based on the percent drop in pressure, additional flows may be available from a hydrant as follows:

- 0 – 10% drop   three times the original flow
- 11 – 15% drop   two times the original flow
- 16 – 25% drop   one time the original flow

NFPA 1002 5.2. *IFPO, 2E:* Chapter 10, page 277.

Question #67

The Iowa State formula can be used for an entire structure or a section of the structure, and is based on volume. NFPA 1002 5.2. *IFPO, 2E:* Chapter 11, pages 292 – 293.

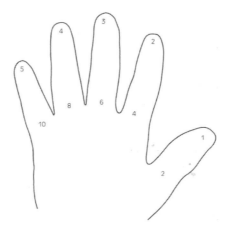

Question #68

The hand method is a fireground method used to estimate friction loss in 100-foot sections of 2 ½-inch hose. Simply select the fingertip representing the gpm flow and then multiply the two figures on the finger for the approximate friction loss pressure for each 100-foot section of 2 ½-inch hose.

   350 gpm = 7 × 3.5 = 24.5 psi per 100-foot sections of 2½-inch hose NFPA 1002 5.2.1, 5.2.2. *IFPO, 2E:* Chapter 10, pages 302 – 303.

Question #69

The drop-ten method simply subtracts 10 from the first 2 numbers of gpm flow. It is not as accurate as other methods, but provide a simple rule of thumb for fireground use.

   150 gpm = 15 – 10 = 5 psi friction loss per 100-foot sections of 2 ½-inch hose

NFPA 1002 5.2.1, 5.2.2. *IFPO, 2E:* Chapter 10, page 303.

Question #70

$$cq^2L$$

$$.2 \times \frac{500}{100} \times \frac{100}{100}$$

$$.2 \times 5 \times 1$$

1 psi friction loss

NFPA 1002 5.2. *IFPO, 2E:* Chapter 11, pages 306 – 307.

Question #71

$$NR = 1.57 \times d^2 \times NP$$

where NR = nozzle reaction
           1.57 = constant
           d = diameter of nozzle orifice in inches
           NP = operating nozzle pressure in psi

NFPA 1002 5.2.4. *IFPO, 2E:* Chapter 10, pages 282 – 283.

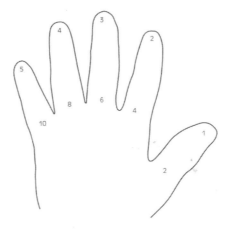

Question #72
The hand method is a fireground method used to estimate friction loss in 100-foot sections of
2 ½-inch hose. Simply select the fingertip representing the gpm flow and then multiply the two
figures on the finger for the approximate friction loss pressure for each 100-foot section of
2 ½-inch hose.

   250 gpm = 4.5 × 9 = 40.5 psi per 100-foot section of 2 ½-inch hose.

NFPA 1002 5.2.1, 5.2.2. *IFPO, 2E:* Chapter 10, pages 302 – 303.

Question #73
The drop-ten method simply subtracts 10 from the first two numbers of gpm flow. It is not as
accurate as other methods, but provides a simple rule of thumb for fireground use. NFPA 1002 5.2.1,
5.2.2. *IFPO, 2E:* Chapter 10, page 303.

Question #74

$2q^2 + q$

$$2 \times \left(\frac{225}{100} \right)^2 + \left(\frac{225}{100} \right)$$

$$2 \times (2.25)^2 + (2.25)$$

$$2 \times 5.06 + 2.25$$

12.37 or 12 psi friction loss

NFPA 1002 5.2. *IFPO, 2E:* Chapter 11, pages 304 – 305.

Question #75

$2q^2 + q$

$2 \times \left(\dfrac{310}{100} \right)^2 + \left(\dfrac{310}{100} \right)$

$2 \times (3.1)^2 + (3.1)$

$2 \times 9.61 + 3.1$

22.32 or 22 psi friction loss

NFPA 1002 5.2. *IFPO, 2E:* Chapter 11, pages 304 – 305.

Question #76

cq^2L

$24 \times \left(\dfrac{125}{100} \right)^2 \times \left(\dfrac{100}{100} \right)$

$24 \times (1.25)^2 \times (1)$

$24 \times 1.56 \times 1$

37.5 psi friction loss

NFPA 1002 5.2. *IFPO, 2E:* Chapter 11, pages 306 – 307.

Question #77

cq^2L

$2 \times \left(\dfrac{300}{100} \right)^2 \times \left(\dfrac{100}{100} \right)$

$2 \times (3)^2 \times (1)$

$2 \times 9 \times 1$

18 psi friction loss

NFPA 1002 5.2. *IFPO, 2E:* Chapter 11, pages 306 – 307.

Question #78

q^2

$\left(\dfrac{350}{100} \right)^2$

$(3.5)^2$

12.25 psi friction loss

NFPA 1002 5.2. *IFPO, 2E:* Chapter 11, pages 306 – 307.

Question #79

q^2

$\left(\dfrac{400}{100} \right)^2$

$(4)^2$

16 psi friction loss

NFPA 1002 5.2. *IFPO, 2E:* Chapter 11, pages 306 – 307.

Question #80

Elevation formula in feet:

$EL = 0.5 \times H$

where EL = the gain or loss of elevation in psi
0.5 = pressure exerted at base of 1-cubic-inch column of water 1 foot high
H = distance in feet above or below the pump

$EL = 0.5 \times 14$
$EL = 7$ psi

NFPA 1002 5.2. *IFPO, 2E:* Chapter 11, page 311.

Question #81

Elevation formula in feet:

$EL = 0.5 \times H$

where EL = the gain or loss of elevation in psi
0.5 = pressure exerted at base of 1-cubic-inch column of water 1 foot high
H = distance in feet above or below the pump

$EL = 0.5 \times -10$
$EL = -5$ psi

NFPA 1002 5.2. *IFPO, 2E:* Chapter 11, page 311.

Question #82

Elevation formula in feet:

$EL = 5 \times H$

where EL = the gain or loss of elevation in psi
5 = gain or loss in pressure for each floor level
H = height in number of floor levels above or below the pump

$EL = 5 \times 9$
$EL = 45$ psi

NFPA 1002 5.2. *IFPO, 2E:* Chapter 11, page 311.

Question #83

Elevation formula by floor level:

$EL = 5 \times H$

where   EL = the gain or loss of elevation in psi
5 = gain or loss in pressure for each floor level
H = height in number of floor levels above or below the pump

EL = 5 × –1
$EL = -5$ psi

NFPA 1002 5.2. *IFPO, 2E:* Chapter 11, page 311.

Question #84

$PDP = NP + FL + AFL \pm EL$

where PDP = pump discharge pressure
NP = nozzle pressure
FL = friction loss in hose
AFL = appliance friction loss
EL = elevation gain or loss

$PDP = 80$ psi + 112 psi
$PDP = 192$ psi

NFPA 1002 5.2.4. *IFPO, 2E:* Chapter 12, pages 317 – 319.

Question #85

$PDP = NP + FL + AFL \pm EL$

where
PDP = pump discharge pressure
NP = nozzle pressure
FL = friction loss in hose
AFL = appliance friction loss
EL = elevation gain or loss

PDP = 100 psi + 36 psi + 10 psi
PDP = 146

NFPA 1002 5.2.4. *IFPO, 2E:* Chapter 12, pages 317 – 319.

Question #86

- *Dry-pipe systems* maintain air or compressed gas under pressure within the system. When a head fuses, water enters the system and discharges through any fused heads.

- *Pre-action systems* are similar to dry-pipe systems in that air or compressed gas is maintained in the system. However, an automatic detection system (smoke, heat, flame, etc.) or manual system (pull box) must operate to allow water to enter the system. At this point, it is similar to a wet-pipe system; when a sprinkler head fuses, water discharges.

- *Deluge systems* maintain all sprinkler heads in an open position. When a detection system operates, water enters the system and is discharged through all the open heads.

- *Wet-pipe systems* maintain water in the system at all times. When a sprinkler head is fused, water immediately discharges through the fused heads.

NFPA 1002 5.2.4. *IFPO, 2E:* Chapter 9, pages 148 – 253.

Question #87

- *Dry-pipe systems* maintain air or compressed gas under pressure within the system. When a head fuses, water enters the system and discharges through any fused heads.

- *Pre-action systems* are similar to dry-pipe systems in that air or compressed gas is maintained in the system. However, an automatic detection system (smoke, heat, flame, etc.) or manual system (pull box) must operate to allow water to enter the system. At this point, it is similar to a wet-pipe system; when a sprinkler head fuses, water discharges.

- *Deluge systems* maintain all sprinkler heads in an open position. When a detection system operates, water enters the system and is discharged through all the open heads.

- *Wet-pipe systems* maintain water in the system at all times. When a sprinkler head is fused, water immediately discharges through the fused heads.

NFPA 1002 5.2.4. *IFPO, 2E:* Chapter 9, pages 148 – 253.

Question #88

$Q = 30 \times d^2 \times \sqrt{NP}$

$Q = 30 \times 0.5^2 \times \sqrt{50}$

$Q = 30 \times .25 \times 7$

Q = 52.5 or 53 gpm

NFPA 1002 5.2. *IFPO, 2E:* Chapter 11, page 296 – 298.

Question #89

$$Q = 30 \times d^2 \times \sqrt{NP}$$

$$Q = 30 \times 1^2 \times \sqrt{70}$$

$$Q = 30 \times 1 \times 8.36$$

$Q = 250.8$ or 251 gpm

NFPA 1002 5.2. *IFPO, 2E:* Chapter 11, page 296 – 298.

Question #90

$$NR = 1.57 \times d^2 \times NP$$

$$NR = 1.57 \times 1.5^2 \times 80$$

$$NR = 1.57 \times 2.25 \times 80$$

$NR = 282.6$ or 283 lb

NFPA 1002 5.2. *IFPO, 2E:* Chapter 10, pages 282 – 283.

Question #91

$$NR = gpm \times \sqrt{NP} \times 0.0505$$

$$NR = 100 \times \sqrt{70} \times 0.0505$$

$$NR = 100 \times 8.37 \times 0.0505$$

$NR = 42.26$ or 42 lb

NFPA 1002 5.2. *IFPO, 2E:* Chapter 10, pages 282 – 283.

Question #92

Centrifugal force is the tendency of an object when moving in a circular pattern to move outward from the center. NFPA 1002 4.3.1, 4.3.6. *IFPO, 2E:* Chapter 3, pages 56 – 57.

Question #93

Total stopping distance is measured from the time a hazard is detected until the vehicle comes to a complete stop. Total stopping distance consists of:

- *Perception distance* (distance apparatus travels from the time the hazard is seen until the brain recognizes it as a hazard)
- *Reaction distance* (distance apparatus travels from the time the brain sends the message to depress the brakes until the brakes are depressed)
- *Braking distance* (distance of travel from the time the brake is depressed until the vehicle comes to a complete stop)

NFPA 1002 4.3. *IFPO, 2E:* Chapter 3, page 55 – 65.

Question #94

The C in the API oil classification indicates use with diesel engines. NFPA 1002 4.2.1, 5.1.1. *IFPO, 2E:* Chapter 2, pages 32 – 33.

Question #95

The purpose of NFPA 1915, Fire Apparatus Preventive Maintenance Program, is to help ensure fire apparatus are maintained in a ready state and in safe operating condition. The standard requires that preventive maintenance inspections be conducted as required by the manufacturer. NFPA 1002 4.2. *IFPO, 2E:* Chapter 2, page 25.

Question #96

Flow meters are increasingly finding their way onto the pump panels because they reduce fireground hydraulic calculations and, because of the use of "smart" electronics, they are easy to calibrate and increase the versatility on pump panel location.

Pressure gauges are still needed to ensure safety from excessive pressures. NFPA 1002 5.2. *IFPO, 2E:* Chapter 5, page 113.

Question #97

Fire department connections are usually a siamese appliance with two 2 ½-inch clapper inlets. NFPA 1002 5.2.4. *IFPO, 2E:* Chapter 9, page 252.

Question #98

A siamese is used to combine two or more lines into one line. NFPA 1002 5.2. *IFPO, 2E:* Chapter 6, page 149.

Question #99

Laminar flow usually causes minimal friction loss, whereas each of the other items causes turbulent flow that increases friction loss. NFPA 1002 5.2. *IFPO, 2E:* Chapter 10, pages 179 – 282.

Question #100

Feathering or gating is the process of slowly opening/closing discharge control valves to adjust discharge settings. NFPA 1002 5.2. *IFPO, 2E:* Chapter 9, pages 237 – 238.

PHASE III

SYNTHESIS & EVALUATION

Referring to Table I-1, the final levels of Bloom's Taxonomy, Cognitive Domain, are covered in section three. Mastery of this section suggests the highest level of understanding of the material. The levels addressed are:

- synthesis

- evaluation

The successful test-taker should be able to rearrange material, modify processes, compare data, and interpret results. For testing at this level, questions will be tied around more of an application process. The student should be able to apply the information learned in the class or in the textbook.

Phase Three, Exam One

1. Although a variety of pump sizes and configurations exist in the fire service today, the general operation of most pumps is basically similar.
   a. True
   b. False

2. If a problem is found during a vehicle inspection and can be readily fixed, there is no need for the pump operator to document the finding.
   a. True
   b. False

3. 100% slippage can occur in centrifugal pumps because they have an open path from the intake to the discharge side of the pump.
   a. True
   b. False

4. Positive-displacement pumps theoretically discharge a varying quantity of water inversely related to each revolution or cycle of the pump.
   a. True
   b. False

5. After arriving on the scene, the driver must safely position the apparatus. One positioning consideration is to never park the apparatus on railroad tracks.
   a. True
   b. False

6. At the scene of a structure fire, you just opened and set the discharge pressures for two lines and then set the pressure-relief valve. Almost immediately, both lines are shut down, and the engine speed slows considerably. You are not concerned because changing engine speed is how the pressure-relief valve limits pressure surges.
   a. True
   b. False

7. As a pump operator, you have several main duties or functions. According to NFPA 1002, these duties consist of driving the apparatus, operating the pump, preventive maintenance, and supervision.
   a. True
   b. False

8. NFPA 1002 requires that driver pump operators be medically fit, licensed to drive, and meet Firefighter I requirements. The standard also includes knowledge and skill requirements for the following duties of the driver pump operator *except*
   a. preventive maintenance.
   b. driving the apparatus.
   c. pump operations.
   d. first-line supervisory responsibilities.

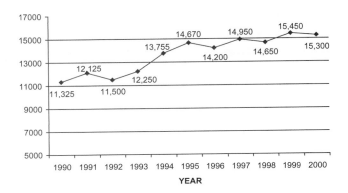

9. The *NFPA Journal* reports on emergency-vehicle accident information, and a summary of the number of accidents is provided in the above graph. Given this information, the trend of vehicle accidents over the next five years will most likely
   a. increase dramatically.
   b. continue to slightly increase or stay about the same.
   c. decline dramatically.
   d. none of the answers is correct.

10. Exemptions extended to the operation of emergency vehicles are usually found in
    a. NFPA 1002.
    b. NFPA 1501.
    c. local and state laws.
    d. federal codes.

11. Your fire department recently submitted its annual budget. The mayor is attempting to reduce its budget and is questioning the need for a preventive maintenance program. All of the following are good justifications for a preventive maintenance program *except:*
    a. The time and money spent on a preventive maintenance program is significantly less than the potential damage likely to occur when preventive maintenance is not conducted.
    b. A poor preventive maintenance program can increase the frequency and cost of repairs and reduce vehicle reliability.
    c. Criminal and civil liability may occur when emergency apparatus are not properly maintained.
    d. Emergency apparatus would not look as good and may cause a negative public relations image.

12. Your department is researching the requirements for a new apparatus. The fire chief and assistant fire chief have been discussing what the requirements should be for quick-buildup air tanks. The chief feels the tanks should reach operating pressure within 30 seconds, while the assistant chief feels at least 2 minutes are required. They have asked you for your opinion. You tell them that, according to NFPA 1901, new apparatus must have a quick-buildup capable of reaching operating pressure within _____.
    a. 30 seconds
    b. 60 seconds
    c. 2 minutes
    d. 4 minutes

13. Refueling of apparatus should occur per department policy or, if no policy exists, when the fuel level reaches the _____ mark.
    a. one-quarter
    b. one-half
    c. three-quarter
    d. one-third

14. A 20-inch steering wheel should have no more than _____ free play in either direction.
    a. 1 inch
    b. 20 degrees
    c. 2 inches
    d. 2 degrees

15. Accidents involving responding apparatus can produce significant outcomes, including
    a. a delay in assistance to those who summoned help.
    b. fire department members and civilians seriously or fatally injured.
    c. the city, department, and driver facing civil and/or criminal proceedings.
    d. each of these is a potential outcome.

16. Most state laws require that emergency-vehicle drivers obey the same laws as other vehicle operators unless specifically exempt from doing so. State laws typically define several conditions that must exist for exemptions to be extended. Which of the following would most likely meet one or more of the conditions?
    a. returning from an incident
    b. responding to an incident with audible and visual warning devices
    c. responding to an incident during good weather (no rain, snow, fog, etc.)
    d. driving 10 mph over the posted speed limit

17. You are the driver of a fire department vehicle responding to a reported structure fire. In which of the following conditions do you not have to bring the apparatus to a complete stop?
    a. at negative right-of-way and blind intersections
    b. at unguarded railroad guard crossing
    c. at school crossings
    d. when a stopped school bus with flashing warning lights is encountered

18. Can a water tanker with a capacity of 2,000 gallons cross over a bridge with a weight limit of 8 tons?
    a. Yes, but only if half the water is drained first.
    b. Yes, the water weighs less than the authorized limit, and the weight of the vehicle would be less than 1 ton.
    c. No, the water alone weighs more than the authorized limit.
    d. No, the water weighs less than the authorized limit but the weight of the tanker exceeds the authorized limit.

19. A new driver seems to have a difficult time backing the apparatus into the fire station. Which of the following driving exercises might help the new driver?
    a. serpentine
    b. alley dock
    c. confined-space turnaround
    d. diminishing clearance

20. Positive-displacement pumps theoretically discharge (displace) a specific quantity of water for each revolution or cycle of the pump. In reality, _____ occurs.
    a. higher pressure
    b. compression
    c. increased speed
    d. slippage

21. With centrifugal pumps, if the speed of the pump is held constant and the flow of water increases, pressure will
    a. increase.
    b. decrease.
    c. remain the same.
    d. none of the answer choices is correct.

22. Which of the following rated-capacity charts is correct for a 1,000-gpm pumper?
    a. 1,000 gpm at 100 psi
        700 gpm at 200 psi
        500 gpm at 250 psi
    b. 1,000 gpm at 150 psi
        700 gpm at 200 psi
        500 gpm at 250 psi
    c. 1,000 gpm at 100 psi
        750 gpm at 200 psi
        500 gpm at 250 psi
    d. 1,000 gpm at 100 psi
        700 gpm at 200 psi
        500 gpm at 300 psi

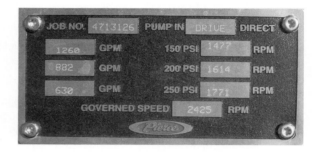

23. The above rated capacity plate is for a _____-gpm pump.
    a. 500
    b. 1,000
    c. 1,250
    d. insufficient information provided

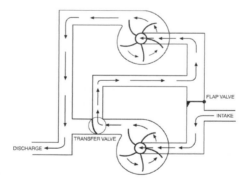

24. The flow of water through the centrifugal pump shown above indicates that it is
    a. in volume mode.
    b. in pressure mode.
    c. a single-stage pump.
    d. none of the answers is correct.

25. A foam system that requires foam to be added to the on-board water tank is called a
    a. pre-mixed system.
    b. in-line eductor system.
    c. around-the-pump proportioning system.
    d. balanced-pressure system.

26. The _____ and _____ control valves can be use together to help keep the pump from overheating by circulating water from the pump to the tank and then back to the pump again.
    a. tank-to-pump, transfer valve
    b. pump-to-tank, intake
    c. intake, discharge
    d. tank-to-pump, pump-to-tank

27. The basic steps for conducting a hose test are provided below. What is the missing step?

    1. Connect water supply and test hose sections
    2. Open discharge lines for test hose sections
    3. Slowly increase discharge pressure to 45 psi
    4. _____
    5. Mark hose near coupling to determine slippage
    6. Increase pressure slowly (NFPA 1962 suggests no faster than 15 psi per second) to test pressure
    7. Maintain pressure for three minutes, periodically checking for leaks
    8. Record results

    a. Check hose for leaks
    b. Clear nonessential personnel from the testing area
    c. Set pressure-relief device
    d. Remove air from within hose

28. When conducting hose testing, the instrumentation used during the test should be calibrated
    a. just prior to the test.
    b. within the past six months.
    c. within the past year.
    d. within the past two years.

29. According to NFPA 1965, a gated wye with a handle that is in a position perpendicular to the hose line is
    a. open
    b. closed.
    c. exactly half open.
    d. not a requirement of the standard.

30. A 5,000-gallon tanker is pumping two lines, each flowing 125 gpm. How long will the on-board water supply last?
    a. 5 minutes
    b. 10 minutes
    c. 15 minutes
    d. 20 minutes

31. When flow-testing hydrants, the next major step after taking a static pressure reading at the test hydrant is to
    a. slowly shut down hydrants one at a time.
    b. record exact interior size, in inches, of each outlet flowed.
    c. open flow hydrants one at a time until a 25% drop in residual pressure is achieved.
    d. continue to flow to clear debris and foreign substances.

32. During a drafting operation, the pressure within the pump is reduced by 7 psi. To what height can the water be lifted/forced?
    a. 14.7 feet
    b. 16 feet
    c. 22.3 feet
    d. none of the answers is correct.

33. Which of the following water supplies is usually the easiest to secure yet often the most limited?
    a. static hydrants
    b. municipal hydrants
    c. on-board tanks
    d. ponds and lakes

34. How far should the second pumper be positioned in a relay if the first pumper is flowing 500 gpm through a 4-inch hose at 185 psi (5 psi friction loss per 100 feet)?
    a. 500 feet
    b. 3,300 feet
    c. 5,000 feet
    d. 5,300 feet

35. Determine the individual shuttle tanker flow for a 1,500-gallon tanker with a 10-minute shuttle cycle time.
    a. 100 gpm
    b. 150 gpm
    c. 500 gpm
    d. 1,000 gpm

36. The same basic steps must be taken to move water from the supply to the discharge point. What is the next step after opening discharge lines?
    a. set transfer valve
    b. throttle to desired pressure
    c. set the pressure regulating device
    d. engage the pump

37. When positioning a pumper at an incident, all of the following should be considered *except*
    a. following department SOPs.
    b. evaluating available water supply.
    c. tactical considerations.
    d. assessing surroundings.
    e. finding easiest location to stop.

38. When engaging a pump utilizing a PTO, the transmission should be in _____ when operating the PTO lever.
    a. neutral
    b. first or lowest gear
    c. fourth or highest gear
    d. reverse

39. When engaging a pump through a split-shaft arrangement, the transmission should
    a. stay in neutral.
    b. be placed in the highest gear after switching the pump shift control from road to pump.
    c. stay in low gear.
    d. be placed in pump drive.

40. The transfer valve should be in the _____ position for a 1,000-gpm pumper expecting to flow 750 gpm at 140 psi.
    a. volume
    b. pressure
    c. off
    d. on

41. After discharge lines are connected, the pump operator starts to increase the pump discharge pressure. When the throttle is slowly increased, a corresponding increase in the main discharge gauge is not noted. What is a common cause of this?
   a. pump is not in gear
   b. pump is not primed
   c. supply line is not open or insufficient
   d. all of these are common causes

42. During a pumping operation, the pump operator attempts to increase the pressure in the only flowing line. As the throttle is increased, a corresponding increase in pressure does not occur. The pump operator then checks the relief device; it is functioning properly, but is not relieving any pressure at this time. What might occur if the pump operator continues to increase the throttle (engine speed) hoping to increase the discharge pressure?
   a. pump may cavitate
   b. may lose prime
   c. intake may collapse
   d. may cause damage to the municipal water main
   e. all of these could occur

43. Place the following on-board water supply procedures in the correct order:
   1. Open "talk-to-pump"
   2. Engage pump
   3. Set pressure regulating device
   4. Increase throttle

   a. 1, 2, 3, 4
   b. 2, 1, 4, 3
   c. 2, 4, 1, 3
   d. 1, 2, 4, 3

44. Selecting a location to position fire apparatus should include evaluating surroundings, such as
   a. the potential exposure to radiant heat and fire extension.
   b. the potential for collapse.
   c. wind direction, power lines, and escape routes.
   d. all are correct.

45. The quickest way to stop a pump from cavitating is to reduce the pump speed. Other ways to stop cavitation include all of the following *except*
   a. increase the supply flow.
   b. remove one or more discharge lines.
   c. add one or more discharge lines.
   d. reduce nozzle gpm settings.

46. When cavitation is occurring within a centrifugal pump, two pressure zones exists. The _____-pressure zone occurs near the center of the impeller, while the _____-pressure zone occurs near the outer edge of the impeller.
   a. high, low
   b. low, high
   c. neutral, high
   d. high, balanced

47. You are conducting pump operations at a large fire. The temperature outside is about 25°F. You should do each of the following *except*
   a. not stop the flow of water within the pump.
   b. stop leaks to reduce the accumulation of ice.
   c. immediately drain water from all hoses when the pumping operation is completed
   d. all are correct.

48. You are the second arriving pumper on scene and have been asked to connect to the fire department connection. The firefighter connecting the discharge lines to the fire department connection informs you that it is marked with "STANDPIPE." This means that
    a. the system is a combination sprinkler and standpipe.
    b. the system is only a standpipe.
    c. the pump operator should pump an additional 25 psi for the standpipe.
    d. both b and c are correct.

49. A Class 2 standpipe system has a minimum flow rate of _____ gpm.
    a. 100
    b. 125
    c. 150
    d. 175

50. All of the following are correct concerning troubleshooting pump problems *except:*
    a. Problems are usually either procedural or mechanical.
    b. Use the manufacturer's troubleshooting guides when available.
    c. Only certified mechanics should attempt to troubleshoot pumping problems.
    d. The best method to troubleshoot is to follow the flow of water from the intake to the discharge while attempting to determine the problem.

51. What is the weight of water in a tank with a volume of 100 cubic feet?
    a. 100 lbs
    b. 834 lbs
    c. 8,340 lbs
    d. 6,234 lbs

52. The nozzle reaction for a ⅛-inch tip smooth-bore hand-line operating at the correct nozzle pressure is
    a. 1 psi.
    b. 5 psi.
    c. 10 psi.
    d. 15 psi.

53. Two hose lines are flowing 125 gpm each into a structure. If the flow is sustained for 10 minutes, how many tons of water were delivered into the structure?
    a. 5 tons
    b. over 10 tons
    c. over 20 tons
    d. not enough information is provided

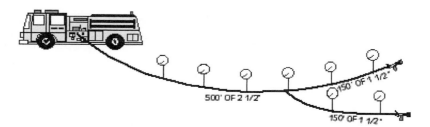

54. The above drawing illustrates which pressure principle? The wyed lines are not flowing, and the pressure readings are all the same.
    a. Pressure of a liquid acting on a surface is perpendicular to that surface.
    b. Pressure at any point in a liquid at rest is equal in every direction.
    c. Pressure exerted at the bottom of a container is independent of the shape or volume of the container.
    d. External pressure applied to a confined liquid (fluid) is transmitted equally throughout the liquid.

55. During a pump operation, an initial static pressure reading of 50 psi was noted. After initiating flow of a 125-gpm pre-connect, the residual pressure was 40 psi. How many more gpm is the hydrant capable of flowing?
    a. 125 gpm
    b. 250 gpm
    c. 375 gpm
    d. 0 gpm

56. Calculate friction loss using the hand method for 300 feet of 2 ½-inch hose flowing 250 gpm.
    a. 3.5 psi
    b. 37.5 psi
    c. 48.5 psi
    d. 58.5 psi

57. Calculate friction loss in 400 feet of 2 ½-inch hose flowing 250 gpm, using the drop-ten method.
    a. 15 psi
    b. 30 psi
    c. 60 psi
    d. 90 psi

58. Calculate friction loss using the hand method for 300 feet of 2 ½-inch hose flowing 200 gpm.
    a. 8 psi
    b. 12 psi
    c. 24 psi
    d. 32 psi

59. The nozzle reaction for a combination nozzle operating at the correct nozzle pressure and flowing 100 gpm is
    a. 51 psi.
    b. 76 psi.
    c. 101 psi.
    d. 151 psi.

60. A 2 ½-inch hose line 200 feet in length is flowing 250 gpm. The friction loss was calculated to be 30 psi. Which fireground friction loss formula was used?
    a. cq^2L
    b. drop-ten
    c. hand method
    d. $2q^2+q$

61. As the pump operator, you are pre-planning the potential fire flow needs for a structure that is 100 feet long by 60 feet wide, and 12 feet tall. Using the Iowa State formula, calculate the needed flow for this structure.
    a. 250 gpm
    b. 500 gpm
    c. 720 gpm
    d. 2,000 gpm

62. You are preplanning the potential fire flow needs for a part of a structure that is 50 feet long by 25 feet wide, and 12 feet tall. Using the NFA formula, calculate the needed flow for this area.
    a. 150 gpm
    b. 417 gpm
    c. 720 gpm
    d. 2,000 gpm

63. You are the pump operator at a small structural fire. The attack crew is setting up 200 feet of 1 ½-inch hose with a 95-gpm combination nozzle. What is the friction loss for this hose lay when using the formula cq^2L?
    a.   32 psi
    b.   43 psi
    c.   53 psi
    d.   98 psi

64. A pump operator has been asked to support a standpipe system within a structure. The interior attack team is heading to the sixth floor, and will connect a high-rise pack consisting of 200 feet of 1 ¾-inch hose with a combination nozzle to the standpipe. The pump operator should increase pump pressure by _____ psi to account for the change in elevation.
    a.   25
    b.   30
    c.   150
    d.   175

65. A relay system operating with three pumpers is moving water from a hydrant to the fire, a total of about 750 feet. The second pumper in the relay is 48 feet above the supply pumper. The supply pump operator should increase pump pressure by _____ psi to account for the change in elevation.
    a.   24
    b.   124
    c.   178
    d.   240

66. You are suppling a pumper in a relay flowing 400 gpm through 500 feet of 3-inch hose. Using the condensed q formula, calculate the friction in the hose line.
    a.   40 psi
    b.   50 psi
    c.   70 psi
    d.   80 psi

67. An attack crew is stretching out 350 feet of 2 ½-inch hose. What is the friction loss in the line if the flow is 200 gpm? Use the formula $2q^2 + q$.
    a.   10 psi
    b.   21 psi
    c.   22 psi
    d.   35 psi

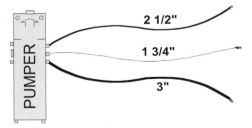

68. For the above pumping operation, if the 2 ½-inch line was 500 feet long flowing 350 gpm from a combination nozzle, then the pump discharge pressure would be _____. Use either the drop-ten method or the formula cq^2L.
    a.   150 psi
    b.   200 psi
    c.   223 psi
    d.   250 psi

69. You are the pump operator at the supply pumper in a relay operation. The hose lay between your pumper and the next pumper consists of 500 feet of 3-inch hose, followed by 250 feet of 2 ½-inch hose. What is the friction loss for the entire hose lay when flowing 325 gpm? Use the formula cq^2L.
    a. 32 psi
    b. 42 psi
    c. 53 psi
    d. 95 psi

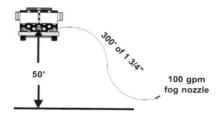

70. For the pumping operation shown above, the pump discharge pressure should be _____ psi.
    a. 122
    b. 147
    c. 177
    d. 222

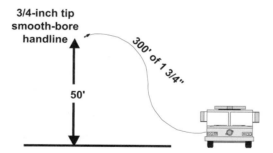

71. For the pumping operation shown above, the pump discharge pressure should be _____ psi.
    a. 115
    b. 140
    c. 187
    d. 222

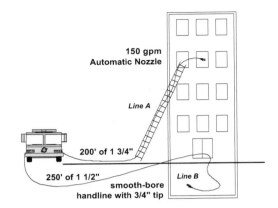

72. The pump discharge pressure for Line A in the figure above should be _____ psi.
    a. 15
    b. 104
    c. 139
    d. 154

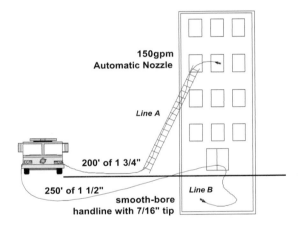

73. The pump discharge pressure for Line B in the figure above should be _____ psi.
    a. 25
    b. 51
    c. 101
    d. 140

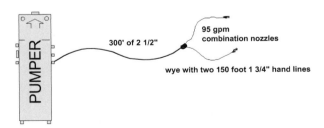

74. For the pumping operation shown above, the pump discharge pressure should be _____ psi.
    a. 100
    b. 153
    c. 177
    d. 222

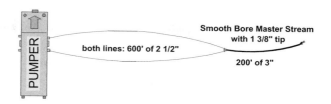

75. For the pumping operation shown above, the pump discharge pressure should be _____ psi.
    a. 100
    b. 153
    c. 186
    d. 220

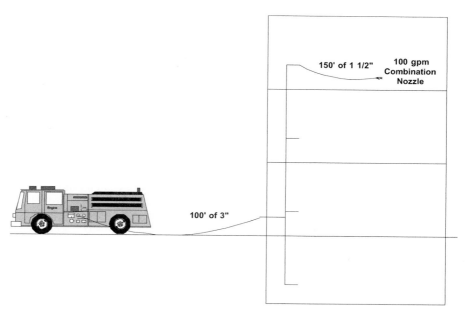

76. For the pumping operation shown above, the pump discharge pressure should be
    a. 150 psi.
    b. 172 psi.
    c. 186 psi.
    d. 220 psi.

77. Determine the PDP for the following hose line configuration:
    150 feet of 1 ¾-inch hose flowing 125 gpm through a combination (fog) nozzle, friction loss
    for the hose line is calculated to be 36 psi.
    a. PDP = 36 psi
    b. PDP = 100 psi
    c. PDP = 136 psi
    d. PDP = 161 psi

78. Determine the PDP for the following hose line configuration:
    300 feet of 2 ½-inch hose with an automatic nozzle following 200 gpm; friction loss for the
    hose line is calculated to be 24 psi. The line is advanced to the second level of the occupancy.
    a. PDP = 124 psi
    b. PDP = 129 psi
    c. PDP = 79 psi
    d. PDP = 109 psi

79. In the pressure mode of a four-stage centrifugal pump, if the total discharge is 500 gpm at 300 psi, then each impeller is discharging
    a. 500 gpm at 150 psi.
    b. 500 gpm at 75 psi.
    c. 250 gpm at 100 psi.
    d. 250 gpm at 75 psi.

80. You are operating the supply pumper in a relay and calculate that your pump discharge pressure should be 196 psi to ensure that 20 psi remains at the next pump. Your supply is a static water source. After priming the pump, you slowly open your discharge and begin flowing water to the next pumper through 4-inch hose. The pumper you are suppling water to is about 126 feet above your pumper, up a winding dirt road. After contacting the pump operator to verify 20 psi at the next pump, you set your pressure-relief device. Do you note any problems with this operation?
    a. No problem, everything is good.
    b. The pressure is too high.
    c. The pressure is too low.
    d. The relief device should be set prior to verifying 20 psi at the next pump.

81. You are the pump operator preparing to initiate flow for three like lines. You determine that one line requires a discharge pressure of 150 psi. For this operation, which of the following is not correct?
    a. To initiate flow, the pump operator can open one line at a time; for each line added, the discharge gate is slowly opened while the throttle is increased to maintain the PDP.
    b. PDP is 150 psi.
    c. To initiate flow, the pump operator can open all three lines at the same time and then increase the throttle to the PDP.
    d. PDP is 450 psi.

82. If the tanker shuttle time is 8 minutes, and the tanker shuttle flow capacity is 250 gpm, what is the size, in gallons, of the tanker?
    a. 250 gallons
    b. 1,250 gallons
    c. 2,000 gallons
    d. not enough information provided

83. How long can a 1,000-gallon on-board water tank sustain a 1 ¾-inch hand-line with a 1-inch tip smooth-bore nozzle?
    a. just under 5 minutes
    b. just over 5 minutes
    c. about 8 minutes
    d. not enough information provided

84. A flow meter on the pump panel reads 100 gpm when flowing water through a discharge line. The hand-line consists of 200 feet of 1 ¾-inch hose with an automatic nozzle operating on the second floor of a structure. How much water is the nozzle discharging?
    a. 100 gpm
    b. 131 gpm
    c. 150 gpm
    d. 164 gpm

85. About how many Btu does it take to raise 1 pound of water from 62°F to 212°F?
    a. 100
    b. 150
    c. 200
    d. 212

86. To test a driver's ability to drive in a straight line, such as on a narrow street or road, the _____ driving exercise should be used.
    a. straight line
    b. alley way and dock
    c. diminishing clearance
    d. confined-space turnaround

87. NFPA 1500 requires that all personnel riding on an apparatus be seated and properly secured with seat belts any time the vehicle is in motion. One exception to this requirement is
    a. for loading hose.
    b. for emergency response.
    c. when donning SCBA.
    d. no exceptions are allowed according to NFPA.

88. You are the driver operator responding to an incident. During the response, you come to an intersection but cannot account for all lanes of traffic. When such instances occur, you should
    a. slow to between 5 to 10 mph and proceed with extreme caution.
    b. come to a complete stop and proceed with caution after lanes and traffic are accounted for.
    c. slow down considerably, but do not stop to ensure the safety of the apparatus.
    d. reduce speed below 20 mph and proceed with caution.

89. If the flow of a centrifugal pump is held constant, and the speed of the pump is increased, then the pressure will
    a. increase.
    b. decrease.
    c. remain the same.
    d. none of the answer choices is correct.

90. You are the pump operator of the supply pumper in a relay operation. While priming the pump, you note a 7-psi pressure reduction on the master intake gauge. You know that the static water supply is over 18 feet below the pump. Will you have any difficulty priming the pump?
    a. No problem in that the general rule of thumb is to attempt to pump no more than 22.5 feet.
    b. Yes, you will have a problem in that 7 psi will only raise water to the height of 16 feet.
    c. No problem in that 7 psi can raise water to a height of 22 feet.
    d. None of the answers is correct.

91. After arriving on scene, you set up a drafting operation and successfully prime the pump. After setting the transfer valve to volume, the next step is to
    a. set transfer valve.
    b. throttle to desire pressure.
    c. set the pressure-regulating device.
    d. open discharge lines.

92. After arriving on scene, you set up a drafting operation and successfully prime the pump. However, after about two minutes, the pump loses it prime. You successfully prime the pump again, only to lose it about two minutes later. Which of the following is most likely *not* the cause?
    a. debris on intake strainer
    b. lift to high
    c. no oil in priming tank
    d. high point in intake line

93. Rotary vane pumps operate based on _____ principles, whereas centrifugal-pump operation is based on _____ principles.
    a. positive-displacement, dynamic
    b. hydrodynamic, hydrostatic
    c. internal pressure, external pressure
    d. hydrostatic, hydrodynamic

94. As the pump operator, you respond to a reported structure fire. Upon arrival, you determine the structure to be a trailer approximately 60 feet long, 30 feet wide, and 10 feet tall. Based on this information, what formula can you use to determine the estimated needed fire flow for the structure?
    a. Iowa State Formula
    b. National Fire Academy (NFA) Formula
    c. Condensed q Formula
    d. both a and c are correct
    e. both a and b are correct

95. You are the pump operator of the second arriving pumper. You are told to secure a water supply from a hydrant that is already supplying the first pumper. This is a strong hydrant in terms of pressure and volume. To secure the supply, you connect a section of hose from the unused intake of the first pumper to the intake of your pump. This pumping configuration is known as
    a. relay pumping.
    b. dual pumping.
    c. tandem pumping.
    d. pressure pumping.

96. The systematic movement of water from a supply source through a pump to a discharge point defines
    a. relay operations.
    b. drafting.
    c. pump operations.
    d. tandem pumping.

97. Conducting preventive maintenance inspections is an important duty of a pump operator. Which of the following standards requires that these inspections be documented?
    a. NFPA 1002
    b. NFPA 1500
    c. NFPA 1911
    d. all are correct

98. While responding to an incident, you notice a hazard in front of you. You travel 35 feet before the brain recognizes the hazard. You travel another 30 feet before your brains sends a signal to your feet and the brake is depressed. It takes another 125 feet before the vehicle comes to a complete stop. What is the total stopping distance?
    a. 65 feet
    b. 155 feet
    c. 160 feet
    d. 190 feet

99. Newton's first and third law of motion can be used to help explain the concept of centrifugal force. Which of the following is correct?
    a. Newton's first law of motion indicates that a moving body travels in a straight line with a constant speed unless affected by an outside force.
    b. Newton's third law of motion states that for every action there is an equal an opposite reaction.
    c. Both a and b are correct.
    d. Neither a nor b are correct.

100. The key to safely navigate a curve is to
    a. make the turn with the brakes slightly pressed.
    b. not turn the steering wheel too abruptly.
    c. maintain traction.
    d. reduce speed on ice.

Phase III, Exam I: Answers to Questions

1. T	26. D	51. D	76. B
2. F	27. D	52. A	77. C
3. T	28. C	53. B	78. B
4. F	29. B	54. D	79. B
5. T	30. D	55. C	80. B
6. F	31. C	56. B	81. D
7. F	32. B	57. C	82. C
8. D	33. C	58. C	83. A
9. B	34. B	59. A	84. A
10. C	35. B	60. B	85. B
11. D	36. B	61. C	86. C
12. B	37. E	62. B	87. A
13. C	38. A	63. B	88. A
14. C	39. B	64. A	89. A
15. D	40. A	65. A	90. B
16. B	41. D	66. D	91. D
17. C	42. E	67. D	92. C
18. C	43. B	68. C	93. D
19. B	44. D	69. D	94. E
20. D	45. C	70. A	95. B
21. B	46. B	71. B	96. C
22. B	47. D	72. D	97. B
23. C	48. B	73. B	98. D
24. A	49. A	74. B	99. C
25. A	50. C	75. D	100. C

Phase III, Exam I:
Rationale & References for Questions

Question #1
Most main pumps on suppression apparatus in the U.S. fire service are centrifugal pumps, while most priming pumps are positive-displacement pumps. Although a variety of sizes and configurations are used, the basic operation of these pumps is similar. NFPA 1002 5.1. *IFPO, 2E:* Chapter 1, page 11.

Question #2
Even though the problem was fixed, the pump operator should document the finding to assist with tracking and trending of problems. NFPA 1002 4.2.2. *IFPO, 2E:* Chapter 2, page 25.

Question #3
Centrifugal pumps have only one moving part, the impeller, and it is open from the intake to the discharge side of the pump. This allows for 100% slippage when all discharges are closed. NFPA 1002 5.2.1, 5.2.2. *IFPO, 2E:* Chapter 4, pages 85 – 90.

Question #4
Actually, positive-displacement pumps theoretically discharge (displace) a *specific* quantity of water for each revolution or cycle of the pump. NFPA 1002 5.2.1, 5.2.2. *IFPO, 2E:* Chapter 4, page 78.

Question #5
Several general considerations for positioning the vehicle include the following:
- Position the vehicle to reduce the likelihood of being struck by traffic
- Use the vehicle to shield emergency personnel from traffic
- Position to enhance emergency operations
- Position away from hazards
- Consider wind direction
- Do not park in the collapse zone
- **Never park on railroad tracks**
- Park the vehicle on the side of the incident

NFPA 1002 4.3. *IFPO, 2E:* Chapter 3, page 65.

Question #6
The pressure-relief valve limits pressure surges by opening and closing a passage between the discharge side of the pump to either the intake or atmosphere. In this case, the engine speed should not slow down. NFPA 1002 5.2.1, 5.2.2, 5.2.4. *IFPO, 2E:* Chapter 5, pages 122 – 123.

Question #7
Although supervision is often a duty or function of a pump operator, NFPA 1002 does not contain supervision requirements. NFPA 1002 identifies the main duties of the pump operator as driving, maintenance, and pump operation. NFPA 1002 4.1. *IFPO, 2E:* Chapter 1, pages 15 – 16.

Question #8
NFPA 1002 requirements for the basic duties of a driver pump operator are as follows:
- Preventive Maintenance (section 4.2 and 5.1.1)
- Driving the apparatus (section 4.3)
- Pump operations (section 5.2)

Supervisory responsibilities are not included in NFPA 1002.

NFPA 1002 4.2, 4.3, 5.1, 5.2. *IFPO, 2E:* Chapter 1, pages 6 – 9.

Question #9
The trend seems to indicate that the number of vehicle accidents has gradually increased from 1995 to 2000. NFPA 1002 4.3. *IFPO, 2E:* Chapter 1, pages 8 – 9.

Question #10
Exemptions extended to the operation of emergency vehicles are most often located in state and local laws. NFPA 1002 4.2, 4.3, 5.1. *IFPO, 2E:* Chapter 1, page 15.

Question #11
Although good public relations is important, increased safety, reduced liability, and costs savings are more important reasons for a well-funded preventive maintenance program. NFPA 1002 4.2.1, 5.1.1. *IFPO, 2E:* Chapter 2, pages 21 – 22.

Question #12
NFPA 1901 requires all new apparatus to have quick pressure buildup capabilities so that operating pressure is reached within 60 seconds. NFPA 1002 4.2.1. *IFPO, 2E:* Chapter 2, page 35.

Question #13
Apparatus fuel tanks should be refilled according to department policy. When no refill policy exists, apparatus fuel tanks should be refilled when the fuel level reaches the three-quarter mark. That is, when the gauge drops from full to three-quarter full. NFPA 1002 4.2.1. *IFPO, 2E:* Chapter 2, page 35.

Question #14
It is not advisable to turn the steering wheel so that the wheels turn. Rather, the steering wheel should be turned until just before the wheels turn. The distance should not exceed 10 degrees in either direction. For a 20-inch steering wheel, that would be approximately 2 inches of movement. NFPA 1002, 4.2.1. *IFPO, 2E:* Chapter 2, pages 35 – 36.

Question #15
The impact of an accident involving responding apparatus can be:

- Delayed assistance to those who summoned help
- Additional units dispatched for the original call as well as for the new accident
- Fire department members and civilians could be seriously or fatally injured
- Fire department and civilian vehicles and property could sustain extensive damage
- The city, department, and driver could face civil and/or criminal proceedings
- Image presented by the accident will last a long time in the mind of the public

NFPA 1002 4.3. *IFPO, 2E:* Chapter 3, page 43.

Question #16
Most state laws require that emergency-vehicle drivers obey the same laws as other vehicle operators unless specifically exempt from doing so. State laws typically define several conditions that must exist for exemptions to be extended and include:

- Only authorized emergency vehicles are covered
- The exemptions are only provided when responding to an emergency
- Audible and visual warning devices must be operating when taking advantage of the exemption.

NFPA 1002 4.3. *IFPO, 2E:* Chapter 3, pages 44 – 45.

Question #17

NFPA 1500 requires that emergency-vehicle drivers come to a complete stop when any intersection hazard is present. Specifically, the vehicle must come to a complete stop when any of the following exists:

- As directed by a law enforcement officer
- At traffic red lights or stop signs
- At negative right-of-way and blind intersections
- When all lanes of traffic in an intersection cannot be accounted for
- When stopped school bus with flashing warning lights is encountered
- At unguarded railroad guard crossing (also for nonemergency)
- When other intersection hazards are present

NFPA 1002 4.3. *IFPO, 2E:* Chapter 3, page 48.

Question #18

The water in the tanker weighs 16,700 lbs (8.35 lbs per gallon × 2,000 gallons), which is equal to 8.35 tons (16,700 lbs ÷ 2,000 lbs/ton). It would be unsafe to cross the bridge because the water alone exceeds the authorized limit. The weight of the vehicle would add even more risk to crossing the bridge. NFPA 1002 4.3. *IFPO, 2E:* Chapter 3, pages 52 – 53.

Question #19

Several driving exercises are typically used to assess the ability of a driver to safely operate and control the vehicle. These include the following:

- *Alley Dock*. Assesses the ability to back the vehicle into a restricted area, such as a fire station or down an alley.
- *Serpentine*. Assesses the ability to drive around obstacles, such as parked cars and tight corners.
- *Confined Space Turnaround*. Assesses the ability to turn the vehicle around within a confined space, such as a narrow street or driveway.
- *Diminishing Clearance*. Assesses the ability to drive the vehicle in a straight line, such as on a narrow street or road.

NFPA 1002 4.3.2, 4.3.3, 4.3.4, 4.3.5. *IFPO, 2E:* Chapter 3, pages 65 – 66.

Question #20

Water actually leaks (or slips) between the surfaces of the internal moving parts of a pump. The greater the slippage, the less efficient the positive-displacement pump. NFPA 1002 5.2.1, 5.2.2. Chapter 4, page 78.

Question #21

Centrifugal pump performance is contingent on three interrelated factors. If one factor remains constant, a change in one of the remaining factors will change the other.

- *Speed*. If the speed of the pump is held constant and the flow of water is increased, pressure will drop. If more water is allowed to flow while the speed of the pump remains the same, pressure will be reduced as less resistance occurs on the discharge side of the pump.
- *Flow*. If the flow of water is held constant and the speed of the pump is increased, pressure will increase. The same amount of water is being discharged, yet the pump is attempting to discharge more water. This results in an increase in pressure.
- *Pressure*. If the pressure is held constant and the speed of the pump is increased, flow will increase. The increased speed of the pump will increase the flow. Constant pressure can be maintained by increasing or reducing the resistance on the discharged side of the pump.

NFPA 1002 5.2.1, 5.2.2. *IFPO, 2E:* Chapter 4, page 86.

Question #22

According to NFPA 1901, a pump must have a rated capacity as follows:

- 100% of its rated capacity at 150 psi
- 70% of its rated capacity at 200 psi
- 50% of its rated capacity at 250 psi

NFPA 1002 5.2.1, 5.2.2. *IFPO, 2E:* Chapter 4, page 91.

Question #23

A pump must deliver 100% of its rated capacity at 150 psi. Note that at 150 psi, the pump can deliver 1,250 gpm. NFPA 1002 5.2.1, 5.2.2. *IFPO, 2E:* Chapter 4, page 91.

Question #24

In volume mode the water enters both impellers from a common intake and leaves from a common discharge. NFPA 1002 5.2.1, 5.2.2, 5.2.4. *IFPO, 2E:* Chapter 4, pages 90 – 91.

Question #25

- A *pre-mixed system* consists of a tank in which foam concentrate and water are added in appropriate proportions. Often, the foam is simply added to the on-board water tank.
- An *in-line eductor system* utilizes eductors to add foam to water in appropriate proportions. This system can have either internal or external controls, and requires an eductor and a foam tank (or supply). Internal systems have a foam concentrate valve and meeting control. Eductors are usually installed between the pump discharge and the discharge outlet.
- The *around-the-pump proportioning system* uses an eductor (located between the intake and discharge sides of the pump) to mix foam with water. Water from the discharge side of the pump picks up foam and delivers it to the intake side of the pump, where it can then be discharged to all outlets.
- A *balanced-pressure system* mixes foam with water by means of pressure. One system uses pressure to force foam from a bladder. The second system uses a separate foam pump (two types are by-pass and demand).
- A *direct injection/compressed-air foam system* (CAFS) consists of a separate foam pump and foam tank. Direct injection moves foam from the tank directly into the discharge lines. A CAFS adds compressed air to create a lightweight foam.

NFPA 1002 5.2.3. *IFPO, 2E:* Chapter 5, pages 126 – 130.

Question #26

- *Tank-to-pump valves* allow water to flow from the on-board water supply to the intake side of the pump.
- *Pump-to-tank valves* (tank fill) allow water to flow from the discharge side of the pump to the tank.
- A *transfer valve* is found on multistage pumps, and it redirects water from the pump between the pressure mode and the volume mode.

NFPA 1002 5.2.1, 5.2.2. *IFPO, 2E:* Chapter 5, page 119.

Question #27

Basic steps for testing hose include:

1. Connect water supply and test hose sections

2. Open discharge lines for test hose sections

3. Slowly increase discharge pressure to 45 psi

4. Remove all air from within the hose

5. Mark hose near coupling to determine slippage

6. Increase pressure slowly (NFPA 1962 suggests no faster than 15 psi per second) to test pressure

7. Maintain pressure for three minutes, periodically checking for leaks

8. Record results

NFPA 1002 5.1. *IFPO, 2E:* Chapter 6, page 147.

Question #28

Testing instrumentation calibration should occur within the past 12 months of the test date. NFPA 1002 5.1. *IFPO, 2E:* Chapter 6, page 147.

Question #29

NFPA 1965 requires that appliances with lever-operated handles must indicate a closed position when the handle is perpendicular to the hose line. NFPA 1002 5.2.1, 5.2.2, 5.2.4. *IFPO, 2E:* Chapter 6, page 148.

Question #30

A pumper with a 5,000-gallon tank flowing 250 gpm (125 gpm × 2 lines) will last about 20 minutes. $250 \text{gallons per minute} \overline{)5000 \text{gallons}}$ = 20 minutes. NFPA 1002 5.2.1. *IFPO, 2E:* Chapter 7, page 167.

Question #31

Basic steps for conducting hydrant flow include:

1. Take static pressure reading at test hydrant (make sure to open hydrant fully and remove air)

2. Open flow hydrants one at a time until a 25% drop in residual pressure is achieved

3. Continue to flow to clear debris and foreign substances

4. Take reading at the same time: residual reading at test hydrant and flow readings at each flow hydrant, record results

5. Slowly shut down hydrants one at a time

6. Record exact interior size, in inches, of each outlet flowed

NFPA 1002 5.2.1. *IFPO, 2E:* Chapter 7, pages 180 – 181.

Question #32

Water will rise approximately 2.3 feet for each 1 psi of pressure.

So, 7 psi × 2.3 feet/psi = 16.1 feet.

NFPA 1002 5.2.1. *IFPO, 2E:* Chapter 7, pages 185 – 186.

Question #33

On-board water supplies are readily available by pulling the tank-to-pump control valve. However, on-board water tanks are limited in quantity. NFPA 1002 5.2.1. *IFPO, 2E:* Chapter 7, pages 166 – 167.

Question #34

The distance between pumpers in a relay operation can be determined as follows:

$(PDP - 20) \times 100/FL$

$(185 - 20) \times 100 / 5$

165×20

3,300 feet

NFPA 1002 5.2.3. *IFPO, 2E:* Chapter 7, page 194.

Question #35

Divide tank volume by the shuttle cycle time.

$1,500 \div 10 = 150$ gpm.

NFPA 1002 5.2.1. *IFPO, 2E:* Chapter 7, page 201.

Question #36

The basic steps for pump operation include:

Step 1 Position apparatus, set parking brake, and let engine return to idle

Step 2 Engage the pump

Step 3 Provide water to intake side of pump (on-board, hydrant, draft)

Step 4 Set transfer valve (if so equipped)

Step 5 Open discharge lines

Step 6 Throttle to desired pressure

Step 7 Set the pressure-regulating device

Step 8 Maintain appropriate flows and pressures

NFPA 1002 5.2.1, 5.2.2, 5.2.4. *IFPO, 2E:* Chapter 8, page 207.

Question #37

Important considerations for positioning pumping apparatus at incidents include:

- If no fire or smoke is visible, park near main entrance

- Be sure to follow department SOPs

- Consider tactical priorities for the incident

- Consider surroundings such as heat from the fire, collapse, overhead lines, escape routes, wind

NFPA 1002 5.2.1, 5.2.2. *IFPO, 2E:* Chapter 8, pages 208 – 209.

Question #38

Basic steps for pump engagement powered through a PTO include:

- Bring apparatus to complete stop, set parking brake, let engine return to idle speed

- Disengage the clutch (push in the clutch pedal)

- Place transmission in neutral

- Operate the PTO lever

- For mobile pumping, place transmission in the proper gear

- For stationary pumping, place transmission in neutral

- Engage the clutch slowly

NFPA 1002 5.2.1, 5.2.2, 5.2.4. *IFPO, 2E:* Chapter 8, pages 211 – 212.

Question #39
Basic steps to engage a pump utilizing a split-shaft arrangement are as follows:

- Bring apparatus to complete stop
- Place transmission in neutral
- Apply parking brake
- Operate pump shift switch from road to pump position
- Ensure the "OK to pump light" comes on
- Shift transmission to pumping gear (usually highest gear)

NFPA 1002 5.2.1, 5.2.2, 5.2.4. *IFPO, 2E:* Chapter 8, pages 211 – 212.

Question #40
Because the flow is greater than 50% of the pump's rated capacity and the expected pump discharge pressure is less than 150 psi, the transfer valve should be in the volume mode. The transfer valve operation rule of thumb is as follows:

- *Volume mode* for flows greater than 50% of a pump's rated capacity and pressures less than 150 psi.
- *Pressure mode* for flows less than 50% of a pump's rated capacity and pressures greater than 150 psi.

NFPA 1002 5.2.1, 5.2.2, 5.2.4. *IFPO, 2E:* Chapter 8, page 214 – 215.

Question #41
A corresponding increase in discharge pressure should be noted when the throttle is increased (engine speed increased). If not, one of the following common causes could be occurring:

- The pump may not be in gear
- The pump may not be primed
- The supply line could be closed or insufficient

NFPA 1002 5.2.1, 5.2.2, 5.2.4. *IFPO, 2E:* Chapter 8, page 216.

Question #42
If the throttle is increased past the point of a corresponding increase in discharge pressure, the following might occur:

- Pump cavitation
- Loss of prime
- Intake line collapse
- Damage to municipal water mains

NFPA 1002 5.2.1, 5.2.2, 5.2.4. *IFPO, 2E:* Chapter 8, page 216.

Question #43

The on-board water supply procedures are as follows:

- Position apparatus, set parking brake
- Engage pump
- Set transfer valve
- Open "tank-to-pump"
- Open discharge control valves
- Increase throttle
- Set pressure-regulating device
- Plan for more water

NFPA 1002 5.2.1, 5.2.2, 5.2.4. *IFPO, 2E:* Chapter 8, pages 220 – 221.

Question #44

Each of the answer selections provided is an important consideration that should be evaluated with regard to selecting a location to position fire apparatus. NFPA 1002 5.2.1, 5.2.2. *IFPO, 2E:* Chapter 8, pages 208 – 209.

Question #45

Cavitation can be stopped by reducing pump pressure through pump speed, reducing discharge flow, or through increasing supply flow. Adding one or more discharge lines would only aggravate a deteriorating system and most likely would cause additional damage. NFPA 1002 5.1.1. *IFPO, 2E:* Chapter 9, pages 245 – 247.

Question #46

The low-pressure zone near the center of the impeller allows vapor pockets to be created. When these vapor pockets pass through the high-pressure zone, they are forced back into water (liquid state) abruptly. NFPA 1002 5.2.1, 5.2.2, 5.2.4. *IFPO, 2E:* Chapter 9, page 256.

Question #47

Cold-weather operations can be grueling. The pump operator must be diligent to stop all leaks, keep water flowing, and drain everything when the pump operation is completed if freezing might occur. NFPA 1002 5.2.1, 5.2.2, 5.2.4. *IFPO, 2E:* Chapter 9, page 247.

Question #48

A fire department connection marked "STANDPIPE" means the system is a standpipe system, not a sprinkler system. NFPA 1002 5.2.4. *IFPO, 2E:* Chapter 9, page 253.

Question #49

Standpipes are classified as follows:

- *Class 1* standpipes provide 2 ½-inch connections for trained firefighters/fire brigades at an initial flow rate of 500 gpm.
- *Class 2* standpipes provide 1 ½-inch connections for initial attack at a minimum flow rate of 100 gpm.
- *Class 3* standpipes provide 1 ½-inch and 2 ½-inch connections for trained firefighters/fire brigades at an initial flow rate of 500 gpm.

NFPA 1002 5.2.4. *IFPO, 2E:* Chapter 9, page 253.

Question #50
Pump operators are expected to be able to troubleshoot pump problems. Several basic troubleshooting considerations include:

- Problems are usually either procedural or mechanical
- When proper procedures are followed, the problem is most likely mechanical
- The best method to troubleshoot is to follow the flow of water from the intake to the discharge while attempting to determine the problem
- Use the manufacturer's troubleshooting guides when available

NFPA 1002 5.2. *IFPO, 2E:* Chapter 9, page 254.

Question #51
To calculate weight, the formula $W = D \times V$ is used.
$W = 62.34$ lb/ft$^3 \times 100$ ft^3
$W = 6{,}234$ lbs (note the cubic feet cancel each other)

NFPA 1002 5.2. *IFPO, 2E:* Chapter 10, page 266.

Question #52
Nozzle reaction calculation
Smooth-bore nozzles:
$NR = 1.57 \times d^2 \times NP$
where NR = nozzle reaction
        1.57 = constant
        d = diameter of nozzle orifice in inches
        NP = operating nozzle pressure in psi

$NR = 1.57 \times 0.125^2 \times 50$

$NR = 1.57 \times .0156 \times 50$

$NR = 1.22$

NFPA 1002 5.2. *IFPO, 2E:* Chapter 10, pages 282 – 283.

Question #53
First, determine the total gpm:
        125 gpm $\times$ 2 = 250 gpm
        250 gpm $\times$ 10 minutes = 2,500 gallons
Next determine the numbers of pounds:
        $W = 8.34$ lb/gal $\times$ 2,500 gallons
        $W = 20{,}850$ lbs
Finally, convert pounds to tons:
        20,850 lbs = 10.4 tons (divide by 2,000)

NFPA 1002 5.2. *IFPO, 2E:* Chapter 10, page 268.

Question #54
Basic pressure principles include:

- Pressure at any point in a liquid at rest is equal in every direction.
- Pressure of a liquid acting on a surface is perpendicular to that surface.
- External pressure applied to a confined liquid (fluid) is transmitted equally throughout the liquid.
- Pressure at any point beneath the surface of a liquid in an open container is directly proportional to its depth.
- Pressure exerted at the bottom of a container is independent of the shape or volume of the container.

NFPA 1002 5.2. *IFPO, 2E:* Chapter 10, pages 271 – 275.

Question #55

Based on the percent drop in pressure, additional flows may be available from a hydrant as follows:

    0 – 10% drop   three times the original flow

    11 – 15% drop   two times the original flow

    16 – 25% drop   one time the original flow

A 10-psi drop in pressure is about 20% of the initial static reading (10 psi /50 psi)

Three times the original flow is 375 psi. NFPA 1002 5.2. *IFPO, 2E:* Chapter 10, page 277.

Question #56

The hand method is a fireground method used to estimate friction loss in 100-foot sections of 2 ½-inch hose. Simply select the fingertip representing the gpm flow and then multiply the 2 figures on the finger for the approximate friction loss pressure for each 100-foot section of 2 ½-inch hose. First, calculate the friction loss in 100-foot sections:

    250 gpm = 2.5 × 5 = 12.5 psi per 100-foot sections of 2 ½-inch hose

Next, multiply the friction loss by the number of 100-foot sections in the line:

    300 feet / 100 psi/feet = 3 psi

    3 psi × 12.5 psi = 37.5 psi friction loss in the line

NFPA 1002 5.2.1, 5.2.2. *IFPO, 2E:* Chapter 10, pages 302 – 303.

Question #57

The drop-ten method simply subtracts 10 from the first 2 numbers of gpm flow. It is not as accurate as other methods, but it provides a simple rule of thumb for fireground use.

    250 gpm = 25 – 10 = 15 psi friction loss per 100-foot sections of 2 ½-inch hose

    15 psi × 4 (4 sections of 100-feet of hose) = 60 psi

NFPA 1002 5.2.1, 5.2.2. *IFPO, 2E:* Chapter 10, page 303.

Question #58

The hand method is a fireground method used to estimate friction loss in 100-foot sections of 2 ½-inch hose. Simply select the fingertip representing the gpm flow and then multiply the 2 figures on the finger for the approximate friction loss pressure for each 100-foot section of 2 ½-inch hose.

FL = 2 × 4 × 3 (number of 100-foot sections of hose)

FL = 24 psi

NFPA 1002 5.2.1, 5.2.2. *IFPO, 2E:* Chapter 10, pages 302 – 303.

Question #59

Nozzle reaction calculation

Combination nozzles:

$$NR = gpm \times \sqrt{NP} \times 0.0505$$

where NR = nozzle reaction
                  0.0505 = constant
                  gpm = flow in gallons per minute
                  NP = operating nozzle pressure in psi

$NR = 100 \times \sqrt{100} \times 0.0505$

$NR = 100 \times 10 \times 0.0505$

NR = 50.5 or 51 psi

NFPA 1002 5.2. *IFPO, 2E:* Chapter 10, pages 282 – 283.

Question #60

The drop-ten method was used:

250 gpm = 25 – 10 = 15 psi friction loss per 100-foot section of 2 ½-inch hose
15 psi × 2 (number of 100-foot sections of hose) = 30 psi

Other fireground formulas field a slightly different result:

cq^2L and hand method = 25 psi
$2q^2 + q$ = 15 psi

NFPA 1002 5.2.1, 5.2.2. *IFPO, 2E:* Chapter 10, page 303.

Question #61

Iowa State formula: $NF = \dfrac{V}{100}$

    where NF = needed flow in gpm
                    V = volume of the area in cubic feet
                    100 = is a constant in ft^3/gpm

$$NF = \frac{100 ft \times 60 ft \times 12 ft}{100 ft^3 / gpm}$$

$$NF = \frac{72,000 ft^3}{100 ft^3 / gpm}$$

$$NF = 720 gpm$$

NFPA 1002 5.2. *IFPO, 2E:* Chapter 11, pages 292 – 293.

Question #62

NFA formula: $NF = \dfrac{A}{3}$

    where NF = needed flow in gpm
                    A = area of the structure in square feet
                    3 = constant in ft^2/gpm

$$NF = \frac{A}{3 ft^2 gpm}$$

$$NF = \frac{50 ft \times 25 ft}{3 ft^2 gpm}$$

$$NF = \frac{1,250 ft^2}{3 ft^2 gpm}$$

$$NF = 416.66 gpm$$

NFPA 1002 5.2. *IFPO, 2E:* Chapter 11, pages 292 – 293.

Question #63

$cq^2 L$
where c for 1 ½-inch hose is 24

$$24 \times \left(\frac{95}{100}\right)^2 \times \frac{200}{100}$$

$$24 \times (.95)^2 \times 2$$

$$24 \times 0.9 \times 2$$

$$43.2 psi$$

NFPA 1002 5.2. *IFPO, 2E:* Chapter 11, pages 306 – 307.

Question #64

Elevation formula by floor level:

$EL = 5 \times H$

where EL = the gain or loss of elevation in psi

5 = gain or loss in pressure for each floor level

H = height in number of floor levels above or below the pump

$EL = 5 \times 5$

$EL = 25 psi$

NFPA 1002 5.2.4. *IFPO, 2E:* Chapter 11, page 311.

Question #65

Elevation formula in feet:

$EL = 0.5 \times H$

where EL = the gain or loss of elevation in psi

0.5 = pressure exerted at base of 1-cubic-inch column of water 1 foot high

H = distance in feet above or below the pump

$EL = 5 \times 48$

$EL = 25 psi$

NFPA 1002 5.2. *IFPO, 2E:* Chapter 11, page 311.

Question #66

q^2

$\left(\dfrac{400}{100}\right)^2$

$(4)^2$

16 psi per 100-foot sections

$FL = 16 \times 5$ (number of 100-foot sections in the hose line)

$FL = 80$ psi

NFPA 1002 5.2. *IFPO, 2E:* Chapter 11, pages 306 – 307.

Question #67

$2q^2 + q$

$2 \times \left(\dfrac{200}{100}\right)^2 + \left(\dfrac{200}{100}\right)$

$2 \times (2)^2 + (2)$

$2 \times 4 + 2$

10 psi friction loss per 100-foot sections of hose

$10 \times 3.5 = 35$ psi friction loss in the line

NFPA 1002 5.2. *IFPO, 2E:* Chapter 11, pages 304 – 305.

Question #68

$PDP = NP + FL + AFL \pm EL$

NP: 100 psi

FL: cq^2L

$$2 \times \left(\frac{350}{100} \right)^2 \times \frac{500}{100}$$

$$2 \times (3.5)^2 \times 5$$

$$2 \times 12.25 \times 5$$

$$122.5psi$$

AFL: 0 psi

$\pm EL$: 0 psi

$PDP = 100 + 123$

$PDP = 223$ psi

NFPA 1002 5.2. *IFPO, 2E:* Chapter 11, pages 317 – 357.

Question #69

cq^2L

where c for 3-inch hose is .8 and for 2 ½-inch hose is 2.

3-inch hose friction loss:

$$.8 \times \left(\frac{325}{100} \right)^2 \times \frac{500}{100}$$

$$.8 \times (3.25)^2 \times 5$$

$$.8 \times 10.56 \times 5$$

$$42.24psi$$

2 ½-inch hose friction loss:

$$2 \times \left(\frac{325}{100} \right)^2 \times \frac{250}{100}$$

$$2 \times (3.25)^2 \times 2.5$$

$$2 \times 10.56 \times 2.5$$

$$52.8psi$$

Total friction loss is 95 psi (42 psi + 53 psi)

NFPA 1002 5.2. *IFPO, 2E:* Chapter 11, pages 306 – 307.

Question #70

$PDP = NP + FL + AFL \pm EL$

NP: 100 psi

FL: cq^2L

$$15.5 \times \left(\frac{100}{100}\right)^2 \times \frac{300}{100}$$

$$15.5 \times (1)^2 \times 3$$

$$15.5 \times 1 \times 3$$

$$46.5psi$$

AFL: 0 psi

$\pm EL$: $0.5 \times H$

$0.5 \times (-50)$

$-25psi$

$PDP = 100 + 47 - 25$

$PDP = 122$ psi

NFPA 1002 5.2. *IFPO, 2E:* Chapter 11, pages 317 – 357.

Question #71

$PDP = NP + FL + AFL \pm EL$

NP: 50 psi

gpm: $30 \times d^2 \times \sqrt{NP}$

$$30 \times (0.75)^2 \times \sqrt{50}$$

$$30 \times 0.56 \times 7$$

$$117.6gpm$$

FL: cq^2L

$$15.5 \times \left(\frac{118}{100}\right)^2 \times \frac{300}{100}$$

$$15.5 \times (1.18)^2 \times 3$$

$$15.5 \times 1.39 \times 3$$

$$64.6psi$$

AFL: 0 psi

$\pm EL$: $0.5 \times H$

$0.5 \times (50)$

$25psi$

$PDP = 50 + 65 + 25$

$PDP = 140$ psi

NFPA 1002 5.2. *IFPO, 2E:* Chapter 11, pages 317 – 357.

Question #72

$PDP = NP + FL + AFL \pm EL$

NP: 100 psi

FL: cq^2L

$$15.5 \times \left(\frac{150}{100}\right)^2 \times \frac{200}{100}$$

$$15.5 \times (1.5)^2 \times 2$$

$$15.5 \times 1.25 \times 2$$

$$38.75 psi$$

AFL: 0 psi

$\pm EL$: $5 \times H$

5×3

$15 psi$

$PDP = 100 + 39 + 15$

$PDP = 154$ psi

NFPA 1002 5.2. *IFPO, 2E:* Chapter 11, pages 317 – 357.

Question #73

$PDP = NP + FL + AFL \pm EL$

NP: 50 psi

gpm: $30 \times d^2 \times \sqrt{NP}$

$$30 \times (0.44)^2 \times \sqrt{50}$$

$$30 \times 0.19 \times 7$$

$$39.9 gpm$$

FL: cq^2L

$$15.5 \times \left(\frac{40}{100}\right)^2 \times \frac{250}{100}$$

$$15.5 \times (.40)^2 \times 2.5$$

$$15.5 \times .16 \times 2.5$$

$$6.2 psi$$

AFL: 0 psi

$\pm EL$: $5 \times H$

$5 \times (-1)$

$-5 psi$

$PDP = 50 + 6 - 5$

$PDP = 51$ psi

NFPA 1002 5.2. *IFPO, 2E:* Chapter 11, pages 317 – 357.

Question #74

First, calculate the friction loss in the 2 ½-inch line, remembering to add the flow through both 1 ¾-inch lines. Next, calculate the friction loss for only one of the 1 ¾-inch lines because they are like lines. Finally, determine the appliance friction loss.

$PDP = NP + FL + AFL \pm EL$

NP: 100 psi

FLs (2 ½-inch line):

$$cq^2L$$

$$2 \times \left(\frac{190}{100}\right)^2 \times \frac{300}{100}$$

$$2 \times (1.9)^2 \times 3$$

$$2 \times 3.61 \times 3$$

$$21.66psi$$

FLa (1 ¾ lines):

$$cq^2L$$

$$15.5 \times \left(\frac{95}{100}\right)^2 \times \frac{150}{100}$$

$$15.5 \times (0.95)^2 \times 1.5$$

$$15.5 \times 0.9 \times 1.5$$

$$20.9psi$$

AFL: 10 psi

$\pm EL$: 0

$PDP = 100 + (22 + 21) + 10$

$PDP = 153$ psi

NFPA 1002 5.2. *IFPO, 2E:* Chapter 12, pages 342 – 344.

Question #75

$PDP = NP + FL + AFL \pm EL$

NP: 80 psi

gpm: $30 \times d^2 \times \sqrt{NP}$

$30 \times (1.375)^2 \times \sqrt{80}$

$30 \times 1.89 \times 9$

$510.3 gpm$

FLs (2 ½-inch lines): (half the flow)

cq^2L

$2 \times \left(\dfrac{255}{100}\right)^2 \times \dfrac{600}{100}$

$2 \times (2.55)^2 \times 6$

$2 \times 6.5 \times 6$

$78 psi$

FLa (3-inch line):

cq^2L

$.8 \times \left(\dfrac{510}{100}\right)^2 \times \dfrac{200}{100}$

$.8 \times (5.1)^2 \times 2$

$.8 \times 26.01 \times 2$

$41.61 psi$

AFL: 5 psi for the siamese and 15 psi for the monitor

$\pm EL$: 0

$PDP = 80 + (78 + 42) + 20$

$PDP = 220$ psi

NFPA 1002 5.2. *IFPO, 2E:* Chapter 12, pages 347 – 348.

Question #76

$PDP = NP + FL + AFL \pm EL$

NP: 100 psi

*FL*s (3-inch lines):

$$cq^2L$$

$$.8 \times \left(\frac{100}{100}\right)^2 \times \frac{100}{100}$$

$$.8 \times (1)^2 \times 1$$

$$.8 \times 1 \times 1$$

$$.8psi$$

*FL*a (1 ½-inch line):

$$cq^2L$$

$$24 \times \left(\frac{100}{100}\right)^2 \times \frac{150}{100}$$

$$24 \times (1)^2 \times 1.5$$

$$24 \times 1 \times 1.5$$

$$36psi$$

AFL: 25 psi for the standpipe

$\pm EL$: 5 psi per level, $5 \times 2 = 10$ psi

$PDP = NP + FL + AFL \pm EL$
$PDP = 100 + (1 + 36) + 25 + 10$
$PDP = 172$ psi

NFPA 1002 5.2.4. *IFPO, 2E:* Chapter 12, pages 349 – 350.

Question #77

The combination fog nozzle requires 100 psi operating pressure.

$PDP = NP + FL + AFL \pm EL$
$PDP = 100 + 36$
$PDP = 136$ psi

NFPA 1002 5.2. *IFPO, 2E:* Chapter 12, pages 319 – 321.

Question #78

$PDP = NP + FL + AFL \pm EL$

$NP = 50$ for smooth-bore on a hand-line
        80 for smooth-bore on a master-stream device
        100 for combination (fog, automatic, selectable flow...)

$EL = 5$ psi added or subtracted for each floor level above/below ground level
        .5 psi added or subtracted for each foot in elevation above/below ground level

$PDP = 100$ (NP) $+ 24$ (FL) $+ 5$ (EL)
$PDP = 129$ psi

NFPA 1002 5.2. *IFPO, 2E:* Chapter 12, pages 319 – 321.

Question #79

In *pressure mode*, each impeller delivers the same quantity to the next impeller, and each impeller adds its pressure before sending to the next impeller. Total discharge then, is the same for each impeller, whereas pressure is divided among each of the impellers. NFPA 1002 5.2.1, 5.2.2, 5.2.4. *IFPO, 2E:* Chapter 4, pages 89 – 91.

Question #80

The pressure is too high because the maximum operating pressure for supply hose is 185 psi. NFPA 1002 5.2.2. *IFPO, 2E:* Chapter 7, page 192.

Question #81

When flowing two or more like lines, the PDP will be the same regardless of how many lines are placed in operation. For each line initiated, the discharge gate is slowly opened while the throttle is increase to maintain the PDP. NFPA 1002 5.2. *IFPO, 2E:* Chapter 12, pages 336 – 337.

Question #82

A tanker shuttle time is determined by dividing the tank volume by the time it takes the tanker to complete one cycle from the dump site, to the fill site, and back to the dump site.

To determine the size of a tanker when the tanker shuttle time and flow is known, simply multiply the shuttle flow capacity by the cycle time.

$250 \times 8 = 2,000$ gallons

NFPA 1002 5.2.1. *IFPO, 2E:* Chapter 7, pages 200 – 201.

Question #83

NP: 50 psi

gpm: $30 \times 1^2 \times \sqrt{50}$

$\qquad 30 \times (1)^2 \times \sqrt{50}$

$\qquad 30 \times 1 \times 7$

$\qquad 210 \, gpm$

$1,000/210 = 4.76$ minutes

NFPA 1002 5.2. *IFPO, 2E:* Chapter 11, pages 296 – 297.

Question #84

Flow meters read the flow within the line. If the flow meter on the pump panel reads 100 gpm, then the nozzle is discharging 100 gpm. The amount of water flow will be the same at any point in the line. NFPA 1002 5.2. *IFPO, 2E:* Chapter 5, pages 113 – 114.

Question #85

The specific heat of water is 1 Btu. It takes 152 Btu to raise 1 pound of water from 62°F to 212°F. NFPA 1002 5.2.1, 5.2.2. *IFPO, 2E:* Chapter 10, page 264.

Question #86

Several driving exercises are typically used to assess the ability of a driver to safely operate and control the vehicle. These include the following:

- *Alley Dock.* Assesses the ability to back the vehicle into a restricted area, such as a fire station or down an alley.
- *Serpentine.* Assesses the ability to drive around obstacles, such as parked cars and tight corners.
- *Confined-Space Turnaround.* Assesses the ability to turn the vehicle around within a confined space, such as a narrow street or driveway.
- *Diminishing Clearance.* Assesses the ability to drive the vehicle in a straight line, such as on a narrow street or road.

NFPA 1002 4.3.2, 4.3.3, 4.3.4, 4.3.5. *IFPO, 2E:* Chapter 3, pages 65 – 66.

Question #87

The only exceptions to not wearing a seat belt in a fire department vehicle are:

- Loading hose
- Tiller driver training
- Patient treatment

NFPA 1002 4.3. *IFPO, 2E:* Chapter 3, page 49.

Question #88

NFPA 1500 requires that emergency-vehicle drivers come to a complete stop when any intersection hazard is present. Specifically, the vehicle must come to a complete stop when any of the following exists:

- As directed by a law enforcement officer
- At traffic red lights or stops signs
- At negative right-of-way and blind intersections
- When all lanes of traffic in an intersection cannot be accounted for
- When stopped school bus with flashing warning lights is encountered
- At unguarded railroad guard crossing (also for nonemergency)
- When other intersection hazards are present

NFPA 1002 4.3. *IFPO, 2E:* Chapter 3, page 48.

Question #89

Centrifugal-pump performance is contingent on three interrelated factors. If one factor remains constant, a change in one of the remaining factors will change the other.

- *Speed.* If the speed of the pump is held constant and the flow of water is increased, pressure will drop. If more water is allowed to flow while the speed of the pump remains the same, pressure will be reduced as less resistance occurs on the discharge side of the pump.
- *Flow.* If the flow of water is held constant and the speed of the pump is increased, pressure will increase. The same amount of water is being discharged, yet the pump is attempting to discharge more water. This results in an increase in pressure.
- *Pressure.* If the pressure is held constant and the speed of the pump is increased, flow will increase. The increased speed of the pump will increase the flow. Constant pressure can be maintained by increasing or reducing the resistance on the discharge side of the pump.

NFPA 1002 5.2.1, 5.2.2. *IFPO, 2E:* Chapter 4, page 86.

Question #90

Water will rise approximately 2.3 feet/psi for each 1 psi of pressure.

So, 7 psi × 2.3 feet/psi = 16.1 feet.

NFPA 1002 5.2.1. *IFPO, 2E:* Chapter 7, pages 185 – 186.

Question #91

The basic steps for pump operation include:

Step 1   Position apparatus, set parking brake, and let engine return to idle

Step 2   Engage the pump

Step 3   Provide water to intake side of pump (on-board, hydrant, draft)

Step 4   Set transfer valve (if so equipped)

Step 5   Open discharge lines

Step 6   Throttle to desired pressure

Step 7   Set the pressure-regulating device

Step 8   Maintain appropriate flows and pressures

NFPA 1002 5.2.1, 5.2.2, 5.2.4. *IFPO, 2E:* Chapter 8, page 207.

Question #92

Each of the following is a possible cause for loss of prime or failure to prime:

- Air leaks

- Debris on intake strainer

- By-pass line open

- No oil in priming tank

- Defective priming valve

- Improper clearance in rotary gear or vane primer pump

- Engine speed too low

- Lift too high

- Primer not operated long enough

Because the pump was primed twice, lack of oil in the priming tank is most likely not the problem. NFPA 1002 5.2. *IFPO, 2E:* Appendix D, page 376.

Question #93

Positive-displacement pumps, such as rotary vane or rotary gear pumps, operate based on hydrostatic principles (liquids at rest).

Dynamic, or centrifugal, pumps operate based on hydrodynamic principles (liquids in motion).

NFPA 1002 5.2. *IFPO, 2E:* Chapter 4, page 76.

Question #94

With the information provided, the needed fire flow can be calculated using either the Iowa State Formula and the National Fire Academy Formula. NFPA 1002 5.2. *IFPO, 2E:* Chapter 11, pages 292 – 293.

Question #95

- *Dual pumping* occurs when one hydrant supplies two pumpers. The second pumper receives water from the intake of the first pumper. In other words, the excess water provided to the first pumper is diverted to the second pumper.

- *Tandem pumping* occurs when one hydrant supplies two pumpers; it is similar to a relay operation. The first pump discharges all its water to the intake of the second pumper, as in a relay. The only significant difference is that a relay is used to move water over extended distances, whereas tandem pumping is used when higher pressures are required than a single pumper can provide.

NFPA 1002 5.2.1. *IFPO, 2E: IFPO, 2E:* Chapter 8, pages 224 – 225.

Question #96
Fire pump operations can be defined as the systematic movement of water from a supply source through a pump to a discharge point. NFPA 1002 5.2. *IFPO, 2E:* Chapter 1, page 8.

Question #97
NFPA 1500 requires that inspections, maintenance, repair, and service records be maintained for all vehicles. NFPA 1002 4.2.2. *IFPO, 2E:* Chapter 2, page 25.

Question #98
Total stopping distance is measured from the time a hazard is detected until the vehicle comes to a complete stop. It includes:

- Perception distance
- Reaction distance
- Braking distance

NFPA 1002 4.3.2, 4.3.3, 4.3.4, 4.3.5. *IFPO, 2E:* Chapter 3, page 55.

Question #99
Newton's first and third laws of motion help explain the concept of centrifugal force. Both Newton's laws are correctly stated. NFPA 1002 5.2.1. *IFPO, 2E:* Chapter 4, page 85.

Question #100
The most correct answer is to maintain traction. Speed, centrifugal force, road conditions, and weight shifts can all affect the ability to maintain traction. When traction is lost, the apparatus will be out of control. NFPA 1002 4.3. *IFPO, 2E:* Chapter 3, page 56.

Phase Three, Exam Two

1. The first few minutes after arrival to the scene can be demanding for pump operators as they connect supply hose, open discharge lines, and set up pressures. After this setup period, pump operator's work can be equally demanding as they continually observe instrumentation, monitor and adjust pressures and flows, and prepare for changing fireground situations.
   a. True
   b. False

2. Preventive maintenance is critical because it helps maintain good public relations through good-looking apparatus.
   a. True
   b. False

3. After starting your shift as a pump operator, you begin to conduct a preventive maintenance inspection on your assigned apparatus. During the inspection, you found three minor problems that you were able to fix immediately, and one problem that required a mechanic to fix. Because you were able to fix the three minor problems, you only need to document the one problem that could not be fixed immediately.
   a. True
   b. False

4. It is a myth that both the driver and the officer can be held accountable for unsafe actions while driving to or returning from an incident.
   a. True
   b. False

5. Typically, municipal water supply systems provide water for both public consumption, such as household and industrial uses, as well as for emergency use, such as supplying hydrants and fixed fire-protection systems.
   a. True
   b. False

6. Upon arriving on the emergency scene, the driver must safely position the apparatus. One important consideration is to not use the vehicle to shield emergency personnel from traffic.
   a. True
   b. False

7. Both the pressure-relief and pressure governor regulating systems limit pressure surges by changing the engine speed.
   a. True
   b. False

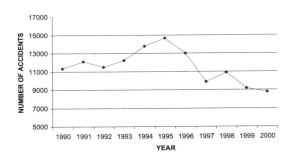

8. Each year the *NFPA Journal* reports on emergency-vehicle accident statistics. Having read several of these reports, what can you say about the graph?
   a. Emergency-vehicle accidents will continue to decrease.
   b. Emergency-vehicle accidents will start to increase or stay about the same.
   c. Emergency-vehicle accidents will most likely increase significantly.
   d. The data in the graph is incorrect, because the number of emergency-vehicle accidents did not decline between 1990 and 2000.

9. Over the years, the title or name used for the position responsible for driving and operating the pump has changed. The tittle used in NFPA 1002 is
   a. driver apparatus operator.
   b. driver/operator – pumper.
   c. engineer.
   d. driver.

10. NFPA 1002 is an example of a
    a. standard.
    b. guidelines.
    c. law.
    d. ordinance.

11. Vehicle inspection components can be grouped into the following general areas:
    1 – Inside cab
    2 – Outside vehicle
    3 – Engine compartment
    4 – Pump and related components

    Using the categories, which of the following is a common sequence used to conduct a vehicle inspection?
    a. 4, 2, 3, 1
    b. 4, 3, 1, 2
    c. 2, 3, 1, 4
    d. 3, 4, 1, 2

12. You just completed a pump test and note that, after pumping at rated capacity for 3 hours, the engine fuel gauge indicates the fuel tank about empty. You know the vehicle was just filled prior to the start of the pump test. Based on this information, which of the following do you know as being correct?
    a. The engine RPM was set too high.
    b. The fuel tank size is too small and does not comply with NFPA 1901.
    c. There is no problem with the fuel tank size as it is in compliance with NFPA 1901.
    d. The pump RPM was set too high.

13. A pump operator on another shift tells you that the 750-gpm pump does not require annual service testing because it is has a rated capacity of less than 1,000 gpm. As the senior pump operator, you know that NFPA _____ requires that pumps with a rated capacity of _____ or greater be service tested at least annually.
    a. 1901, 500 gpm
    b. 1901, 500 gpm
    c. 1911, 250 gpm
    d. 1915, 500 gpm

14. As the pump operator at the beginning of a new shift, your first step to begin a preventive maintenance inspection on your assigned apparatus is to
    a. start with the engine compartment.
    b. look over the pump and related components.
    c. conduct an inventory of equipment.
    d. check the previous inspection reports.

15. While responding to a reported structure fire, you are involved in an accident. In addition to the accident itself, another outcome includes
    a. a delay in assistance to those who summoned help.
    b. additional units must be dispatched for the original call as well as for the new accident.
    c. the city, department, and driver could face civil and/or criminal proceedings.
    d. all of these as potential outcomes.

16. You are the driver of a pumper returning from a working structure fire. Under which of the following conditions would you *not* have to bring the apparatus to a complete stop?
    a. at traffic red lights or stop signs
    b. as directed by a law enforcement officer
    c. when all lanes of traffic in an intersection cannot be accounted for
    d. at guarded railroad grade crossings

17. A firefighter training to be a pump operator is having a difficult time with a certain driving skill. The instructor has the firefighter practice the confined-space turnaround driving exercise. With this information, you know the firefighter must be having trouble with
    a. backing the vehicle into a restricted area, such as a fire station.
    b. driving around obstacles, such as parked cars and tight corners.
    c. turning the vehicle around within a small space, such as a narrow street or driveway.
    d. driving the vehicle in a straight line, such as on a narrow street or road.

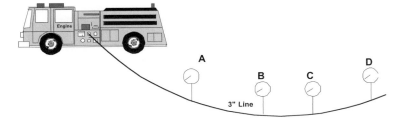

18. During a pumping operation, you initiate flow water through a hose line. After setting the correct pump discharge pressure, the nozzle is closed. Assuming the pressure-regulating device is working, which of the gauges will have the highest pressure reading?
    a. A
    b. D
    c. All will have the same pressure.
    d. There is not enough information provided to answer the question.

19. If you increase the speed of a positive-displacement pump, then the flow will
    a. increase.
    b. decrease.
    c. not change.
    d. increase inversely with pump speed.

20. With centrifugal pumps, if the flow of water is held constant and the speed of the pump is increased, pressure will
    a. increase.
    b. decrease.
    c. remain the same.
    d. increase inversely.

21. Which of the following rated-capacity charts is correct for a 1,500-gpm pumper?
    a. 1,500 gpm at 100 psi
        1,000 gpm at 200 psi
        750 gpm at 250 psi
    b. 1,500 gpm at 150 psi
        1,050 gpm at 200 psi
        750 gpm at 250 psi
    c. 1,500 gpm at 100 psi
        1,050 gpm at 250 psi
        750 gpm at 3000 psi
    d. 1,500 gpm at 150 psi
        1,250 gpm at 200 psi
        750 gpm at 250 psi

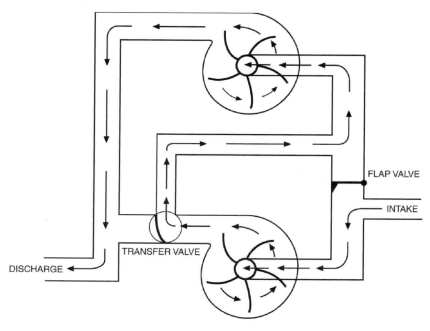

22. The flow of water through this centrifugal pump indicates that it is
    a. in volume mode.
    b. in pressure mode.
    c. a single-stage pump.
    d. none of the answers is correct.

23. The void space created by the impeller being mounted eccentric to the pump casing is called the
    a. centrifuge.
    b. impeller shroud.
    c. positive-displacement void (pdv).
    d. volute.

24. A foam system that mixes foam with water by means of pressure is called a(n)
    a. pre-mixed system.
    b. in-line eductor system.
    c. around-the-pump proportioning system.
    d. balanced pressure system.

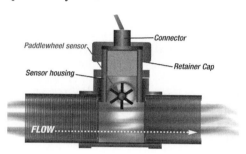

25. This device is a _____, which measures _____.
    a. sensor, pressure
    b. meter, flow
    c. relief device, excessive pressure
    d. sensor, negative pressure

26. NFPA 1961 requires that each hose section be indelibly marked on both ends using 1-inch letters with the following information, *except*
    a. month and year of manufacture.
    b. manufacturer's identification.
    c. statement "Service Test to _____ per NFPA 1962."
    d. testing date and pressure.

27. The basic steps for conducting a hose test are provided below. What is the missing step?
    1 – Connect water supply and test hose sections
    2 – Open discharge lines for test hose sections
    3 – Slowly increase discharge pressure to 45 psi
    4 – Remove all air from within the hose
    5 – _____
    6 – Increase pressure slowly (NFPA 1962 suggests no faster than 15 psi per second) to test pressure
    7 – Maintain pressure for three minutes, while periodically checking for leaks
    8 – Record results

    a. Mark hose near coupling to determine slippage
    b. Clear nonessential personnel from testing area
    c. Check hose for leaks
    d. Set pressure relief device

28. When conducting hose testing, a maximum of _____ -foot sections should be tested.
    a. 100
    b. 300
    c. 500
    d. 1,000

29. A manifold, according to NFPA 1965, with a handle that is in a position perpendicular to the hose line is
    a. open.
    b. closed.
    c. exactly half open.
    d. not a requirement of the standard.

30. A 5,000-gallon tanker is dumping its water into a portable tank through two 2 ½-inch lines, each flowing 200 gpm. How long will it take to empty the tanker?
    a. 5 minutes
    b. 12.5 minutes
    c. 15 minutes
    d. 20.5 minutes

31. When flow testing hydrants, the next major step after recording residual readings at the test hydrant and flow readings at each flow hydrant is to
    a. slowly shut down hydrants one at a time.
    b. record exact interior size, in inches, of each outlet flowed.
    c. open flow hydrants one at a time until a 25% drop in residual pressure is achieved.
    d. continue to flow to clear debris and foreign substances.

32. If the pressure within the pump is reduced by 4 psi, water can be raised to a height of
    _____.
    a. 7 feet.
    b. 9 feet.
    c. 14.7 feet.
    d. none of the answers is correct.

33. All of the following are factors that affect static water supply reliability *except*
    a. hydrant and water main conditions.
    b. the pump and drafting-related components.
    c. environmental and seasonal conditions.
    d. the pump operator's drafting knowledge and skill.

34. In a relay operation, what is the maximum distance for the next pumper of a 1,000-gpm pump discharging 50% of its rated capacity through a 3-inch line if the friction is 20 psi per 100-foot sections of hose?
    a. 500 feet
    b. 1,150 feet
    c. 2,150 feet
    d. 3,300 feet

35. Your captain has asked you to calculate the individual shuttle flow for your 5,000-gallon tanker. If the shuttle cycle time is 20 minutes, the individual shuttle tanker flow would be
    _____.
    a. 250 gpm
    b. 500 gpm
    c. 750 gpm
    d. 1,000 gpm

36. During pump operations, what is the next step after setting discharge pressures and flow?
    a. Set transfer valve.
    b. Open discharge lines.
    c. Set the pressure-regulating device.
    d. Engage the pump.

37. All of the following are important considerations when positioning a pumper at an incident, *except*
    a. analyzing available water supply.
    b. following department SOPs.
    c. reviewing tactical considerations.
    d. being mindful of surroundings.
    e. each of these should be considered when positioning a pumper.

38. When engaging a pump utilizing a PTO, the transmission should be in _____ when the apparatus will be stationary.
    a. neutral
    b. first or lowest gear
    c. fourth or highest gear
    d. reverse

39. When engaging a pump through a split-shaft arraignment, the "OK to Pump" light will come on after
    a. the transmission is placed in neutral.
    b. switching the pump shift control from pump to road.
    c. switching the pump shift control from road to pump.
    d. increasing the discharge pressure by 20 psi.

40. You are the pump operator on the first-arriving apparatus, a 1,250-gpm pumper, to a reported structure fire. Smoke is visible upon arrival. The needed fire flow is estimated to be about 500 gpm. Because of the size and length of the attached lines, you feel the pump discharge pressure will be approximately 215 psi. As a result, you should set the transfer valve to the _____ position.
    a. volume
    b. pressure
    c. off
    d. on

41. When the engine throttle is slowly increased, a corresponding increase in the main discharge gauge should be noted. If this does not occur, what might be the problem?
    a. The transfer valve is in the wrong position.
    b. The pressure regulator is off.
    c. The supply line is closed.
    d. All of these could be the problem.

42. During a pumping operation, the pump operator attempts to increase the pressure in one of the two flowing lines. As the throttle is increased, a corresponding increase in pressure does not occur. The pump operator then checks the relief device; it is functioning properly, but is not relieving any pressure at this time. What can the pump operator do to increase pressure in the one line?
    a. Increase the water supply.
    b. Shut down the other line.
    c. Reset the pressure-regulating device.
    d. Only A and C are correct.
    e. Only A and B are correct.

43. Place the following on-board water supply procedures in the correct order:
    1. Open "tank-to-pump."
    2. Engage pump.
    3. Set pressure-regulating device.
    4. Position apparatus.

    a. 4, 2, 1, 3
    b. 2, 4, 3, 1
    c. 4, 2, 3, 1
    d. 4, 1, 2, 3

44. Positioning fire apparatus should include the evaluation of tactical priorities. For example, if the fire is small and/or a very quick attack is needed, the apparatus should
    a. drop required equipment and hose at the scene and then proceed directly to the nearest water supply.
    b. always position near a strong water supply.
    c. catch the closest hydrant and then advance to the incident.
    d. be positioned so that pre-connects can be pulled and the on-board water supply used.

45. All of the following are ways to stop cavitation in a pump *except*
    a. reducing pump speed.
    b. increasing the water supply.
    c. adding one or more discharge lines.
    d. reducing nozzle gpm settings.

46. During cavitation a _____ pressure zone occurs near the center of the impeller, while a _____ pressure zone occurs near the outer edge of the impeller.
    a. low, high
    b. high, low
    c. neutral, high
    d. high, balanced

47. A Class 3 standpipe system has a minimum flow rate of _____ gpm.
    a. 100
    b. 150
    c. 250
    d. 500

48. When troubleshooting pump problems, all of the following are correct *except*
    a. when proper procedures are followed, the problem is most likely mechanical.
    b. use the manufacturer's troubleshooting guides when available.
    c. only certified mechanics should attempt to troubleshoot pumping problems.
    d. the best method to troubleshoot is to follow the flow of water from the intake to the discharge while attempting to determine the problem.

49. What is the weight of water in a tank measuring 10 ft × 50 ft × 25 ft?
    a. 31,170 lbs
    b. 104,250 lbs
    c. 779,250 lbs
    d. 10,425 lbs

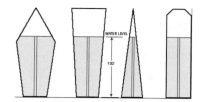

50. The pressure at the base of each vessel is the same and illustrates what pressure principle?
    a. Pressure exerted at the bottom of a container is independent of the shape or volume of the container.
    b. Pressure at any point beneath the surface of a liquid in an open container is directly proportional to its depth.
    c. External pressure applied to a confined liquid (fluid) is transmitted equally throughout the liquid.
    d. Pressure at any point in a liquid at rest is equal in every direction.

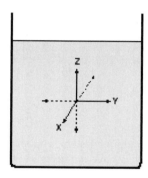

51. The drawing illustrates which pressure principle?
    a. Pressure of a liquid acting on a surface is perpendicular to that surface.
    b. Pressure at any point in a liquid at rest is equal in every direction.
    c. Pressure exerted at the bottom of a container is independent of the shape or volume of the container.
    d. External pressure applied to a confined liquid (fluid) is transmitted equally throughout the liquid.

52. A hose line is flowing 225 gpm into a structure. If the flow is sustained for 3 minutes, how many gallons of water will be delivered into the structure?
    a. 5,629 lbs
    b. 2.8 tons
    c. 675 gallons
    d. Not enough information is provided

53. During a pump operation, an initial static pressure reading of 30 psi was noted. After initiating flow of two 95-gpm pre-connects, the residual pressure at the hydrant was 25 psi. How many more gpm is the hydrant capable of flowing?
    a. 0 gpm
    b. 95 gpm
    c. 190 gpm
    d. 285 gpm

54. A smooth-bore nozzle master-stream device operating at the correct nozzle pressure with a 2-inch tip will have a nozzle reaction of
    a. 80 psi.
    b. 180 psi.
    c. 202 psi.
    d. 502 psi.

55. A combination nozzle master-stream device operating at the correct nozzle pressure flowing 300 gpm will have a nozzle reaction of
    a. 100 psi.
    b. 180 psi.
    c. 152 psi.
    d. 512 psi.

56. A hose lay consisting of 550 feet of 2 ½-inch is flowing 250 gpm. Calculate the friction loss in this hose lay using the hand method.
    a. 69 psi
    b. 124 psi
    c. 139 psi
    d. 159 psi

57. What is the friction loss in 250 feet of 2 ½-inch hose flowing 450 gpm, using the hand method?
    a. 101 psi
    b. 121 psi
    c. 131 psi
    d. 151 psi

58. While supporting a standpipe system, you are informed that the attack crew has engaged the fire on the 9th floor. They have connected to the standpipe with two 1 ½-inch lines with combination nozzles. You should increase pump pressure by _____ psi to account for elevation.
    a. 25
    b. 40
    c. 45
    d. 140

59. During a relay operation, you are informed that the next pumper in line is 63 feet below your pumper. As you begin to supply this pumper, you should decrease pump pressure by _____ psi to account for the change in elevation.
    a. 20
    b. 25
    c. 32
    d. 52

60. During a structure fire, a master-stream device is being set up with a 2-inch tip supplied by two sections of 3-inch hose that is 500 feet long. If the master-stream is to flow 502 gpm at 80 psi nozzle pressure, the friction loss is _____ psi using the formula cq^2L.
    a. 25
    b. 50
    c. 55
    d. 76

61. You are suppling a 2 ½-inch exposure line that is 400 feet and flowing 225 gpm. Calculate the friction loss for this line using the drop-ten method.
    a. 12 psi
    b. 22 psi
    c. 48 psi
    d. 65 psi

62. A 2 ½-inch hose line 200 feet in length is flowing 250 gpm. The friction loss was calculated to be 25 psi. Which fireground friction loss formula was used?
    a. cq^2L
    b. drop-ten
    c. hand method
    d. both a and c are correct

63. Using the NFA formula, calculate the needed flow for part of a structure that is 35 feet long by 15 feet wide, and 16 feet tall.
    a. 84 gpm
    b. 175 gpm
    c. 525 gpm
    d. 550 gpm

64. You are pre-planning the potential fire flow needs for a structure. The structure is 75 feet long, 55 feet wide, and 11 feet tall. Using the Iowa State formula, determine if a hydrant that is capable of flowing 500 gpm can provide the estimated needed fire flow.
    a. Yes, it can.
    b. No, it cannot.
    c. It depends on how long the flow is needed.
    d. Not enough information is provided.

65. As part of a relay operation, you are supplying a 500-foot hose line flowing 250 gpm through 3-inch hose. Using the condensed q formula, calculate the friction in the hose line.
    a. 31 psi
    b. 41 psi
    c. 70 psi
    d. 80 psi

66. A crew is stretching out 450 feet of 2 ½-inch hose for an exposure line. What is the friction loss in the line if the flow is 225 gpm? Use the formula $2q^2 + q$.
    a. 26 psi
    b. 54 psi
    c. 72 psi
    d. 85 psi

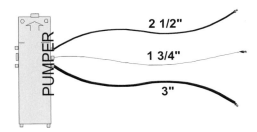

67. For this pumping operation, if the 3-inch line was 500 feet long flowing 350 gpm through a combination nozzle, then the pump discharge pressure would be _____ psi.
    a. 149
    b. 204
    c. 223
    d. 250

68. You are the pump operator supplying an exposure line consisting of 300 feet of 3-inch hose, followed by 300 feet of 2 ½-inch hose. What is the friction loss for the entire hose lay when flowing 225 gpm? Use the formula cq^2L.
    a. 30 psi
    b. 42 psi
    c. 53 psi
    d. 95 psi

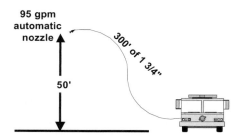

69. For this pumping operation, the pump discharge pressure should be _____.
    a. 115 psi
    b. 140 psi
    c. 167 psi
    d. 222 psi

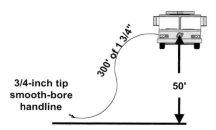

70. For this pumping operation, the pump discharge pressure should be _____.
    a. 25 psi
    b. 75 psi
    c. 118 psi
    d. 140 psi

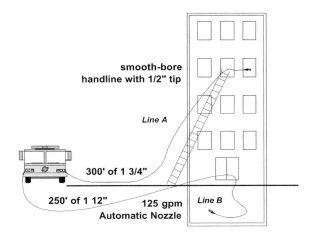

71. The pump discharge pressure for Line A should be _____.
    a.   46 psi
    b.   58 psi
    c.   101 psi
    d.   140 psi

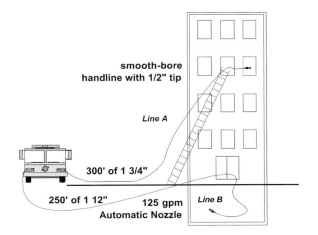

72. The pump discharge pressure for Line B should be _____.
    a.   46 psi
    b.   58 psi
    c.   101 psi
    d.   189 psi

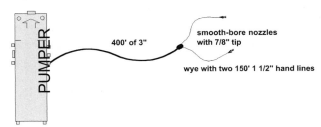

73. For this pumping operation, the pump discharge pressure should be _____.
    a.   100 psi
    b.   153 psi
    c.   186 psi
    d.   222 psi

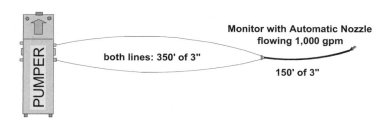

74. For this pumping operation, the pump discharge pressure should be _____.
   a. 100 psi
   b. 253 psi
   c. 310 psi
   d. 575 psi

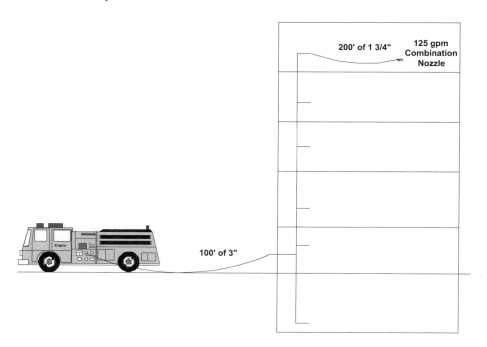

75. For this pumping operation, the pump discharge pressure should be _____.
   a. 150 psi
   b. 172 psi
   c. 186 psi
   d. 214 psi

76. Determine the PDP for the following hose line configuration:

   300 feet of 1 ¾-inch hose with a combination nozzle flowing 100 gpm, friction loss for the
   hose line is calculated to be 47 psi. The line is taken down one level to the basement.
   a. PDP = 92 psi
   b. PDP = 97 psi
   c. PDP = 142 psi
   d. PDP = 147 psi

77. Determine the PDP for the following hose line configuration:

    150 feet of 1 ¾-inch hand-line equipped with a ¾-inch tip smooth-bore nozzle flowing
    118 gpm, friction loss for the hose line is calculated to be 32 psi.
    a. PDP = 32 psi
    b. PDP = 50 psi
    c. PDP = 82 psi
    d. PDP = 132 psi

78. You are determining the PDP and initiating flow from two like lines. One of the discharge lines
    has an operating pressure of 130 psi. Based on this information, which of the following is
    incorrect?
    a. PDP is 260 psi.
    b. PDP is 130 psi.
    c. To initiate flow, the pump operator can open both lines at the same time and then increase
        the throttle to the correct PDP.
    d. To initiate flow, the pump operator can open one line at a time and then increase the
        throttle to maintain the correct PDP.

79. In the volume mode of a four-stage centrifugal pump, if the total discharge is 1,000 gpm at
    100 psi, then each impeller is discharging
    a. 500 gpm at 100 psi.
    b. 500 gpm at 50 psi.
    c. 250 gpm at 100 psi.
    d. 250 gpm at 25 psi.

80. You are operating the last pumper in a relay and you calculate that your pump discharge
    pressure should be 325 psi to supply three attack lines and one exposure line. Your supply from
    the previous pumper seems adequate, with a residual intake pressure of 23 psi. You slowly
    open your discharges and then set your pressure-relief device. Do you note any problems with
    this operation?
    a. There is no problem; everything is good.
    b. The discharge pressure is too high.
    c. The intake pressure is too low.
    d. The relief device should be set prior to verifying 20 psi at the next pump.

81. A tanker operating in a shuttle has a cycle time of 10 minutes and a flow capacity of 150 gpm.
    What is the size of the tanker?
    a. 250 gallons
    b. 1,500 gallons
    c. 2,000 gallons
    d. not enough information provided

82. A 1 ½-inch hand-line of 400 feet with a 1-inch tip smooth-bore nozzle is being supplied by a
    750-gallon on-board water tank. How long can the hand-line be sustained?
    a. about 3 ½ minutes
    b. just over 5 minutes
    c. under 10 minutes
    d. not enough information provided

83. As the pump operator, you just initiated flow through hand-line consisting of 150 feet of
    1 ½-inch hose with a ¾-inch tip smooth-bore nozzle. The residual pressure from the hydrant
    reads 58 psi on the intake, and the discharge flow meter is 118 gpm. What is the flow from the
    smooth-bore nozzle if the hand-line is operating in the basement of a structure?
    a. 50 gpm
    b. 80 gpm
    c. 118 gpm
    d. not enough information provided.

84. It takes about _____ Btu to raise the temperature of 1 gallon of water from 62°F to 212°F.
    a. 100
    b. 150
    c. 212
    d. 1,268

85. To test a driver's ability to drive around obstacles such as parked cars and tight corners, the _____ driving exercise should be used.
    a. alley dock
    b. serpentine
    c. diminishing clearance
    d. confined-space turnaround

86. Prior to moving a fire department vehicle, all personnel riding on an apparatus must be seated and properly secured with seat belts according to NFPA 1500. One exception to this requirement is
    a. when donning PPE.
    b. for emergency response.
    c. when conducting tiller training.
    d. no exceptions are allowed according to NFPA.

87. While responding to an incident, you come to an unguarded railroad guard crossing. According to NFPA 1500, you should
    a. proceed over the railroad tracks at a slow rate of speed.
    b. move quickly across the tracks to reduce the likelihood of a train striking the apparatus.
    c. slow down considerably, but do not stop to ensure the safety of the apparatus.
    d. come to a complete stop before proceeding across the tracks.

88. If the pressure in a centrifugal pump is held constant while the speed of the pump is increased, flow will
    a. increase.
    b. decrease.
    c. remain the same.
    d. none of the answer choices is correct.

89. While priming the pump, you note a 4-psi pressure reduction on the master intake gauge. You know that the static water supply is just over 8 feet below the pump. Will you have any difficulty priming the pump?
    a. There is no problem in that the general rule of thumb is to attempt to pump no more than 22.5 feet.
    b. Yes, you will have a problem in that 4 psi will only raise water to the height of 6.8 feet.
    c. There is no problem in that 4 psi can raise water to a height of 9 feet.
    d. None of the answers is correct.

90. You arrived on scene to a small residential fire. After catching a hydrant, you position the apparatus and set the parking brake. What should you do next?
    a. Set transfer valve.
    b. Engage the pump.
    c. Set the pressure-regulating device.
    d. Provide water to the pump.

91. As the pump operator, you set up a drafting operation and successfully prime the pump. However, before you are able to open discharge lines, the pump loses its prime. You successfully prime the pump again, only to lose it about 3 minutes later. Which of the following is most likely the cause?
    a. air leaks
    b. end of intake hose not submerged deep enough
    c. engine speed too low
    d. both a and b are correct
    e. all are correct

92. Positive-displacement pumps operate based on _____ principles, while dynamic (centrifugal) pumps operate based on _____ principles.
    a. static, dynamic
    b. hydrodynamic, hydrostatic
    c. internal pressure, external pressure
    d. hydrostatic, hydrodynamic

93. You are the pump operator arriving at the scene of a reported structural fire. You estimate the structure to be approximately 130 feet long and 200 feet wide. Based on this information, what formula can you use to determine the estimated needed fire flow for the structure?
    a. Iowa State formula
    b. National Fire Academy (NFA) Formula
    c. condensed q formula
    d. both a and c are correct
    e. both a and b are correct

94. Which of the following is *not* a duty or function of the pump operator as listed in NFPA 1002?
    a. driving apparatus
    b. operating apparatus
    c. supervision
    d. preventive maintenance

95. You are the pump operator of the second-arriving pumper. You are told to secure a water supply and to support a high-rise standpipe system. The first-arriving pumper has the only accessible hydrant but cannot provide the required pressure to support the standpipe. You decide to have the first pumper discharge all water directly to your pump so that you can increase the pressure to the required level. This pumping configuration is known as
    a. relay pumping.
    b. dual pumping.
    c. tandem pumping.
    d. this is not a recommended pumping operation.

96. The three interdependent activities of pump operations include
    a. securing a water supply, pump procedures, and discharge maintenance.
    b. preventive maintenance, driving the apparatus, and operating the pump.
    c. drafting, relay operations, and shuttle operations.
    d. both a and b are correct.

97. Conducting preventive maintenance inspections is an important duty of a pump operator. Which of the following standards includes requirements for the frequency or schedule of inspections?
    a. NFPA 1002
    b. NFPA 1500
    c. NFPA 1915
    d. both b and c are correct

98. While responding to an incident, you notice a hazard in front of you. You travel 35 feet before the brain recognizes the hazard. It takes another 125 feet before the vehicle comes to a complete stop. What is the total stopping distance?
    a. 35 feet
    b. 125 feet
    c. 160 feet
    d. not enough information provided

99. Newton's first and third laws of motion can be used to help explain the concept of centrifugal force. Which of the following is correct?
    a. Newton's first law of motion indicates that a moving body travels in a straight line with a constant speed reduction unless affected by an outside force.
    b. Newton's third law of motion states that for every action there is an equal and opposite reaction.
    c. Both a and b are correct.
    d. Neither a nor b are correct.

100. While responding to an incident, you begin to navigate a curve. As a knowledgeable pump operator, you know that the key to safely navigating a curve is to
    a. make the turn with the brakes slightly pressed.
    b. maintain traction.
    c. slightly decrease your speed.
    d. slightly increase your speed.

Phase III, Exam II: Answers to Questions

1.	T	26.	D	51.	B	76.	C
2.	F	27.	A	52.	C	77.	C
3.	F	28.	B	53.	C	78.	A
4.	F	29.	B	54.	D	79.	C
5.	T	30.	B	55.	C	80.	B
6.	F	31.	A	56.	A	81.	B
7.	F	32.	B	57.	A	82.	A
8.	D	33.	A	58.	B	83.	C
9.	B	34.	B	59.	C	84.	D
10.	A	35.	A	60.	A	85.	B
11.	C	36.	C	61.	C	86.	C
12.	C	37.	E	62.	D	87.	D
13.	C	38.	A	63.	B	88.	A
14.	D	39.	C	64.	A	89.	C
15.	D	40.	B	65.	A	90.	B
16.	D	41.	C	66.	B	91.	E
17.	C	42.	E	67.	A	92.	D
18.	C	43.	A	68.	B	93.	B
19.	A	44.	D	69.	C	94.	C
20.	A	45.	C	70.	D	95.	C
21.	B	46.	A	71.	B	96.	A
22.	B	47.	D	72.	D	97.	D
23.	D	48.	C	73.	C	98.	D
24.	D	49.	C	74.	C	99.	B
25.	B	50.	A	75.	D	100.	B

Phase III, Exam II:
Rationale & References for Questions

Question #1
It is a misperception about fire pump operations that most of the activities related to fire pump operations occur in the first few minutes upon arrival to the scene. After initial operations are set up, pump operators must:

- Continually observe instrumentation.
- Adjust flows and pressure as appropriate for safety of personnel and equipment.
- Be prepared to readily adapt to changing fireground situations.
- Monitor and plan for water supply needs and long-term operations.
- Maintain a constant vigilance to safety.

NFPA 1002 5.1. *IFPO, 2E:* Chapter 1, page 4.

Question #2
Although good public relations is important, increased safety, reduced liability, and costs savings are more important reasons for a well-funded preventive maintenance program. NFPA 1002 4.2.1, 5.1.1. *IFPO, 2E:* Chapter 2, pages 21 – 22.

Question #3
Even though a problem can be fixed immediately, the pump operator should document all findings to assist with tracking and trending of problems. NFPA 1002 4.2.2. *IFPO, 2E:* Chapter 2, page 25.

Question #4
The relationship between the driver and officer is both legal and practical. Both can be held accountable for unsafe actions. NFPA 1002 4.3. *IFPO, 2E:* Chapter 3, pages 53 – 54.

Question #5
Municipal water supplies commonly provide water for both normal public consumption and for emergency use. NFPA 1002 5.2.1. *IFPO, 2E:* Chapter 7, pages 169 – 170.

Question #6
Several general considerations for positioning the vehicle include the following:

- Position the vehicle to reduce the likelihood of being struck by traffic.
- Use the vehicle to shield emergency personnel from traffic.
- Position to enhance emergency operations.
- Position a way from hazards.
- Consider wind direction.
- Do not park in the collapse zone.
- Never park on railroad tracks.
- Park the vehicle on the side of the incident.

NFPA 1002 4.3. *IFPO, 2E:* Chapter 3, page 65.

Question #7
Only the pressure governor changes the engine speed. The pressure-relief valve limits pressure surges by opening and closing a passage between the discharge side of the pump to either the intake or atmosphere. NFPA 1002 5.2.1, 5.2.2, 5.2.4. *IFPO, 2E:* Chapter 5, pages 122 – 123.

Question #8
The actual trend indicates that the number of vehicle accidents has gradually increased from 1995 to 2000. NFPA 4.3. *IFPO, 2E:* Chapter 1, pages 8 – 9.

Question #9

According to NFPA 1002 (5.1), "driver/operator – pumper" is the term used to describe the individual who has met the requirements of Chapter 5 for driving and operating apparatus equipped with fire pumps. NFPA 1002 5.1. *IFPO, 2E:* Chapter 1, page 15.

Question #10

Laws are rules that are legally binding and enforceable. Standards are guidelines that are not legally binding or enforceable. NFPA 1002 is an example of a standard. The full title is "NFPA 1002 Standard on Fire Apparatus Driver/Operator Professional Qualifications." NFPA 1002 4.2, 4.3, 5.1. *IFPO, 2E:* Chapter 1, page 15.

Question #11

A common sequence for conducting a vehicle inspection is to start with a walk around the outside of the vehicle, followed by the engine compartment. Next, the inside of the cab is inspected, when the engine is started. With the engine running, the final step is to complete the inspection of the pump and related components. NFPA 1002 4.2.1, 5.1.1. *IFPO, 2E:* Chapter 2, pages 27 – 28.

Question #12

According to NFPA 1901, the fuel tank must be sufficient in size to drive the pump for at least 2 ½ hours at its rated capacity when pumping at draft. In this case, the fuel tank surpassed this requirement by pumping for 3 hours. NFPA 1002 4.2.1. *IFPO, 2E:* Chapter 2, page 35.

Question #13

NFPA 1911 requires that pumps with a rated capacity of 250 gpm or greater be service tested at least annually and after any major repair or modification. NFPA 1002 4.2.1, 5.1.1. *IFPO, 2E:* Chapter 2, page 24.

Question #14

Although no specific inspection sequence is mandated, it is a good idea to review previous vehicle inspection reports before starting a preventive maintenance inspection. NFPA 1002 4.2.1. *IFPO, 2E:* Chapter 2, page 28.

Question #15

The impact of an accident involving responding apparatus can be:

- Delayed assistance to those who summoned help.
- Additional units must be dispatched for the original call as well as for the new accident.
- Fire department members and civilians could be seriously or fatally injured.
- Fire department and civilian vehicles and property could sustain extensive damage.
- The city, department, and driver could face civil and/or criminal proceedings.
- The image presented by the accident will last a long time in the mind of the public.

NFPA 1002 4.3. *IFPO, 2E:* Chapter 3, page 43.

Question #16

NFPA 1500 requires that emergency-vehicle drivers come to a complete stop when any intersection hazard is present. Specifically, the vehicle must come to a complete stop when any of the following exists:

- As directed by a law enforcement officer
- At traffic red lights or stop signs
- At negative right-of-way and blind intersections
- When all lanes of traffic in an intersection cannot be accounted for
- When a stopped school bus with flashing warning lights is encountered
- At unguarded railroad guard crossings (also for nonemergency)
- When other intersection hazards are present

NFPA 1002 4.3. *IFPO, 2E:* Chapter 3, page 48.

Question #17

Several driving exercises are typically used to assess the driver's ability to safely operate and control the vehicle. These include the following:

- *Alley Dock.* Assesses the ability to back the vehicle into a restricted area, such as a fire station or down an alley.
- *Serpentine.* Assesses the ability to drive around obstacles, such as parked cars and tight corners.
- *Confined-Space Turnaround.* Assesses the ability to turn the vehicle around within a confined space, such as a narrow street or driveway.
- *Diminishing Clearance.* Assesses the ability to drive the vehicle in a straight line, such as on a narrow street or road.

NFPA 1002 4.3.2, 4.3.3, 4.3.4, 4.3.5. *IFPO, 2E:* Chapter 3, pages 65 – 66.

Question #18

When no water is flowing, the pressure at any point in a hose line will be the same. Each of the gauges would have the same pressure reading, and the regulating device would have no effect on the pressure readings among the gauges. NFPA 1002 5.2.1, 5.2.2. *IFPO, 2E:* Chapter 4, page 78.

Question #19

Each revolution of a positive-displacement pump will yield (displace) a specific quantity of water. If the speed of the pump increases, then the quantity of water displaced increases. NFPA 1001 5.2.1, 5.2.2. *IFPO, 2E:* Chapter 4, page 78.

Question #20

Centrifugal-pump performance is contingent on three interrelated factors. If one factor remains constant, a change in one of the remaining factors will change the other.

- *Speed.* If the speed of the pump is held constant and the flow of water increases, pressure will drop. If more water is allowed to flow while the speed of the pump remains the same, pressure will be reduced, because less resistance occurs on the discharge side of the pump.
- *Flow.* If the flow of water is held constant and the speed of the pump is increased, pressure will increase. The same amount of water is being discharged, yet the pump is attempting to discharge more water. This results in an increase in pressure.
- *Pressure.* If the pressure is held constant and the speed of the pump is increased, flow will increase. The increased speed of the pump will increase the flow. Pressure can be maintained constant by increasing or reducing the resistance on the discharge side of the pump.

NFPA 1002 5.2.1, 5.2.2. *IFPO, 2E:* Chapter 4, page 86.

Question #21

According to NFPA 1901, a pump must have a rated capacity as follows:

- 100% of its rated capacity at 150 psi
- 70% of its rated capacity at 200 psi
- 50% of its rated capacity at 250 psi

NFPA 1002 5.2.1, 5.2.2. *IFPO, 2E:* Chapter 4, page 91.

Question #22

This line drawing indicates the pump is in pressure mode because the transfer valve redirects the discharge from one impeller to the intake of the second impeller. NFPA 1002 5.2.1, 5.2.2, 5.2.4. *IFPO, 2E:* Chapter 4, pages 90 – 91.

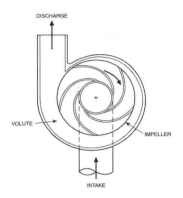

Question #23
As noted in the diagram, the void space is called the volute. NPFA 1002 5.2.1, 5.2.2. *IFPO, 2E:* Chapter 4, pages 86 – 89.

Question #24
- A *pre-mixed system* consists of a tank in which foam concentrate and water are added in appropriate proportions. Often, the foam is simply added to the on-board water tank.
- An *in-line eductor system* utilizes eductors to add foam to water in appropriate proportions. This system can have either internal or external controls, and requires an eductor and a foam tank (or supply). Internal systems have a foam concentrate valve and meeting control. Eductors are usually installed between the pump discharge and the discharge outlet.
- An *around-the-pump proportioning system* uses an eductor (located between the intake and discharge sides of the pump) to mix foam with water. Water from the discharge side of the pump picks up foam and delivers to the intake side of the pump, where it can then be discharged to all outlets.
- A *balanced-pressure system* mixes foam with water by means of pressure. One system uses pressure to force foam from a bladder. The second system uses a separate foam pump (two types are by-pass and demand).
- A *direct injection/compressed-air foam system* (CAFS) consists of a separate foam pump and foam tank. Direct injection moves foam from the tank directly into the discharge lines. CAFS adds compressed air to create a lightweight foam.

NFPA 1002 5.2.3. *IFPO, 2E:* Chapter 5, pages 126 – 103.

Question #25
Flow meters often use paddle wheels to measure the flow within a line. NFPA 1002 5.2.1, 5.2.2. *IFPO, 2E:* Chapter 5, pages 113 – 114.

Question #26
The test date and pressure is usually stenciled on the hose, as opposed to indelibly marked. NFPA 1002 5.1.1. *IFPO, 2E:* Chapter 6, page 146.

Question #27
Basic steps for testing hose include:
1. Connect water supply and test hose sections
2. Open discharge lines for test hose sections
3. Slowly increase discharge pressure to 45 psi
4. Remove all air from within the hose
5. Mark hose near coupling to determine slippage
6. Increase pressure slowly (NFPA 1962 suggests no faster than 15 psi per second) to test pressure

7. Maintain pressure for three minutes, while periodically checking for leaks

8. Record results

NFPA 1002 5.1. *IFPO, 2E:* Chapter 6, page 147.

Question #28

A maximum of 300-foot hose sections should be tested. NFPA 1002 5.1. *IFPO, 2E:* Chapter 6, page 147.

Question #29

NFPA 1965 requires that appliances with lever-operated handles must indicate a closed position when the handle is perpendicular to the hose line. NFPA 1002 5.2.1, 5.2.2, 5.2.4. *IFPO, 2E:* Chapter 6, page 148.

Question #30

A pumper with a 5,000-gallon tanker flowing 400 gpm (200 gpm × 2 lines) will last about 12.5 minutes. 400 gallons per minute ⟩ 5000 gallons = 12.5 minutes. NFPA 1002 5.2.1. *IFPO, 2E:* Chapter 7, page 167.

Question #31

Basic steps for conducting hydrant flow include:

1. Take static pressure reading at test hydrant (make sure to open hydrant fully and remove air).

2. Open flow hydrants one at a time until a 25% drop in residual pressure is achieved.

3. Continue to flow to clear debris and foreign substances.

4. Take readings at the same time: residual reading at test hydrant and flow readings at each flow hydrant, and record results.

5. Slowly shut down hydrants one at a time.

6. Record exact interior size, in inches, of each outlet flowed.

NFPA 1002 5.2.1. *IFPO, 2E:* Chapter 7, pages 180 – 181.

Question #32

Water will rise approximately 2.3 feet/psi for each 1 psi of pressure.
So, 4 psi × 2.3 feet/psi = 9.2 feet. NFPA 1002 5.2.1. *IFPO, 2E:* Chapter 7, pages 185 – 186.

Question #33

Hydrants and water mains are not part of a static water supply. Rather, they are integral parts of a municipal water supply system. NFPA 1002 5.2.1. *IFPO, 2E:* Chapter 7, page 184.

Question #34

The distance can be calculated as follows:

- (PDP – 20) × 100/FL
- (250 – 20) × 100 / 20
- 230 × 5
- 1,150 feet

NFPA 1002 5.2.3. *IFPO, 2E:* Chapter 7, page 194.

Question #35

Divide tank volume by the shuttle cycle time.

5000 ÷ 20 = 250 gpm.

NFPA 1002 5.2.1. *IFPO, 2E:* Chapter 7, page 201.

Question #36

The basic steps for pump operation include:

- Step 1   Position apparatus, set parking brake, and let engine return to idle.
- Step 2   Engage the pump.
- Step 3   Provide water to intake side of pump (on-board, hydrant, draft).
- Step 4   Set transfer valve (if so equipped).
- Step 5   Open discharge lines.
- Step 6   Throttle to desired pressure.
- Step 7   Set the pressure-regulating device.
- Step 8   Maintain appropriate flows and pressures.

NFPA 1002 5.2.1, 5.2.2, 5.2.4. *IFPO, 2E:* Chapter 8, page 207.

Question #37

Important considerations for positioning pumping apparatus at incidents include:

- If no fire or smoke is visible, park near main entrance.
- Be sure to follow department SOPs.
- Consider tactical priorities for the incident.
- Consider surroundings such as heat from the fire, collapse, overhead lines, escape routes, and wind.

NFPA 1002 5.2.1, 5.2.2. *IFPO, 2E:* Chapter 8, pages 208 – 209.

Question #38

Basic steps for pump engagement powered through a PTO include:

- Bring apparatus to complete stop, set parking brake, and let engine return to idle speed.
- Disengage the clutch (push in the clutch pedal).
- Place transmission in neutral.
- Operate the PTO lever.
- For mobile pumping, place transmission in the proper gear (low).
- For stationary pumping, place transmission in neutral.
- Engage the clutch slowly.

NFPA 1002 5.2.1, 5.2.2, 5.2.4. *IFPO, 2E:* Chapter 8, pages 210 – 211.

Question #39

Basic steps to engage a pump utilizing a split-shaft arrangement are as follows:

- Bring apparatus to complete stop.
- Place transmission in neutral.
- Apply parking brake.
- Operate pump shift switch from road to pump position.
- Ensure the "OK to pump" light comes on.
- Shift transmission to pumping gear (usually highest gear).

NFPA 1002 5.2.1, 5.2.2, 5.2.4. *IFPO, 2E:* Chapter 8, pages 211 – 212.

Question #40
Transfer valve operation rule of thumb:

- *Volume mode* should be used when expected flows are greater than 50% of a pump's rated capacity with pressures less than 150 psi.

- *Pressure mode* should be used when expected flows are less than 50% of a pump's rated capacity with pressures greater than 150 psi.

NFPA 1002 5.2.1, 5.2.2, 5.2.4. *IFPO, 2E:* Chapter 8, pages 214 – 215.

Question #41
A corresponding increase in discharge pressure should be noted when the throttle is increased (engine speed increased). If not, one of the following common causes could be occurring:

- The pump may not be in gear.

- The pump may not be primed.

- The supply line could be closed or insufficient.

NFPA 1002 5.2.1, 5.2.2, 5.2.4. *IFPO, 2E:* Chapter 8, page 216.

Question #42
Resetting the pressure regulating device will not enable the pump operator to increase pressure because the device is not currently relieving pressure. NFPA 1002 5.2.1, 5.2.2, 5.2.4. *IFPO, 2E:* Chapter 8, page 216.

Question #43
The basic on-board water supply procedures are:

- Position apparatus, set parking brake.

- Engage pump.

- Set transfer valve.

- Open "tank-to-pump."

- Open discharge control valves.

- Increase throttle.

- Set pressure-regulating device.

- Plan for more water.

NFPA 1002 5.2.1, 5.2.2, 5.2.4. *IFPO, 2E:* Chapter 8, pages 220 – 221.

Question #44
The fastest attack will most often be pulling directly to the scene and advancing pre-connected lines supplied by on-board water. If this occurs, the pump operator should immediately start planning for an alternate water supply. NFPA 1002 5.2.1, 5.2.2. *IFPO, 2E:* Chapter 8, pages 208 – 209.

Question #45
Cavitation can be stopped by reducing pump pressure through pump speed, reduced discharge flow, or increased supply flow. Adding one or more discharge lines would only aggravate a deteriorating system and most likely would cause additional damage. NFPA 1002 5.1.1. *IFPO, 2E:* Chapter 9, pages 245 – 247.

Question #46
The low-pressure zone near the center of the impeller allows vapor pockets to be created. When these vapor pockets pass through the high-pressure zone, they are forced back into water (liquid state) abruptly. NFPA 1002 5.2.1, 5.2.2, 5.2.4. *IFPO, 2E:* Chapter 9, page 256.

Question #47
- *Class 1* standpipes provide 2 ½-inch connections for trained firefighters/fire brigades, at an initial flow rate of 500 gpm.
- *Class 2* standpipes provide 1 ½-inch connections for initial attack, at a minimum flow rate of 100 gpm.
- *Class 3* standpipes provide 1 ½-inch and 2 ½-inch connections for trained firefighters/fire brigades, at an initial flow rate of 500 gpm.

NFPA 1002 5.2.4. *IFPO, 2E:* Chapter 9, page 253.

Question #48
Pump operators are expected to be able to troubleshoot pump problems. Several basic troubleshooting considerations include:
- Problems are usually either procedural or mechanical.
- When proper procedures are followed, the problem is most likely mechanical.
- The best method to troubleshoot is to follow the flow of water from the intake to the discharge while attempting to determine the problem.
- Use the manufacturer's troubleshooting guides when available.

NFPA 1002 5.2. *IFPO, 2E:* Chapter 9, page 254.

Question #49
To calculate weight, the formula $W = D \times V$ is used (V = L × W × H).
- W = 62.34 lb/ft^3 × (10 ft × 50 ft × 25 ft)
- W = 62.34 lb/ft^3 × 12,500
- W = 779,250 lbs (note the cubic feet cancel each other)

NFPA 1002 5.2. *IFPO, 2E:* Chapter 10, page 266.

Question #50
Pressure at any point in a liquid at rest is equal in every direction.
- Pressure of a liquid acting on a surface is perpendicular to that surface.
- External pressure applied to a confined liquid (fluid) is transmitted equally throughout the liquid.
- Pressure at any point beneath the surface of a liquid in an open container is directly proportional to its depth.
- Pressure exerted at the bottom of a container is independent of the shape or volume of the container.

NFPA 1002 5.2. *IFPO, 2E:* Chapter 10, pages 271 – 275.

Question #51
Pressure at any point in a liquid at rest is equal in every direction.
- Pressure of a liquid acting on a surface is perpendicular to that surface.
- External pressure applied to a confined liquid (fluid) is transmitted equally throughout the liquid.
- Pressure at any point beneath the surface of a liquid in an open container is directly proportional to its depth.
- Pressure exerted at the bottom of a container is independent of the shape or volume of the container.

NFPA 1002 5.2. *IFPO, 2E:* Chapter 10, pages 271 – 275.

Question #52

225 gpm × 3 minutes = 675 gallons

NFPA 1002 5.2. *IFPO, 2E:* Chapter 10, page 268.

Question #53

Based on the percent drop in pressure, additional flows may be available from a hydrant has follows:

- 0 – 10% drop   three times the original flow
- 11 – 15% drop   two times the original flow
- 16 – 25% drop   one times the original flow

A 5-psi drop in pressure is about 17% of the initial static reading (5 psi /30 psi).
One times the original flow is 190 psi (remember, two lines were flowing). NFPA 1002 5.2.
IFPO, 2E: Chapter 10, page 277.

Question #54

Nozzle reaction calculation
Smooth-bore nozzles:

$$NR = 1.57 \times d^2 \times NP$$

where NR = nozzle reaction
        1.57 = constant
        d = diameter of nozzle orifice in inches
        NP = operating nozzle pressure in psi

$$NR = 1.57 \times 2^2 \times 80$$

$$NR = 1.57 \times 4 \times 80$$

$$NR = 502.4$$

NFPA 1002 5.2. *IFPO, 2E:* Chapter 10, pages 282 – 283.

Question #55

Nozzle reaction calculation
Combination nozzles:

$$NR = gpm \times \sqrt{NP} \times 0.0505$$

where NR = nozzle reaction
        0.0505 = constant
        gpm = flow in gallons per minute
        NP = operating nozzle pressure in psi

$$NR = 300 \times \sqrt{100} \times 0.0505$$

$$NR = 300 \times 10 \times 0.0505$$

$$NR = 151.5$$

NFPA 1002 5.2. *IFPO, 2E:* Chapter 10, pages 282 – 283.

Question #56

The hand method is a fireground method used to estimate friction loss in 100-foot sections of 2 ½-inch hose. Simply select the fingertip representing the gpm flow and then multiply the two figures on the finger for the approximate friction loss pressure for each 100-foot section of 2 ½-inch hose.

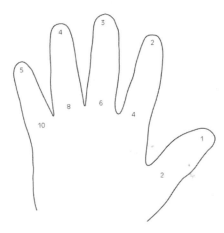

FL = 2.5 × 5 × 5.5 (number of 100-foot sections of hose)

FL = 68.75 or 69 psi

NFPA 1002 5.2.1, 5.2.2. *IFPO, 2E:* Chapter 10, pages 302 – 303.

Question #57

The hand method is a fireground method used to estimate friction loss in 100-foot sections of 2 ½-inch hose. Simply select the fingertip representing the gpm flow and then multiply the two figures on the finger for the approximate friction loss pressure for each 100-foot section of 2 ½-inch hose.

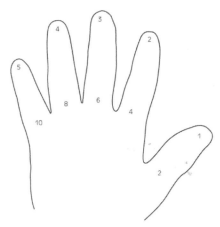

FL = 4.5 × 9 × 2.5 (number of 100-foot sections of hose)

FL = 101.25 psi

NFPA 1002 5.2.1, 5.2.2. *IFPO, 2E:* Chapter 10, pages 302 – 303.

Question #58
Elevation formula by floor level

$EL = 5 \times H$

where EL = the gain or loss of elevation in psi
        5 = gain or loss in pressure for each floor level
        H = height in number of floor levels above or below the pump

$EL = 5 \times 8$

$EL = 40psi$

NFPA 1002 5.2.4. *IFPO, 2E:* Chapter 11, page 311.

Question #59
Elevation formula in feet:

$EL = 0.5 \times H$

where EL = the gain or loss of elevation in psi
        0.5 = pressure exerted at base of 1-cubic-inch column of water 1 foot high
        H = distance in feet above or below the pump

$EL = 0.5 \times 63$

$EL = 31.5psi$

NFPA 1002 5.2. *IFPO, 2E:* Chapter 11, page 311.

Question #60
cq^2L, where c for 3-inch hose is .8
Each of the two lines will flow half of the total gpm, in this case 251 gpm through each line.

$$.8 \times \left(\frac{251}{100} \right)^2 \times \frac{500}{100}$$

$.8 \times 6.2 \times 5$

$25.2psi$

NFPA 1002 5.2. *IFPO, 2E:* Chapter 11, pages 306 – 307.

Question #61
The drop-ten method simply subtracts 10 from the first two numbers of gpm flow. It is not as accurate as other methods, but provides a simple rule of thumb for fireground use.

- 225 gpm = 22 – 10 = 12 psi friction loss per 100-foot sections of 2½-inch hose
- 12 psi × 4 (4 sections of 100-feet of hose) = 48 psi

NFPA 1002 5.2.1, 5.2.2. *IFPO, 2E:* Chapter 10, page 303.

Question #62
Drop-ten method was used:

- 250 gpm = 25 – 10 = 15 psi friction loss per 100-foot sections of 2½"
- 15 psi × 2 (2 sections of 100-feet of hose) = 30 psi

Other fireground formulas field a slightly different result:

- cq^2L and hand method= 25 psi
- $2q^2 + q = 15$ psi

NFPA 1002 5.2.1, 5.2.2. *IFPO, 2E:* Chapter 10, page 303.

Question #63

NFA formula: $NF = \dfrac{A}{3}$

where NF = needed flow in gpm
A = area of the structure in square feet
3 = constant in ft^2/gpm

$NF = \dfrac{A}{3 ft^2 gpm}$

$NF = \dfrac{35 ft \times 15 ft}{3 ft^2 gpm}$

$NF = \dfrac{525 ft^2}{3 ft^2 gpm}$

$NF = 175 gpm$

NFPA 1002 5.2. *IFPO, 2E:* Chapter 11, pages 292 – 293.

Question #64

Iowa State formula: $NF = \dfrac{V}{100}$

where NF = needed flow in gpm
V = volume of the area in cubic feet
100 = is a constant in ft^3/gpm

$NF = \dfrac{75 ft \times 55 ft \times 11 ft}{100 ft^3 / gpm}$

$NF = \dfrac{45,375 ft^3}{100 ft^3 / gpm}$

$NF = 453.75 gpm$

Yes, the hydrant capable of flowing 500 gpm and can supply the needed fire flow of 454 gpm for the structure. NFPA 1002 5.2. *IFPO, 2E:* Chapter 11, pages 292 – 293.

Question #65

q^2

$\left(\dfrac{250}{100}\right)^2$

$(2.5)^2$

6.25 psi per 100-foot sections
$FL = 6.25 \times 5$ (number of 100-foot sections in the hose line)
$FL = 31.25$ psi

NFPA 1002 5.2. *IFPO, 2E:* Chapter 11, pages 306 – 307.

Question #66

$2q^2 + q$

$2 \times \left(\dfrac{225}{100}\right)^2 + \left(\dfrac{225}{100}\right)$

$2 \times (2.25)^2 + (2.25)$

$2 \times 5.06 + 2.25$

12.37 or 12 psi friction loss per 100-foot sections of hose
$12 \times 4.5 = 54$ psi

NFPA 1002 5.2. *IFPO, 2E:* Chapter 11, pages 304 – 305.

Question #67

$PDP = NP + FL + AFL \pm EL$

NP: 100 psi

FL: cq^2L

$\qquad .8 \times \left(\dfrac{350}{100}\right)^2 \times \dfrac{500}{100}$

$\qquad .8 \times (3.5)^2 \times 5$

$\qquad .8 \times 12.25 \times 5$

$\qquad 49psi$

AFL: 0 psi

$\pm EL$:   0 psi

$PDP = 100 \ + \ 49$

$PDP = 149$ psi

NFPA 1002 5.2. *IFPO, 2E:* Chapter 11, pages 317 – 357.

Question #68

cq^2L

where c for 3-inch hose is .8 and for 2 ½-inch hose is 2

3-inch hose friction loss

$.8 \times \left(\dfrac{225}{100}\right)^2 \times \dfrac{300}{100}$

$.8 \times (2.25)^2 \times 3$

$.8 \times 5.06 \times 3$

$12.14psi$

2 ½-inch hose friction loss

$2 \times \left(\dfrac{225}{100}\right)^2 \times \dfrac{300}{100}$

$2 \times (2.25)^2 \times 3$

$2 \times 5.06 \times 3$

$30.36psi$

Total friction loss is 42 psi (12 psi + 30 psi)

NFPA 1002 5.2. *IFPO, 2E:* Chapter 11, pages 306 – 307.

Question #69

$PDP = NP + FL + AFL \pm EL$

NP: 100 psi

FL: cq^2L

$$15.5 \times \left(\frac{95}{100}\right)^2 \times \frac{300}{100}$$

$$15.5 \times (.95)^2 \times 3$$

$$15.5 \times .9 \times 3$$

$41.85psi$

AFL: 0 psi

$\pm EL$ $0.5 \times H$

$0.5 \times (50)$

$25psi$

$PDP = 100 \ + \ 42 \ + \ 25$

$PDP = 167$ psi

NFPA 1002 5.2. *IFPO, 2E:* Chapter 11, pages 317 – 357.

Question #70

$PDP = NP + FL + AFL \pm EL$

NP: 50 psi

gpm: $30 \times d^2 \times \sqrt{NP}$

$$30 \times (0.75)^2 \times \sqrt{50}$$

$$30 \times 0.56 \times 7$$

$117.6gpm$

FL: cq^2L

$$15.5 \times \left(\frac{118}{100}\right)^2 \times \frac{300}{100}$$

$$15.5 \times (1.18)^2 \times 3$$

$$15.5 \times 1.39 \times 3$$

$64.6psi$

AFL: 0 psi

$\pm EL$ $0.5 \times H$

$0.5 \times (-50)$

$-25psi$

$PDP = 50 \ + \ 65 - 25$

$PDP = 140$ psi

NFPA 1002 5.2. *IFPO, 2E:* Chapter 11, pages 317 – 357.

Question #71

$PDP = NP + FL + AFL \pm EL$

NP: 50 psi

gpm: $30 \times d^2 \times \sqrt{NP}$

         $30 \times (0.5)^2 \times \sqrt{50}$

         $30 \times 0.25 \times 7$

         $52.5 gpm$

FL: $cq^2 L$

         $15.5 \times \left(\dfrac{53}{100}\right)^2 \times \dfrac{300}{100}$

         $15.5 \times (0.53)^2 \times 3$

         $15.5 \times 0.28 \times 3$

         $13 psi$

AFL: 0 psi

$\pm EL$ $5 \times H$

         $5 \times (-1)$

         $-5 psi$

$PDP = 50 \ + \ 13 - 5$

$PDP = 58$ psi

NFPA 1002 5.2. *IFPO, 2E:* Chapter 11, pages 317 – 357.

Question #72

$PDP = NP + FL + AFL \pm EL$

NP: 100 psi

FL: $cq^2 L$

         $24 \times \left(\dfrac{125}{100}\right)^2 \times \dfrac{250}{100}$

         $24 \times (1.25)^2 \times 2.5$

         $24 \times 1.56 \times 2.5$

         $93.6 psi$

AFL: 0 psi

$\pm EL$ $5 \times H$

         $5 \times (-1)$

         $-5 psi$

$PDP = 100 \ + \ 94 - 5$

$PDP = 189$ psi

NFPA 1002 5.2. *IFPO, 2E:* Chapter 11, pages 317 – 357.

Question #73

First, calculate the friction loss in the 3-inch line, remembering to add the flow through both 1 ½-inch lines. Next, calculate the friction loss for only one of the 1 ½-inch lines because they are like lines. Finally, determine the appliance friction loss.

$PDP = NP + FL + AFL \pm EL$

NP: 50 psi

gpm: $30 \times d^2 \times \sqrt{NP}$

$30 \times (0.875)^2 \times \sqrt{50}$

$30 \times 0.765 \times 7$

$160.65\, gpm$

FLs (3-inch line):

cq^2L

$.8 \times \left(\dfrac{322}{100}\right)^2 \times \dfrac{400}{100}$

$.8 \times (3.22)^2 \times 4$

$.8 \times 10.37 \times 4$

$33.18\, psi$

FLa (1 ½ lines):

cq^2L

$24 \times \left(\dfrac{161}{100}\right)^2 \times \dfrac{150}{100}$

$24 \times (1.61)^2 \times 1.5$

$24 \times 2.59 \times 1.5$

$93.24\, psi$

AFL: 10 psi

$\pm EL$:   0

$PDP = 50 \; + \; (33 + 93) + 10$

$PDP = 186$ psi

NFPA 1002 5.2. *IFPO, 2E:* Chapter 12, pages 342 – 344.

Question #74

$PDP = NP + FL + AFL \pm EL$

NP: 100 psi

FLs (two 3-inch lines):

cq^2L

$.8 \times \left(\dfrac{500}{100}\right)^2 \times \dfrac{350}{100}$

$.8 \times (5)^2 \times 3.5$

$.8 \times 25 \times 3.5$

$70\, psi$

FLa (3-inch line):

$$cq^2L$$

$$.8 \times \left(\frac{1,000}{100} \right)^2 \times \frac{150}{100}$$

$$.8 \times (10)^2 \times 1.5$$

$$.8 \times 100 \times 1.5$$

$$120psi$$

AFL: 5 psi for the siamese and 15 psi for the monitor
$\pm EL$:   0
$PDP = 100 \ + \ (70 + 120) + 20$
$PDP = 310$ psi

NFPA 1002 5.2. *IFPO, 2E:* Chapter 12, pages 34 – 348.

Question #75
$PDP = NP + FL + AFL \pm EL$
NP: 100 psi
FLs (3-inch lines):

$$cq^2L$$

$$.8 \times \left(\frac{125}{100} \right)^2 \times \frac{100}{100}$$

$$.8 \times (1.25)^2 \times 1$$

$$.8 \times 1.56 \times 1$$

$$1.24psi$$

FLa (1 ¾-inch line):

$$cq^2L$$

$$15.5 \times \left(\frac{125}{100} \right)^2 \times \frac{200}{100}$$

$$15.5 \times (1.25)^2 \times 2$$

$$15.5 \times 1.56 \times 2$$

$$48.36psi$$

AFL: 25 psi for the standpipe
$\pm EL$:   5 psi per level, $5 \times 4 = 40$ psi
$PDP = 100 \ + \ (1 + \ 48) \ + \ 25 \ + \ 40$
$PDP = 214$ psi

NFPA 1002 5.2.4. *IFPO, 2E:* Chapter 12, pages 349 – 350.

Question #76

300 feet of 1 ¾-inch hose with a combination nozzle flowing 100 gpm, friction loss for the hose line is calculated to be 47 psi. The line is taken down one level to the basement.

$PDP = NP + FL + AFL \pm EL$

$NP =$ 50 for smooth-bore on a hand-line

80 for smooth-bore on a master-stream device

100 for combination (fog, automatic, selectable flow, etc.)

$EL =$ 5 psi added or subtracted for each floor level above/below ground level

.5 psi added or subtracted for each foot in elevation above/below ground level

$PDP = 100\ (NP)\ +\ 47\ (FL) - 5\ (EL)$

$PDP = 142$ psi

NFPA 1002 5.2.4. *IFPO, 2E:* Chapter 12, pages 319 – 321.

Question #77

A smooth-bore nozzle on a hand-line requires 50 psi operating pressure.

$PDP = NP\ +\ FL\ +\ AFL \pm EL$

$PDP = 50\ +\ 32$

$PDP = 82$ psi

NFPA 1002 5.2.4. *IFPO, 2E:* Chapter 12, pages 319 – 321.

Question #78

When flowing two or more like lines, the PDP will be the same regardless of how many lines are placed in operation. For each line initiated, the discharge gate is slowly opened while the throttle is increased to maintain the PDP. NFPA 1002 5.2.4. *IFPO, 2E:* Chapter 12, pages 336 – 337.

Question #79

In *volume mode*, impellers spin at the same speed, and each provides the same quantity of water. Total discharge then, is divided among the number of impellers while the pressure each produces is the same. NFPA 1002 5.2.1, 5.2.2, 5.2.4. *IFPO, 2E:* Chapter 4, pages 89 – 91.

Question #80

The pressure is too high because the maximum operating pressure of the attack hose is 275 psi. NFPA 1002 5.2.2. *IFPO, 2E:* Chapter 7, page 192.

Question #81

A tanker shuttle time is determined by dividing the tank volume by the time it takes the tanker to complete one cycle from the dump site, to the fill site, and back to the dump site.

To determine the size of a tanker when the tanker shuttle time and flow is known, simply multiply the shuttle flow capacity by the cycle time.

$150 \times 10 = 1,500$ gallons

NFPA 1002 5.2.1. *IFPO, 2E:* Chapter 7, pages 200 – 201.

Question #82

NP: 50 psi

gpm: $30 \times 1^2 \times \sqrt{50}$

$30 \times (1)^2 \times \sqrt{50}$

$30 \times 1 \times 7$

$210\,gpm$

$750/210 = 3.57$ minutes

NFPA 1002 5.2. *IFPO, 2E:* Chapter 11, pages 296 – 297.

Question #83

If the flow meter on the pump panel reads 118 gpm, then the nozzle is discharging 118 gpm. The water flow will be the same at any point in the line. NFPA 1002 5.2. *IFPO, 2E:* Chapter 5, pages 113 – 114.

Question #84

The specific heat of water is 1 Btu. It takes 152 Btu to raise 1 pound of water from 62°F to 212°F.

It takes 1,268 Btu to raise the temperature of 1 gallon of water from 62°F to 212°F.

NFPA 1002 5.2.1, 5.2.2. *IFPO, 2E:* Chapter 10, page 264.

Question #85

Several driving exercises are typically used to assess the driver's ability to safely operate and control the vehicle. These include the following:

- *Alley Dock*. Assesses the ability to back the vehicle into a restricted area, such as a fire station or down an alley.
- *Serpentine*. Assesses the ability to drive around obstacles, such as parked cars and tight corners.
- *Confined-Space Turnaround*. Assesses the ability to turn the vehicle around within a confined space, such as a narrow street or driveway.
- *Diminishing Clearance*. Assesses the ability to drive the vehicle in a straight line, such as on a narrow street or road.

NFPA 1002 4.3.2, 4.3.3, 4.3.4, 4.3.5. *IFPO, 2E:* Chapter 3, pages 65 – 66.

Question #86

The only exceptions to not wearing a seat belt in a fire department vehicle are:

- Loading hose
- Tiller driver training
- Patient treatment

NFPA 1002 4.3. *IFPO, 2E:* Chapter 3, page 49.

Question #87

NFPA 1500 requires that emergency-vehicle drivers come to a complete stop when any intersection hazard is present. Specifically, the vehicle must come to a complete stop when any of the following exists:

- As directed by a law enforcement officer
- At traffic red lights or stop signs
- At negative right-of-way and blind intersections
- When all lanes of traffic in an intersection cannot be accounted for
- When stopped school bus with flashing warning lights is encountered
- At unguarded railroad guard crossing (also for nonemergency)
- When other intersection hazards are present

NFPA 1002 4.3. *IFPO, 2E:* Chapter 3, page 48.

Question #88

Centrifugal pump performance is contingent on three interrelated factors. If one factor remains constant, a change in one of the remaining factors will change the other.

- *Speed.* If the speed of the pump is held constant and the flow of water increases, pressure will drop. If more water is allowed to flow while the speed of the pump remains the same, pressure will be reduced, because less resistance occurs on the discharge side of the pump.

- *Flow.* If the flow of water is held constant and the speed of the pump is increased, pressure will increase. The same amount of water is being discharged, yet the pump is attempting to discharge more water. This results in an increase in pressure.

- *Pressure.* If the pressure is held constant and the speed of the pump is increased, flow will increase. The increased speed of the pump will increase the flow. Pressure can be maintained constant by increasing or reducing the resistance on the discharge side of the pump.

NFPA 1002 5.2.1, 5.2.2. *IFPO, 2E:* Chapter 4, page 86.

Question #89

Water will rise approximately 2.3 feet/psi for each 1 psi of pressure.

So, 7 psi × 2.3 feet/psi = 16.1 feet.

NFPA 1002 5.2.1. *IFPO, 2E:* Chapter 7, pages 185 – 186.

Question #90

The basic steps for pump operation include:

Step 1	Position apparatus, set parking brake, and let engine return to idle.
Step 2	Engage the pump.
Step 3	Provide water to intake side of pump (on-board, hydrant, draft).
Step 4	Set transfer valve (if so equipped).
Step 5	Open discharge lines.
Step 6	Throttle to desired pressure.
Step 7	Set the pressure-regulating device.
Step 8	Maintain appropriate flows and pressures.

NFPA 1002 5.2.1, 5.2.2, 5.2.4. *IFPO, 2E:* Chapter 8, page 207.

Question #91

Each of the following are possible causes for loss of prime or failure to prime:

- Air leaks
- Debris on intake strainer
- By-pass line open
- No oil in priming tank
- Defective priming valve
- Improper clearance in rotary gear or vane primer pump
- End of intake hose not submerged deep enough
- Engine speed too low
- Lift too high
- Primer not operated long enough

Because the pump was primed twice, lack of oil in the priming tank is most likely not the problem. NFPA 1002 5.2. *IFPO, 2E:* Appendix D, page 376.

Question #92
Positive-displacement pumps, such as rotary vane or rotary gear pumps, operate based on hydrostatic principles (liquids at rest).
Dynamic, or centrifugal, pumps operate based on hydrodynamic principles (liquids in motion). NFPA 1002 5.2. *IFPO, 2E:* Chapter 4, page 76.

Question #93
You would need to use the National Fire Academy Formula because it is based on the area of the structure, and that is the only information you have. To use the Iowa State formula, you would need to be able to calculate volume. NFPA 1002 5.2. *IFPO, 2E:* Chapter 11, pages 292 – 293.

Question #94
Although supervision is often a duty or function of a pump operator, NFPA 1002 does not contain supervision requirements. NFPA 1002 identifies the main duties of the pump operator as driving, maintenance, and pump operation. NFPA 1002 1.4. *IFPO, 2E:* Chapter 1, pages 15 – 16.

Question #95
- *Dual pumping* occurs when one hydrant supplies two pumpers. The second pumper receives water from the intake of the first pumper. In other words, the excess water provided to the first pumper is diverted to the second pumper.

- *Tandem pumping* occurs when one hydrant supplies two pumpers; it is similar to a relay operation. The first pump discharges all its water to the intake of the second pumper, as in a relay. The only significant difference is that a relay is used to move water over extended distances, whereas tandem pumping is used when higher pressures are required than a single pumper can provide.

NFPA 1002 5.2.1. *IFPO, 2E:* Chapter 8, pages 224 – 225.

Question #96
Fire pump operations can be defined as the systematic movement of water from a supply source through a pump to a discharge point. This definition identifies three interdependent activities of 1) securing a water supply, 2) operating the pump, and 3) discharge maintenance. NFPA 1002 5.2. *IFPO, 2E:* Chapter 1, page 8.

Question #97
NFPA 1500 requires that inspections, maintenance, repair, and service records be maintained for all vehicles. NFPA 1915 requires that vehicle inspections be performed according to manufacturer's recommended intervals. NFPA 1002 4.2.2. *IFPO, 2E:* Chapter 2, page 25.

Question #98
Total stopping distance is measured from the time a hazard is detected until the vehicle comes to a complete stop and includes:

- Perception distance – 35 feet

- Reaction distance – ?

- Braking distance – 125 feet

NFPA 1002 4.3.2, 4.3.3, 4.3.4, 4.3.5. *IFPO, 2E:* Chapter 3, page 55.

Question #99
Newton's first law of motion indicates that a moving body travels in a straight line with a constant speed *(not a constant speed reduction)* unless affected by an outside force. NFPA 1002 5.2.1. *IFPO, 2E:* Chapter 4, page 85.

Question #100
The most correct answer is to maintain traction. Speed, centrifugal force, road conditions, and weight shifts can all affect the ability to maintain traction. When traction is lost, the apparatus will be out of control. NFPA 1002 4.3. *IFPO, 2E:* Chapter 3, page 56.

Phase Three, Exam Three

1. Over the years, the title or name used for the position responsible for driving and operating the pump has changed. The title used in NFPA 1002 is apparatus driver/engineer.
   a. True
   b. False

2. Pump operators should never add oil to the engine. This is a preventive maintenance activity for certified mechanics only.
   a. True
   b. False

3. Problems found during preventive maintenance inspections should be documented even if the problem is readily fixed.
   a. True
   b. False

4. Supply reliability is the extent to which the water supply will consistently provide water.
   a. True
   b. False

5. Nozzle reaction for all nozzles used in the fire service can be calculated using the formula $NR = 1.57 \times d^2 \times NP$.
   a. True
   b. False

6. One important consideration when positioning a vehicle at an emergency scene is to position the vehicle to reduce the likelihood of being struck by traffic and to enhance emergency operations.
   a. True
   b. False

7. As the pump operator, you just secured a water supply, opened and set the discharge pressures for three lines, and then set the pressure governor. Almost immediately, two lines are shut down and the engine speed slows considerably. You are not concerned because changing engine speed is how the pressure governor limits pressure surges.
   a. True
   b. False

8. The *NFPA Journal* reports on emergency-vehicle accident data on an annual basis. Based on this, which of the following is correct?
   a. The number of accidents will most likely continue to decrease or stay about the same.
   b. The number of accidents will most likely stay at the same level or slightly increase as in the past several years.
   c. The number of accidents will most likely increase significantly.
   d. No such data exists to answer the question.

9. Limitations and immunity extended to the operation of emergency vehicles are usually found in
   a. standards.
   b. regulations.
   c. local and state laws.
   d. federal codes.

10. The city council has asked the fire department to consider dropping its vehicle preventive maintenance program as a means to help reduce the budget. Which of the following is the weakest reason or argument to justify keeping the preventive maintenance program?
    a. Manufacturers and insurance companies as well as current standards require a department to have a preventive maintenance program.
    b. Criminal and civil liability may occur when emergency apparatus are not properly maintained.
    c. Emergency apparatus would not look as good and may cause a negative public relations image.
    d. The time and money spent on a preventive maintenance program is significantly less than the potential damage likely to occur when preventive maintenance is not conducted.

11. The sequence for conducting a preventive maintenance inspection can vary from department to department. However, inspection components can be grouped in general areas as follows:
    1 – Inside cab
    2 – Outside vehicle
    3 – Engine compartment
    4 – Pump and related components

    Using these categories, which of the following is a common sequence used to conduct a vehicle inspection?
    a. 1, 2, 3, 4
    b. 4, 3, 2, 1
    c. 2, 3, 1, 4
    d. 3, 4, 1, 2

12. According to NFPA 1901, new apparatus must have quick-buildup capabilities that can reach operating pressure within _____.
    a. 1 minute.
    b. 2 minutes.
    c. 4 minutes.
    d. none of the answers is correct.

13. When no fuel refill policy exists, apparatus should be refilled when the fuel level reaches the _____ mark.
    a. one-quarter
    b. one-half
    c. three-quarter
    d. one-third

14. NFPA 1911 requires that pumps with a rated capacity of _____ gpm or greater be service tested at least annually.
    a. 250
    b. 500
    c. 750
    d. 1,000

15. State laws typically define several conditions that must exist for exemptions from normal driving requirements to be extended to emergency vehicles. Which of the following would most likely meet one or more of the conditions?
    a. Only authorized emergency vehicles are covered.
    b. Exemptions are provided only when responding to an emergency.
    c. Audible and visual warning devices must be operating.
    d. All of these conditions must be met.

16. All of the following conditions require the emergency-vehicle driver to bring the apparatus to a complete stop *except*
    a. at school crossings.
    b. when all lanes of traffic in an intersection cannot be accounted for.
    c. at unguarded railroad guard crossings.
    d. when a stopped school bus with flashing warning lights is encountered.

Sign reads: Weight Limit Per Axle 32000 LBS

17. Can a mobile water tanker with a capacity of 1,000 gallons cross this bridge?
    a. Yes, but only if half the water is drained first.
    b. Yes, the water weighs less than the authorized limit and it can cross as long as the tanker itself weighs less than 26,650 lbs.
    c. No, the water weighs more than the authorized limit.
    d. No, the water weighs less than the authorized limit but the weight of the tanker exceeds the authorized limit.

18. A driver has several near-miss accidents. Because of this, the company officer and training officer decide that the driver should practice the serpentine driving exercise. The driver has most likely has had difficulty with
    a. backing the vehicle into a restricted area, such as a fire station.
    b. driving around obstacles, such as parked cars and tight corners.
    c. turning the vehicle around within a confined space, such as a narrow street or driveway.
    d. driving the vehicle in a straight line, such as on a narrow street or road.

19. With positive-displacement pumps, when pressure is increased on the intake side of the pump
    a. discharge pressure will not be affected.
    b. discharge pressure will increase.
    c. discharge pressure will decrease.
    d. none of the answers is correct.

20. With centrifugal pumps, if the pressure is held constant and the speed of the pump is increased, flow will
    a. increase.
    b. decrease.
    c. remain the same.
    d. none of the answers is correct.

21. Which of the following rated capacity charts is correct for a 500-gpm pumper?
    a. 500 gpm at 150 psi
       350 gpm at 200 psi
       250 gpm at 250 psi
    b. 500 gpm at 100 psi
       350 gpm at 250 psi
       250 gpm at 3900 psi
    c. 500 gpm at 100 psi
       350 gpm at 200 psi
       250 gpm at 250 psi
    d. 500 gpm at 100 psi
       250 gpm at 200 psi
       150 gpm at 250 psi

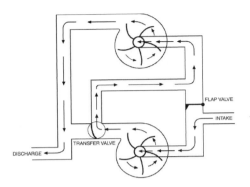

22. The flow of water through this centrifugal pump indicates that it is
    a. in volume mode.
    b. a single-stage pump.
    c. pumping at rated capacity.
    d. none of the answers is correct.

23. Which of the following is incorrect concerning positive-displacement pumps?
    a. The pressure on the intake side of the pump does not affect the discharge.
    b. Each revolution of the pump yields a specific quantity of water.
    c. The quantity of water discharged per revolution will increase as pump speed increases.
    d. The water being pumped is independent of discharge pressure.

24. A foam system that uses water from the discharge side of the pump to pick up foam and deliver to the intake side of the pump is called a(n)
    a. pre-mixed system.
    b. in-line eductor system.
    c. around-the-pump proportioning system.
    d. balanced pressure system.

25. While pre-planning an industrial complex, it is decided that a foam system is needed that will provide foam from all discharges on the apparatus at the same time. The best foam system on an apparatus would be a(n)
    a. pre-mixed system.
    b. in-line eductor system.
    c. around-the-pump proportioning system.
    d. balanced pressure system.

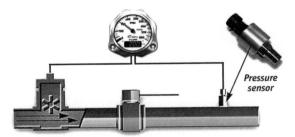

Pressure sensor

26. This device is unique because it
    a. can regulate pressure between the two points.
    b. can calculate friction loss between the two points.
    c. measures and reports both flow and pressure on one gauge.
    d. can accurately inject foam at the appropriate concentration.

27. Fire departments should establish and maintain an accurate record of hose
    a. to help determine cost-effectiveness of hose.
    b. to analyze hose repair trends.
    c. per NFPA 1962.
    d. all are correct answers.

28. The basic steps for conducting a hose test is provided below. What is the missing step?

    1. Connect water supply and test hose sections.
    2. Open discharge lines for test hose sections.
    3. Slowly increase discharge pressure to 45 psi.
    4. Remove all air from within the hose.
    5. Mark hose near coupling to determine slippage.
    6. Increase pressure slowly (NFPA 1962 suggests no faster than 15 psi per second) to test pressure.
    7. _____
    8. Record results.

    a. Reduce pressure at the three-minute mark, then increase pressure to complete the test.
    b. Clear nonessential personnel from testing area.
    c. Maintain pressure for three minutes, periodically checking for leaks.
    d. Identify the correct test pressure for the hose.

29. According to NFPA 1965, a gated wye with a handle that is in a position perpendicular to the hose is
    a. open.
    b. closed.
    c. exactly half open.
    d. not a requirement of the standard.

30. When flow testing hydrants, the next major step after opening flow hydrants one at a time until a 25% drop in residual pressure is achieved is to
    a. slowly shut down hydrants one at a time.
    b. record exact interior size, in inches, of each outlet flowed.
    c. conduct a static pressure reading.
    d. continue to flow to clear debris and foreign substances and then take residual reading at test hydrant and flow readings at each flow hydrant.

31. During a drafting operation, if the pressure within the pump is reduced by 8.5 psi, then water can be raised to a height of _____.
    a. 8.5 feet
    b. 19.5 feet
    c. 22.5 feet
    d. just over 23 feet

32. A pump operator is conducting a drafting operation and knows that the internal pressure of the pump can be reduced by 12 psi given the current configuration and condition of the pump and related components. If the pump operator were to attempt to lift the maximum height for this reduced pressure, would the general rule of thumb of not attempting to pump more than two-thirds of the theoretical lift be violated?
    a. Yes.
    b. No such rule of thumb exists.
    c. No.
    d. Not enough information is provided.

33. One of the major difficulties in setting up and maintaining relay operations is
    a. having enough pumpers to use in the relay.
    b. walking from one pumper to the next.
    c. how to control changes in pressures and flows within the system.
    d. ensuring that an adequate water supply is selected.

34. What is the maximum distance for the next pumper in a relay if the supply pumper is providing 650 gpm at 200 psi through 4-inch hose (8 psi friction loss per 100 feet)?
    a. 500 feet
    b. 2,150 feet
    c. 3,150 feet
    d. 4,500 feet

35. Determine the shuttle flow capacity for the following tanker shuttle:
    1,000-gallon tanker with a 10-minute shuttle cycle
    750-gallon tanker with a 5-minute shuttle cycle
    2,500-gallon tanker with a 10-minute cycle
    a. 250 gpm
    b. 500 gpm
    c. 750 gpm
    d. 1,000 gpm

36. During pump operations, what is the next step after engaging the pump?
    a. Secure a water supply.
    b. Throttle to desired pressure.
    c. Set the pressure-regulating device.
    d. Open discharge lines.

37. When positioning apparatus at an incident, the pump operator should consider all of the following *except*
    a. following department SOPs.
    b. analyzing available water supply.
    c. reviewing tactical considerations.
    d. being mindful of surroundings.
    e. each of these should be considered when positioning a pumper.

38. When engaging a pump utilizing a PTO, the transmission should be in _____ for mobile pump operations.
    a. neutral
    b. first or lowest gear
    c. fourth or highest gear
    d. reverse

39. When engaging a pump through a split-shaft arraignment, the next step after placing the transmission in neutral is to
    a. apply the parking brake.
    b. operate the pump shift switch.
    c. increase engine speed slightly.
    d. increase the discharge pressure by 20 psi.

40. The transfer valve should be in the _____ position for a 1,000-gpm pumper expecting to flow 500 gpm at 150 psi.
    a. volume
    b. pressure
    c. either volume or pressure
    d. none of the answers is correct

41. What might be the problem if a corresponding increase in pump discharge pressure is not noted when the engine throttle is slowly increased?
    a. There is no problem; this is a common occurrence.
    b. The transfer valve is in the wrong position.
    c. The pump may not be in gear.
    d. The pressure regulator is off.

42. During a pumping operation, the pump operator attempts to increase the pressure in one of the two flowing lines. As the throttle is increased, a corresponding increase in pressure does not occur. The pump operator then checks the relief device; it is functioning properly, but is not relieving any pressure at this time. What might occur if the pump operator continues to increase the throttle (engine speed), hoping to increase the discharge pressure?
    a. Pump may cavitate
    b. May lose prime
    c. Intake may collapse
    d. May cause damage to the municipal water main
    e. All of these could occur

43. Place the following on-board water supply procedures in the correct order:
    1. Open "tank-to-pump"
    2. Increase throttle
    3. Set transfer valve
    4. Engage pump
    a. 4, 1, 2, 3
    b. 4, 2, 3, 1
    c. 4, 2, 1, 3
    d. 4, 3, 1, 2

44. The following are ways to stop cavitation in a pump *except*
    a. reducing nozzle gpm settings.
    b. reducing pump speed.
    c. increasing the number of discharge lines.
    d. increasing pump speed.

45. The _____ pressure zone occurs near the center of the impeller, while the _____ pressure zone occurs near the outer edge of the impeller during cavitation within a centrifugal pump.
    a. high, low
    b. low, high
    c. neutral, high
    d. high, balanced

46. Each of the following are correct actions to take during cold-weather operations *except*
    a. stopping leaks to reduce the accumulation of ice.
    b. not stopping the flow of water within the pump.
    c. immediately draining water from all hose when the pumping operation is completed.
    d. all of the following are correct actions during cold-weather operations.

47. A fire department connection marked with "STANDPIPE" means
    a. the system is only a standpipe.
    b. the system is a combination sprinkler and standpipe.
    c. the pump operator should pump an additional 25 psi for the standpipe.
    d. both b and c are correct.

48. A Class 1 standpipe system has a minimum flow rate of _____ gpm.
    a. 100
    b. 150
    c. 250
    d. 500

49. Good troubleshooting is important to pump operations. All of the following are correct concerning troubleshooting pump problems *except*
    a. using the manufacturer's troubleshooting guides when available.
    b. the concept that problems are usually either procedural or mechanical.
    c. that the best method to troubleshoot is to follow the flow of water from the intake to the discharge while attempting to determine the problem.
    d. only certified mechanics should attempt to troubleshoot pumping problems.

50. What is the weight of water in a tank with a volume of 50 cubic feet?
    a. 417 lbs
    b. 4,170 lbs
    c. 3,117 lbs
    d. 30,117 lbs

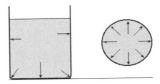

51. The drawing illustrates what pressure principle?
    a. Pressure exerted at the bottom of a container is independent of the shape or volume of the container.
    b. Pressure at any point beneath the surface of a liquid in an open container is directly proportional to its depth.
    c. Pressure of a liquid acting on a surface is perpendicular to that surface.
    d. Pressure at any point in a liquid at rest is equal in every direction.

52. A hose line is flowing 225 gpm into a structure. If the flow is sustained for 3 minutes, how many tons of water will be delivered into the structure?
    a. less than 1 ton
    b. just under 3 tons
    c. just under 5 tons
    d. over 5 tons

53. An initial static pressure reading of 30 psi was noted and, after initiating flow of a 100-gpm pre-connect, the residual pressure was 20 psi. How many more similar lines is the hydrant capable of flowing?
    a. no additional lines of 100 gpm
    b. one additional like line
    c. two additional like lines
    d. three additional like lines

54. When setting up a pump test, the pump operator initiates a test flow and records a nozzle pressure of 60 psi from a 1-inch tip smooth-bore nozzle. What is the nozzle reaction?
    a. 79 lbs.
    b. 84 lbs.
    c. 94 lbs.
    d. 109 lbs.

55. A combination nozzle on a hand-line operating at 75 psi nozzle pressure flowing 95 gpm will have a nozzle reaction of
    a. 42 psi
    b. 75 psi
    c. 82 psi
    d. 100 psi

56. A pump operator is supplying a master-stream device with two 2 ½-inch hose lines of 300 feet each. The master-stream device is flowing 500 gpm. Calculate the friction loss in one of the hose lines using the hand method.
    a. 18 psi
    b. 28 psi
    c. 38 psi
    d. 48 psi

57. A pump operator is supplying a master-stream device with four 2 ½-inch hose lines of 500 feet each. The master-stream device is flowing 1,000 gpm. Calculate the friction loss in one of the hose lines using the hand method.
    a. 24 psi
    b. 34 psi
    c. 43 psi
    d. 63 psi

58. During a relay operation, you are suppling a 2 ½-inch line that is 400 feet and flowing 300 gpm. Using the drop-ten method, determine the friction loss in the hose line.
    a. 12 psi
    b. 22 psi
    c. 60 psi
    d. 80 psi

59. Using the drop-ten method, determine the friction loss in a hose line that is 450 feet of 2 ½-inch hose flowing 325 gpm.
    a. 25 psi
    b. 32 psi
    c. 99 psi
    d. 125 psi

60. A structure that is 120 feet long, 60 feet wide, and 11 feet tall will have a needed fire flow of _____ gpm when calculated using the Iowa State formula.
    a. 792
    b. 1,242
    c. 1,989
    d. 2,400

61. Using the NFA formula, calculate the needed flow for a part of a room that is 25 feet long, 20 feet wide, and 16 feet tall.
    a. 80 gpm
    b. 167 gpm
    c. 187 gpm
    d. 196 gpm

62. For a pump operation in which the attack line is advanced to the second floor of a structure, you should increase pump pressure by _____ psi to account for elevation.
    a. 0
    b. 5
    c. 10
    d. 15

63. You are supplying a master-stream device locate about 500 feet down an incline of about 42 feet. As you begin to support this master-stream device, you should change the pump pressure by _____ psi to account for the change in elevation.
    a. 21
    b. 42
    c. 50
    d. −21

64. You are setting up a relay operation and will be supplying the next pumper through 1,000 feet of 4-inch hose. If the flow through the line is 650 gpm, then the friction loss for the entire line is _____ psi using the formula cq^2L.
    a. 25
    b. 50
    c. 55
    d. 85

65. You are suppling an exposure line that is 350 feet of 3-inch hose flowing 300 gpm. Using the condensed q formula calculate the friction in the hose line.
    a. 32 psi
    b. 42 psi
    c. 53 psi
    d. 78 psi

66. A crew is stretching out 200 feet of 2 ½-inch hose for an attack line. What is the friction loss in the line if the flow is 100 gpm? Use the formula $2q^2 + q$.
    a. 6 psi
    b. 12 psi
    c. 42 psi
    d. 65 psi

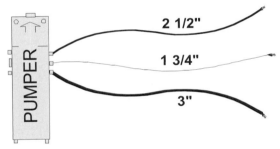

67. For this pumping operation, if the 1 ¾-inch line was 500 feet long flowing 125 gpm from a combination nozzle, then the pump discharge pressure would be _____ psi.
    a. 149
    b. 204
    c. 221
    d. 250

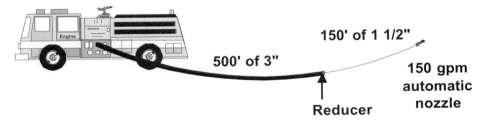

68. Calculate the friction loss for the hose line configuration using the formula cq^2L.
    a. 81 psi
    b. 90 psi
    c. 176 psi
    d. 231 psi

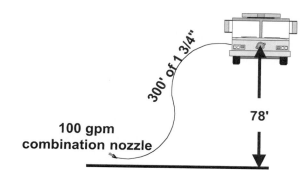

69. The pump discharge pressure should be _____ for this pumping operation.
    a. 108 psi
    b. 147 psi
    c. 186 psi
    d. 208 psi

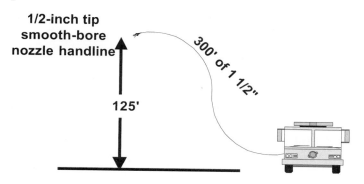

70. For this pumping operation, the pump discharge pressure should be _____.
    a. 94 psi
    b. 140 psi
    c. 187 psi
    d. 222 psi

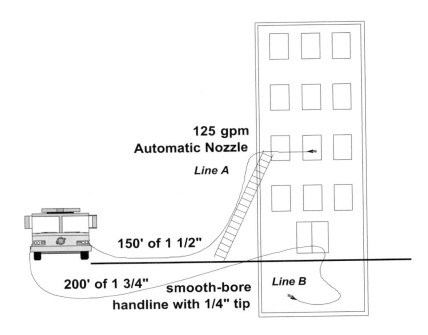

71. The pump discharge pressure for Line A should be _____.
    a. 15 psi
    b. 104 psi
    c. 139 psi
    d. 166 psi

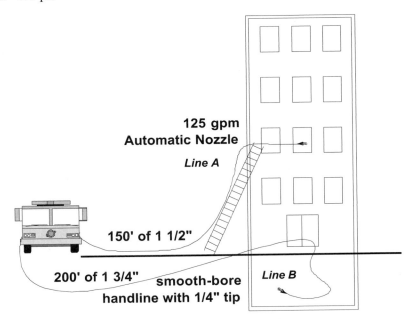

72. The pump discharge pressure for Line B should be _____.
    a. 46 psi
    b. 61 psi
    c. 101 psi
    d. 140 psi

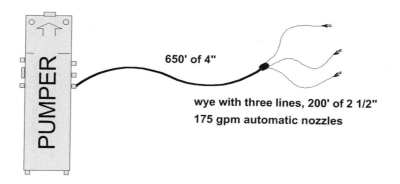

73. For this pumping operation, the pump discharge pressure should be _____.
    a. 123 psi
    b. 163 psi
    c. 183 psi
    d. 222 psi

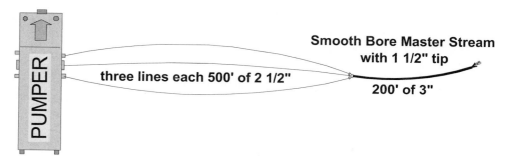

74. For this pumping operation, the pump discharge pressure should be _____.
    a. 100 psi
    b. 153 psi
    c. 186 psi
    d. 215 psi

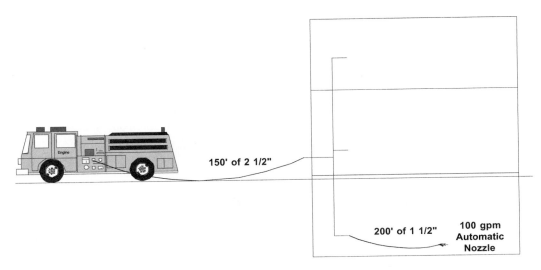

75. For this pumping operation, the pump discharge pressure should be _____.
    a. 150 psi
    b. 171 psi
    c. 186 psi
    d. 214 psi

76. Determine the PDP for the following hose line configuration:
    200 feet of 1 ½-inch hose with a ¾-inch tip smooth-bore nozzle following 118 gpm, friction loss for the hose line is calculated to be 67 psi. The line is taken to the third floor.
    a. PDP = 127 psi
    b. PDP = 117 psi
    c. PDP = 162 psi
    d. PDP = 167 psi

77. Determine the PDP for the following hose line configuration:
    Two like lines that are each 300 feet of 2 ½-inch hose with automatic nozzles following 200 gpm, friction loss for one of the hose lines is calculated to be 24 psi.
    a. PDP = 124 psi
    b. PDP = 148 psi
    c. PDP = 74 psi
    d. PDP = 98 psi

78. When determining PDP and initiating flow for two like lines (each line requires 130 psi), all of the flowing are correct *except*
    a. PDP is 260 psi.
    b. PDP is 130 psi.
    c. to initiate flow, the pump operator can either open all both lines at the same time and then increase the throttle to the PDP or open one line at a time, for each line added, the discharge gate is slowly opened while the throttle is increased to maintain the PDP.
    d. both a and c are correct.

79. In the volume mode of a four-stage centrifugal pump, if the total discharge is 1,250 gpm at 100 psi, then each impeller is discharging about
    a. 625 gpm at 100 psi.
    b. 625 gpm at 50 psi.
    c. 313 gpm at 100 psi.
    d. 313 gpm at 25 psi.

80. You are operating the supply pumper in a relay and calculate that your pump discharge pressure should be 199 psi to ensure that 20 psi remains at the next pump. Your supply is a weak hydrant at a dead-end main. After securing the hydrant supply, you slowly open your discharge and begin flowing water to the next pumper through 4-inch hose. The pumper you are supplying is about 126 feet above your pumper, up a winding dirt road. After contacting the pump operator to verify 20 psi at the next pump, you set your pressure-relief device. Do you note any problems with this operation?
   a. There is no problem; everything is good.
   b. The pressure is too high.
   c. You should not be using a weak hydrant in a relay operation.
   d. The relief device should be set prior to verifying 20 psi at the next pump.

81. A _____-gallon tanker operating in a shuttle has a cycle time of 14 minutes and a flow capacity of 250 gpm.
   a. 250
   b. 2,500
   c. 3,500
   d. 4,500

82. How long can a 1,000-gallon on-board water tank sustain a 1 ¾-inch hand-line with an automatic nozzle?
   a. about 10.5 minutes
   b. about 10 minutes
   c. about 8 minutes
   d. not enough information provided

83. You just initiated flow through two 3-inch lines, 350 feet each, supplying a smooth-bore master-stream device with a 1 ⅜-inch tip. The residual hydrant pressure is 45 psi at the intake, and each discharge flow meter reading is 420. What is the total flow through the smooth-bore monitor?
   a. 420 gpm
   b. 560 gpm
   c. 640 gpm
   d. 840 gpm

84. Five Btu is equivalent to about _____ calorie(s).
   a. 1
   b. 152
   c. 252
   d. 1,268

85. To test a driver's ability to back the vehicle into a restricted area such as a fire station or down an alley, the _____ driving exercise should be used.
   a. straight-line
   b. alley dock
   c. diminishing clearance
   d. confined-space turnaround

86. According to NFPA 1500, prior to moving a fire department vehicle, all personnel must be seated and properly secured with seat belts. One exception to this requirement is
   a. when donning PPE.
   b. for patient treatment.
   c. when conducting tiller training.
   d. both b and c are correct.

87. During an emergency response, you come to an intersection with a traffic red light. According to NFPA 1500, you should
    a. proceed through the intersection at a reduced rate of speed.
    b. move quickly through the intersection to reduce the likelihood of being hit by another vehicle.
    c. slow down considerably, but do not stop to ensure the safety of the apparatus.
    d. come to a complete stop before proceeding through the intersection.

88. If the flow of a centrifugal pump is held constant while the speed of the pump is increased, then the pressure will
    a. increase.
    b. decrease.
    c. remain the same.
    d. none of the answer choices is correct.

89. You just successfully primed the pump during a drafting operation. You know the static water supply is about 16 feet below the pump. Because of this, you know that you reduced the pressure in the pump at least _____ psi.
    a. 5
    b. 7
    c. 16
    d. 22.5

90. You are the pump operator for the last pump in a long relay. After securing water from the preceding pumper, you open the discharge lines, and throttle to the desired pressure. What should you do next?
    a. Set the transfer valve.
    b. Maintain appropriate flows and pressures.
    c. Set the pressure-regulating device.
    d. Provide water to the pump.

91. During a drafting operation you successfully primed the pump. However, before you are able to open discharge lines, the pump lost its prime. You successfully primed the pump again, only to lose it about three minutes later. Which of the following is most likely *not* the cause?
    a. The by-pass line open.
    b. The end of the intake hose is not submerged deep enough.
    c. There is an air leak some place.
    d. The transfer valve is in pressure mode instead of volume mode.

92. A centrifugal pump operates based on _____ principles, whereas rotary gear pump operation is based on _____ principles.
    a. positive-displacement, dynamic
    b. hydrodynamic, hydrostatic
    c. internal pressure, external pressure
    d. hydrostatic, hydrodynamic

93. A structure has a volume of 48,000 cubic feet (100 feet wide, 40 feet long, and 12 feet high). Based on this information, what formula can you use to determine the estimated needed fire flow for the structure?
    a. Iowa State formula
    b. National Fire Academy (NFA) Formula
    c. Condensed q formula
    d. both a and c are correct
    e. both a and b are correct

94. As a pump operator, you have several main duties or functions. Which of the following is not one of those duties as required by NFPA 1002?
    a. driving apparatus
    b. operating apparatus
    c. supervision
    d. preventive maintenance

95. Two special pump operations are tandem and dual pumping. Dual pumping operations are similar to _____, whereas tandem pumping is similar to _____.
    a. relay pumping, tanker shuttle operations
    b. volume mode, pressure mode
    c. tanker shuttle operations, relay pumping
    d. pressure mode, volume mode

96. _____ can be defined as the systematic movement of water from a supply source through a pump to a discharge point.
    a. Relay operations
    b. Drafting
    c. Pump operations
    d. Tandem pumping

97. Conducting preventive maintenance inspections is an important duty of a pump operator. Which of the following standards requires that these inspections be documented?
    a. NFPA 1002
    b. NFPA 1500
    c. NFPA 1901
    d. All are correct

98. While responding to an incident, you notice a hazard in front of you. You travel 35 feet before your brain recognizes the hazard. You travel another 30 feet before your brains sends a signal and the brake is depressed. It takes another 125 feet before the vehicle comes to a complete stop. What is the braking distance?
    a. 65 feet
    b. 125 feet
    c. 160 feet
    d. 190 feet

99. Newton's first and third laws of motion can be used to help explain the concept of centrifugal force. Which of the following is correct?
    a. Newton's first law of motion indicates that a moving body travels in a straight line with a constant speed unless affected by an outside force.
    b. Newton's third law of motion states that for every action there is a relational opposite reaction.
    c. Both a and b are correct.
    d. Neither a nor b are correct.

100. As a knowledgeable pump operator, you know that speed, road conditions, and weight shifts affect your ability to maintain safe control of the vehicle. You also know that the key to safely navigating a curve is to
    a. maintain traction.
    b. make the turn with the brakes slightly pressed.
    c. slightly decrease your speed.
    d. slightly increase your speed.

Phase III, Exam III: Answers to Questions

1. F	26. C	51. C	76. A
2. F	27. D	52. B	77. A
3. T	28. C	53. A	78. A
4. T	29. A	54. C	79. C
5. F	30. D	55. A	80. B
6. T	31. B	56. C	81. C
7. T	32. A	57. D	82. D
8. B	33. C	58. D	83. D
9. C	34. D	59. C	84. C
10. C	35. B	60. A	85. B
11. C	36. A	61. B	86. D
12. A	37. E	62. B	87. D
13. C	38. B	63. D	88. A
14. A	39. A	64. D	89. B
15. D	40. C	65. A	90. B
16. A	41. C	66. A	91. D
17. B	42. E	67. C	92. B
18. B	43. D	68. C	93. E
19. A	44. C	69. A	94. C
20. A	45. B	70. A	95. B
21. A	46. D	71. D	96. C
22. D	47. A	72. A	97. B
23. C	48. D	73. B	98. B
24. C	49. D	74. D	99. D
25. C	50. C	75. B	100. A

Phase III, Exam III:
Rationale & References for Questions

Question #1
According to NFPA 1002 (5.1), "driver/operator – pumper" is the term used to describe the individual that has met the requirements of Chapter 5 for driving and operating apparatus equipped with fire pumps. NFPA 1002 5.1. *IFPO, 2E:* Chapter 1, page 15.

Question #2
The specific preventive maintenance activities conducted by pump operators and mechanics depend on the level of training and the type of preventive maintenance activity being conducted. In general, certified mechanics conduct those activities that require apparatus to be taken out, require several hours to complete, or are detailed and complicated repairs. Checking and adding engine oil or battery fluid may be conducted by the pump operator, whereas changing the engine oil or replacing the battery is most likely performed by a mechanic. NFPA 1002 4.2.1, 5.1.1. *IFPO, 2E:* Chapter 2, page 22.

Question #3
Even though the problem was fixed, the pump operator should document the finding to assist with tracking and trending of problems. NFPA 1002 4.2.2. *IFPO, 2E:* Chapter 2, page 25.

Question #4
The supply reliability relates to the extent to which the supply fluctuates in flow, pressure, and quantity. NFPA 1002 5.2.1. *IFPO, 2E:* Chapter 7, page 165.

Question #5
This formula only works with smooth-bore nozzles. For combination nozzles the formula $NR = gpm \times \sqrt{NP} \times .0505$ must be used. NFPA 1002 5.2.1, 5.2.2, 5.2.4. *IFPO, 2E:* Chapter 10, pages 282 – 283.

Question #6
Several general considerations for positioning the vehicle include the following:

- Position the vehicle to reduce the likelihood of being struck by traffic.
- Use the vehicle to shield emergency personnel from traffic.
- Position to enhance emergency operations.
- Position way from hazards.
- Consider wind direction.
- Do not park in the collapse zone.
- Never park on railroad tracks.
- Park the vehicle on the side of the incident.

NFPA 1002 4.3. *IFPO, 2E:* Chapter 3, page 65.

Question #7
The pressure governor limits pressure surges by changing engine speed while the pressure-relief valve limits pressure surges by opening and closing a passage between the discharge side of the pump to either the intake or atmosphere. NFPA 1002 5.2.1, 5.2.2, 5.2.4. *IFPO, 2E:* Chapter 5, pages 122 – 123.

Question #8
The trend seems to indicate that the number of vehicle accidents has gradually increased from 1995 to 2000. NFPA 4.3. *IFPO, 2E:* Chapter 1, pages 8 – 9.

Question #9
Laws are rules that are legally binding and enforceable. Exemptions extended to the operation of emergency vehicles are most often located in state and local laws. Standards are guidelines that are not legally binding or enforceable. NFPA 1002 4.2, 4.3, 5.1. *IFPO, 2E:* Chapter 1, page 15.

Question #10
Although good public relations is important, increased safety, reduced liability, and costs savings are more important reasons for a well-funded preventive maintenance program. NFPA 1002 4.2.1, 5.1.1. *IFPO, 2E:* Chapter 2, pages 21 – 22.

Question #11
A common sequence for conducting a vehicle inspection is to start with a walk around the outside of the vehicle, followed by an inspection of the engine compartment. Next, the inside of the cab is inspected and the engine is started. With the engine running, the final step is to complete the inspection of the pump and related components. NFPA 1002 4.2.1, 5.1.1. *IFPO, 2E:* Chapter 2, pages 27 – 28.

Question #12
NFPA 1901 requires all new apparatus to have quick pressure build-up capabilities so that operating pressure is reached within 60 seconds. NFPA 1002 4.2.1. *IFPO, 2E:* Chapter 2, page 35.

Question #13
Apparatus fuel tanks should be refilled according to department policy. When no refill policy exists, apparatus fuel tanks should be refilled when the fuel level reaches the three-quarter mark. That is, when the gauge drops from full to three-quarter full. NFPA 1002 4.2.1. *IFPO, 2E:* Chapter 2, page 35.

Question #14
NFPA 1911 requires that pumps with a rated capacity of 250 gpm or greater be service tested at least annually and after any major repair or modification. NFPA 1002 4.2.1, 5.1.1. *IFPO, 2E:* Chapter 2, page 24.

Question #15
Most state laws require that emergency-vehicle drivers obey the same laws as other vehicle operators unless specifically exempt from doing so. State laws typically define several conditions that must exist for exemptions to be extended and include:

- Only authorized emergency vehicles are covered.
- The exemptions are only provided when responding to an emergency.
- Audible and visual warning devices must be operating when taking advantage of the exemption.

NFPA 1002 4.3. *IFPO, 2E:* Chapter 3, pages 44 – 45.

Question #16
NFPA 1500 requires that emergency-vehicle drivers come to a complete stop when any intersection hazard is present. Specifically, the vehicle must come to a complete stop when any of the following exists:

- As directed by a law enforcement officer
- At traffic red lights or stop signs
- At negative right-of-way and blind intersections
- When all lanes of traffic in an intersection cannot be accounted for
- When stopped school bus with flashing warning lights is encountered
- At unguarded railroad guard crossing (also for nonemergency)
- When other intersection hazards are present

NFPA 1002 4.3. *IFPO, 2E:* Chapter 3, page 48.

Question #17

The water in the tanker weighs 8,350 lbs (8.35 lbs per gallon × 1,000 gallons). As long as the vehicle itself weighs less than 26,650 lbs (32,000 lb – 8,350 lbs limit), it would be safe to cross over the bridge. The weight would not exceed 32,000 lbs per axle. NFPA 1002 4.3. *IFPO, 2E:* Chapter 3, pages 52 – 53.

Question #18

Several driving exercises are typically used to assess the ability to safely operate and control the vehicle. These include the following:

- *Alley Dock*. Assesses the ability to back the vehicle into a restricted area, such as a fire station or down an alley.

- *Serpentine*. Assesses the ability to drive around obstacles, such as parked cars and tight corners.

- *Confined-Space Turnaround*. Assesses the ability to turn the vehicle around within a confined space, such as a narrow street or driveway.

- *Diminishing Clearance*. Assesses the ability to drive the vehicle in a straight line, such as on a narrow street or road.

NFPA 1002 4.3.2, 4.3.3, 4.3.4, 4.3.5. *IFPO, 2E:* Chapter 3, pages 65 – 66.

Question #19

Each revolution of a positive-displacement pump will yield (displace) a specific quantity of water. Consequently, the pressure on the intake side of the pump is irrelevant. No matter what the intake pressure is, the pump will only discharge its specific quantity of water per revolution. NFPA 1002 5.2.1, 5.2.1. *IFPO, 2E:* Chapter 4, page 78.

Question #20

Centrifugal pump performance is contingent on three interrelated factors. If one factor remains constant, a change in one of the remaining factors will change the other.

- *Speed*. If the speed of the pump is held constant and the flow of water increases, pressure will drop. If more water is allowed to flow while the speed of the pump remains the same, pressure will be reduced, because less resistance occurs on the discharge side of the pump.

- *Flow*. If the flow of water is held constant and the speed of the pump is increased, pressure will increase. The same amount of water is being discharged, yet the pump is attempting to discharge more water. This results in an increase in pressure.

- *Pressure*. If the pressure is held constant and the speed of the pump is increased, flow will increase. The increased speed of the pump will increase the flow. Pressure can be maintained constant by increasing or reducing the resistance on the discharge side of the pump.

NFPA 1002 5.2.1, 5.2.2. *IFPO, 2E:* Chapter 4, page 86.

Question #21

According to NFPA 1901, a pump must have a rated capacity as follows:

- 100% of its rated capacity at 150 psi

- 70% of its rated capacity at 200 psi

- 50% of its rated capacity at 250 psi

NFPA 1002 5.2.1, 5.2.2. *IFPO, 2E:* Chapter 4, page 91.

Question #22

- In *pressure mode*, the transfer valve redirects the discharge from one impeller to the intake of the second impeller.

- In *volume mode*, water enters both impellers from a common intake and leaves from a common discharge.

NFPA 1002 5.2.1, 5.2.2. *IFPO, 2E:* Chapter 4, pages 89 – 90.

Question #23
Total discharge rate will increase when the pump speed is increased. However, the quantity of water discharged per revolution will never change. In other words, the quantity per revolution stays the same regardless of how slow or fast the pump is operating. NFPA 1002 5.2.1, 5.2.2. *IFPO, 2E:* Chapter 4, pages 78 – 79.

Question #24
- A *pre-mixed system* consists of a tank in which foam concentrate and water are added in appropriate proportions. Often, the foam is simply added to the on-board water tank.

- An *in-line eductor system* utilizes eductors to add foam to water in appropriate proportions. This system can have either internal or external controls, and requires an eductor and a foam tank (or supply). Internal systems have a foam concentrate valve and meeting control. Eductors are usually installed between the pump discharge and the discharge outlet.

- An *around-the-pump proportioning system* uses an eductor (located between the intake and discharge sides of the pump) to mix foam with water. Water from the discharge side of the pump picks up foam and delivers to the intake side of the pump, where it can then be discharged to all outlets.

- A *balanced-pressure system* mixes foam with water by means of pressure. One system uses pressure to force foam from a bladder. The second system uses a separate foam pump (two types are by-pass and demand).

- A *direct injection/compressed-air foam system* (CAFS) consists of a separate foam pump and foam tank. Direct injection moves foam from the tank directly into the discharge lines. CAFS adds compressed air to create a lightweight foam.

NFPA 1002 5.2.3. *IFPO, 2E:* Chapter 5, pages 126 – 130.

Question #25
- A *pre-mixed system* consists of a tank in which foam concentrate and water are added in appropriate proportions. Often, the foam is simply added to the on-board water tank.

- An *in-line eductor system* utilizes eductors to add foam to water in appropriate proportions. This system can have either internal or external controls, and requires an eductor and a foam tank (or supply). Internal systems have a foam concentrate valve and meeting control. Eductors are usually installed between the pump discharge and the discharge outlet.

- An *around-the-pump proportioning system* uses an eductor (located between the intake and discharge sides of the pump) to mix foam with water. Water from the discharge side of the pump picks up foam and delivers to the intake side of the pump, where it can then be discharged to all outlets.

- A *balanced-pressure system* mixes foam with water by means of pressure. One system uses pressure to force foam from a bladder. The second system uses a separate foam pump (two types are by-pass and demand).

- A *direct injection/compressed-air foam system* (CAFS) consists of a separate foam pump and foam tank. Direct injection moves foam from the tank directly into the discharge lines. CAFS adds compressed air to create a lightweight foam.

NFPA 1002 5.2.3. *IFPO, 2E:* Chapter 5, pages 126 – 130.

Question #26
Flow meters often use paddle wheels to measure the flow within a line while pressure gauges use either a Bourdon tube or, in this case, an electronic pressure sensor. Both pressure and flow are provided to a single gauge. NFPA 1002 5.2.1, 5.2.2. *IFPO, 2E:* Chapter 5, pages 113 – 114.

Question #27
NFPA 1962 Standard for the Inspection, Care, and Use of Fire Hose, Couplings, and Nozzles and the Service Testing of Hose requires hose records be maintained. Each of the stated reasons is important. NFPA 1002 5.1.1. *IFPO, 2E:* Chapter 6, page 146.

Question #28
Basic steps for testing hose include:

1. Connect water supply and test hose sections.

2. Open discharge lines for test hose sections.

3. Slowly increase discharge pressure to 45 psi.

4. Remove all air from within the hose.

5. Mark hose near coupling to determine slippage.

6. Increase pressure slowly (NFPA 1962 suggests no faster than 15 psi per second) to test pressure.

7. Maintain pressure for three minutes, periodically checking for leaks.

8. Record results.

NFPA 1002 5.1. *IFPO, 2E:* Chapter 6, page 147.

Question #29
NFPA 1965 requires that appliances with lever-operated handles must indicate a closed position when the handle is perpendicular to the hose line. NFPA 1002 5.2.1, 5.2.2, 5.2.4. *IFPO, 2E:* Chapter 6, page 148.

Question #30
Basic steps for conducting hydrant flow include:

1. Take static pressure reading at test hydrant (make sure open hydrant fully and remove air).

2. Open flow hydrants one at a time until a 25% drop in residual pressure is achieved.

3. Continue to flow to clear debris and foreign substances.

4. Take reading at the same time; residual reading at test hydrant and flow readings at each flow hydrant, and record results.

5. Slowly shut down hydrants one at a time.

6. Record exact interior size, in inches, of each outlet flowed.

NFPA 1002 5.2.1. *IFPO, 2E:* Chapter 7, page 180 – 181.

Question #31
Water will rise approximately 2.3 feet/psi for each 1 psi of pressure.
So, 8.5 psi × 2.3 feet/psi = 19.5 feet. NFPA 1002 5.2.1. *IFPO, 2E:* Chapter 7, pages 185 – 186.

Question #32
Theoretical lift is 14.7 × 2.3 = 33.81 feet
⅔ of this = 22.5
The maximum height for a reduced pressure of 12 psi is 27.6 (12 × 2.3).
So, the rule of thumb is violated. NFPA 1002 5.2.1. *IFPO, 2E:* Chapter 7, pages 185 – 186.

Question #33
Changes in pressure and flows within a relay are difficult to deal with. Changes in any part of the relay can affect the remaining elements. For example, if an attack line is shut down, pressure can increase throughout the system. NFPA 1002 5.2.2. *IFPO, 2E:* Chapter 7, page 190.

Question #34
The distance between pumpers in a relay can be calculated as follows:
(PDP − 20) × 100/FL
(200 − 20) × 100 / 4
180 × 25
4,500 feet
NFPA 1002 5.2.3. *IFPO, 2E:* Chapter 7, page 194.

Question #35
First determine individual shuttle flow capacities and then add them up.
Divide tank volume by the shuttle cycle time.
1,000 ÷ 10 = 100 gpm
750 ÷ 5 = 150 gpm
2,500 ÷ 10 = 250 gpm
Shuttle flow capacity is 500 gpm (100 + 150 + 250). NFPA 1002 5.2.1. *IFPO, 2E:* Chapter 7, page 201.

Question #36
The basic steps for pump operation include:

Step 1	Position apparatus, set parking brake, and let engine return to idle.
Step 2	Engage the pump.
Step 3	Provide water to intake side of pump (on-board, hydrant, draft).
Step 4	Set transfer valve (if so equipped).
Step 5	Open discharge lines.
Step 6	Throttle to desired pressure.
Step 7	Set the pressure-regulating device.
Step 8	Maintain appropriate flows and pressures.

NFPA 1002 5.2.1, 5.2.2, 5.2.4. *IFPO, 2E:* Chapter 8, page 207.

Question #37
Important considerations for positioning pumping apparatus at incidents include:

- If no fire or smoke is visible, park near main entrance.
- Be sure to follow department SOPs.
- Consider tactical priorities for the incident.
- Consider surroundings such as heat from the fire, collapse, overhead lines, escape routes, and wind.

NFPA 1002 5.2.1, 5.2.2. *IFPO, 2E:* Chapter 8, pages 208 − 209.

Question #38
Basic steps for pump engagement powered through a PTO include:

- Bring apparatus to complete stop, set parking brake, and let engine return to idle speed.
- Disengage the clutch (push in the clutch pedal).
- Place transmission in neutral.
- Operate the PTO lever.
- For mobile pumping, place transmission in the proper gear (low).
- For stationary pumping, place transmission in neutral.
- Engage the clutch slowly.

NFPA 1002 5.2.1, 5.2.2, 5.2.4. *IFPO, 2E:* Chapter 8, pages 210 − 211.

Question #39
Basic steps to engage a pump utilizing a split-shaft arrangement are as follows:

- Bring apparatus to complete stop.
- Place transmission in neutral.
- Apply parking brake.
- Operate pump shift switch from road to pump position.
- Ensure "OK to pump" light comes on.
- Shift transmission to pumping gear (usually highest gear).

NFPA 1002 5.2.1, 5.2.2, 5.2.4. *IFPO, 2E:* Chapter 8, pages 211 – 212.

Question #40
Transfer valve operation rule of thumb.

- Use *volume mode* if flows are greater than 50% of a pump's rated capacity, and pressures are less than 150 psi.
- Use *pressure mode* if flows are less than 50% of a pump's rated capacity, and pressures are greater than 150 psi.

NFPA 1002 5.2.1, 5.2.2, 5.2.4. *IFPO, 2E:* Chapter 8, page 214 – 215.

Question #41
A corresponding increase in discharge pressure should be noted when the throttle is increased (engine speed increased). If not, one of the following common causes could be occurring:

- The pump may not be in gear.
- The pump may not be primed.
- The supply line could be closed or insufficient.

NFPA 1002 5.2.1, 5.2.2, 5.2.4. *IFPO, 2E:* Chapter 8, page 216.

Question #42
If the throttle is increased past the point of a corresponding increase in discharge pressure, the following might occur:

- Pump cavitation
- Loss of prime
- Intake line collapse
- Damage to municipal water mains

NFPA 1002 5.2.1, 5.2.2, 5.2.4. *IFPO, 2E:* Chapter 8, page 216.

Question #43
On-board water supply procedures include:

- Position apparatus, set parking brake
- Engage pump
- Set transfer valve
- Open "tank-to-pump"
- Open discharge control valves
- Increase throttle
- Set pressure-regulating device
- Plan for more water

NFPA 1002 5.2.1, 5.2.2, 5.2.4. *IFPO, 2E:* Chapter 8, page 220 – 221.

Question #44

Cavitation can be stopped by reducing pump pressure through pump speed, reducing discharge flow, or through increased supply flow. Adding one or more discharge lines would only aggravate a deteriorating system and most likely would cause additional damage. NFPA 1002 5.1.1. *IFPO, 2E:* Chapter 9, pages 245 – 247.

Question #45

The low-pressure zone near the center of the impeller allows vapor pockets to be created. When these vapor pockets pass through the high-pressure zone, they are forced back into water (liquid state) abruptly. NFPA 1002 5.2.1, 5.2.2, 5.2.4. *IFPO, 2E:* Chapter 9, page 256.

Question #46

Cold-weather operations can be grueling. The pump operator must be diligent to stop all leaks, keep water flowing, and drain everything when the pump operation is completed if freezing might occur. NFPA 1002 5.2.1, 5.2.2, 5.2.4. *IFPO, 2E:* Chapter 9, page 247.

Question #47

A fire department connection marked "STANDPIPE" means the system is a standpipe system, not a sprinkler system. NFPA 1002 5.2.4. *IFPO, 2E:* Chapter 9, page 253.

Question #48

- *Class 1* standpipes provide 2 ½-inch connections for trained firefighters/fire brigades, at an initial flow rate of 500 gpm.
- *Class 2* standpipes provide 1 ½-inch connections for initial attack, at a minimum flow rate of 100 gpm.
- *Class 3* standpipes provide 1 ½-inch and 2 ½-inch connections for trained firefighters/fire brigades, at an initial flow rate of 500 gpm.

NFPA 1002 5.2.4. *IFPO, 2E:* Chapter 9, page 253.

Question #49

Pump operators must be able to perform basic troubleshooting for pump problems. Basic troubleshooting considerations include:

- Problems are usually either procedural or mechanical.
- When proper procedures are followed, the problem is most likely mechanical.
- Best method to troubleshoot is to follow the flow of water from the intake to the discharge while attempting to determine the problem.
- Use the manufacturer's troubleshooting guides when available.

NFPA 1002 5.2. *IFPO, 2E:* Chapter 9, page 254.

Question #50

To calculate weight, the formula $W = D = V$ is used.

$W = 62.34 \text{ lb/ft}^3 \times 50 \text{ ft}^3$

$W = 3,117$ lbs (note the cubic feet cancel each other)

NFPA 1002 5.2. *IFPO, 2E:* Chapter 10, page 266.

Question #51

Although the sum of all forces acting on a molecule within a body of liquid is zero, the force created by the pressure acting on a surface area does have direction. Specifically, this force is perpendicular to any surface it acts upon. NFPA 1002 5.2. *IFPO, 2E:* Chapter 10, pages 271 – 275.

Question #52

225 gpm × 3 minutes = 675 gallons
W = 8.34 lb/gal × 675 gallons
W = 5,629.5 lbs
5,629.5 lbs = 2.8 tons (divide by 2,000)

NFPA 1002 5.2. *IFPO, 2E:* Chapter 10, page 268.

Question #53

Based on the percent drop in pressure, additional flows may be available from a hydrant has follows:

- 0 – 10% drop   three times the original flow
- 11 – 15% drop   two times the original flow
- 16 – 25% drop   on time the original flow

A 10 psi drop in pressure is about 33% of the initial static reading (10 psi /30 psi). The hydrant will most likely not be able to support an additional line of 100 gpm.

NFPA 1002 5.2. *IFPO, 2E:* Chapter 10, page 277.

Question #54

Nozzle reaction calculation
Smooth-bore nozzles:

$$NR = 1.57 \times d^2 \times NP$$

where NR = nozzle reaction
    1.57 = constant
    d = diameter of nozzle orifice in inches
    NP = operating nozzle pressure in psi

$NR = 1.57 \times d^2 \times NP$
$NR = 1.57 \times 1^2 \times 60$
$NR = 1.57 \times 1 \times 60$
$NR = 94.2$

NFPA 1002 5.2. *IFPO, 2E:* Chapter 10, pages 282 – 283.

Question #55

Nozzle reaction calculation
Combination nozzles:

$$NR = gpm \times \sqrt{NP} \times 0.0505$$

where NR = nozzle reaction
    0.0505 = constant
    gpm = flow in gallons per minute
    NP = operating nozzle pressure in psi

$NR = 95 \times \sqrt{75} \times 0.0505$

$NR = 95 \times 8.66 \times 0.0505$

$NR = 41.5$

NFPA 1002 5.2. *IFPO, 2E:* Chapter 10, pages 282 – 283.

Question #56

The hand method is a fireground method used to estimate friction loss in 100-foot sections of 2 ½-inch hose. Simply select the fingertip representing the gpm flow and then multiply the two figures on the finger for the approximate friction loss pressure for each 100-foot section of 2 ½-inch hose.

Each line will be flowing half of the total flow from the master-stream.

FL = 2.5 × 5 × 3 (number of 100-foot sections of hose)

FL = 37.5 or 38 psi

NFPA 1002 5.2.1, 5.2.2. *IFPO, 2E:* Chapter 10, pages 302 – 303.

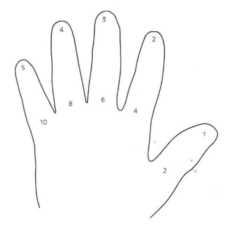

Question #57

The hand method is a fireground method used to estimate friction loss in 100-foot sections of 2 ½-inch hose. Simply select the fingertip representing the gpm flow and then multiply the two figures on the finger for the approximate friction loss pressure for each 100-foot section of 2 ½-inch hose.

Each line will be flowing half of the total flow from the master-stream.

FL = 2.5 × 5 × 5 (number of 100-foot sections of hose)

FL = 62.5 or 63 psi

NFPA 1002 5.2.1, 5.2.2. *IFPO, 2E:* Chapter 10, pages 302 – 303.

Question #58

The drop-ten method simply subtracts 10 from the first two numbers of gpm flow. It is not as accurate as other methods, but it provides a simple rule of thumb for fireground use.

300 gpm = 30 – 10 = 20 psi friction loss per 100-foot sections of 2 ½-inch hose

20 psi × 4 (4 sections of 100-feet of hose) = 80 psi

NFPA 1002 5.2.1, 5.2.2. *IFPO, 2E:* Chapter 10, page 303.

Question #59
The drop-ten method simply subtracts 10 from the first two numbers of gpm flow. It is not as accurate as other methods, but it provides a simple rule of thumb for fireground use.
325 gpm = 32 – 10 = 22 psi friction loss per 100-foot sections of 2 ½-inch hose
22 psi × 4.5 (4.5 sections of 100-feet of hose) = 99 psi

NFPA 1002 5.2.1, 5.2.2. *IFPO, 2E:* Chapter 10, page 303.

Question #60
Iowa State formula: $NF = \dfrac{V}{100}$

where NF = needed flow in gpm
V = volume of the area in cubic feet
100 = is a constant in ft^3/gpm

$$NF = \frac{120 ft \times 60 ft \times 11 ft}{100 ft^3 / gpm}$$

$$NF = \frac{79,200 ft^3}{100 ft^3 / gpm}$$

$NF = 792 gpm$

NFPA 1002 5.2. *IFPO, 2E:* Chapter 11, pages 292 – 293.

Question #61
NFA formula: $NF = \dfrac{A}{3}$

where NF = needed flow in gpm
A = area of the structure in square feet
3 = constant in ft^2/gpm

$$NF = \frac{A}{3 ft^2 gpm}$$

$$NF = \frac{25 ft \times 20 ft}{3 ft^2 gpm}$$

$$NF = \frac{500 ft^2}{3 ft^2 gpm}$$

$NF = 166.66 gpm$

NFPA 1002 5.2. *IFPO, 2E:* Chapter 11, pages 292 – 293.

Question #62
Elevation formula by floor level:

$EL = 5 \times H$
where EL = the gain or loss of elevation in psi
5 = gain or loss in pressure for each floor level
H = height in number of floor levels above or below the pump

$EL = 5 \times 1$

$EL = 5 psi$

NFPA 1002 5.2.4. *IFPO, 2E:* Chapter 11, page 311.

Question #63

Elevation formula in feet:

$EL = 0.5 \times H$

where EL = the gain or loss of elevation in psi

0.5 = pressure exerted at base of 1-cubic-inch column of water 1 foot high

H = distance in feet above or below the pump

$EL = 0.5 \times -42$

$EL = -21 psi$

NFPA 1002 5.2. *IFPO, 2E:* Chapter 11, page 311.

Question #64

$cq^2 L$ (c for 4-inch hose is .2)

$.2 \times \left(\dfrac{650}{100} \right)^2 \times \dfrac{1,000}{100}$

$.2 \times (6.5)^2 \times 10$

$.2 \times 42.25 \times 10$

$84.5 psi$

NFPA 1002 5.2. *IFPO, 2E:* Chapter 11, pages 306 – 307.

Question #65

q^2

$\left(\dfrac{300}{100} \right)^2$

$(3)^2$

9 psi per 100-foot sections

$FL = 9 \times 3.5$ (number of 100-foot sections in the hose line)

$FL = 31.5$ psi NFPA 1002 5.2. *IFPO, 2E:* Chapter 11, pages 306 – 307.

Question #66

$2q^2 + q$

$2 \times \left(\dfrac{100}{100} \right)^2 + \left(\dfrac{100}{100} \right)$

$2 \times (1)^2 + (1)$

$2 \times 1 + 1$

3 psi friction loss per 100-foot sections of hose

$3 \times 2 = 6$ psi

NFPA 1002 5.2. *IFPO, 2E:* Chapter 11, pages 304 – 305.

Question #67

$PDP = NP + FL + AFL\ EL$

NP: 100 psi

FL: cq^2L

$$15.5 \times \left(\frac{125}{100}\right)^2 \times \frac{500}{100}$$

$$15.5 \times (1.25)^2 \times 5$$

$$15.5 \times 1.56 \times 5$$

$$120.9psi$$

AFL: 0 psi

EL: 0 psi

$PDP = 100 + 121$

$PDP = 221$ psi

NFPA 1002 5.2. *IFPO, 2E:* Chapter 11, pages 317 – 357.

Question #68

cq^2L

c for 3-inch hose is .8 and for 1 ½-inch hose is 24

3-inch hose friction loss

$$.8 \times \left(\frac{150}{100}\right)^2 \times \frac{500}{100}$$

$$.8 \times (1.5)^2 \times 5$$

$$.8 \times 2.25 \times 5$$

$$90psi$$

1 ½-inch hose friction loss

$$24 \times \left(\frac{150}{100}\right)^2 \times \frac{150}{100}$$

$$24 \times (1.5)^2 \times 1.5$$

$$24 \times 2.25 \times 1.5$$

$$81psi$$

Reducer = 5 psi

Total friction loss is 176 psi (90 psi + 81 psi + 5 psi)

NFPA 1002 5.2. *IFPO, 2E:* Chapter 11, pages 306 – 307.

Question #69

$PDP = NP + FL + AFL \pm EL$

NP: 100 psi

FL: cq^2L

$$15.5 \times \left(\frac{100}{100}\right)^2 \times \frac{300}{100}$$

$$15.5 \times (1)^2 \times 3$$

$$15.5 \times 1 \times 3$$

$$46.5 psi$$

AFL: 0 psi

$\pm EL$: $0.5 \times H$

$$0.5 \times (-78)$$

$$-39 psi$$

$PDP = 100 + 47 - 39$

$= 108$ psi

NFPA 1002 5.2. *IFPO, 2E:* Chapter 11, pages 317 – 357.

Question #70

$PDP = NP + FL + AFL \pm EL$

NP: 50 psi

gpm: $30 \times d^2 \times \sqrt{NP}$

$$30 \times (0.5)^2 \times \sqrt{50}$$

$$30 \times 0.25 \times 7$$

$$52.5 gpm$$

FL: cq^2L

$$24 \times \left(\frac{53}{100}\right)^2 \times \frac{300}{100}$$

$$24 \times (.52)^2 \times 3$$

$$24 \times .27 \times 3$$

$$19.44 \, psi$$

AFL: 0 psi

$\pm EL$: $0.5 \times H$

$$0.5 \times (50)$$

$$25 psi$$

$PDP = 50 + 19 + 25$

$PDP = 94$ psi

NFPA 1002 5.2. *IFPO, 2E:* Chapter 11, pages 317 – 357.

Question #71

$PDP = NP + FL + AFL \pm EL$

NP: 100 psi

FL: cq^2L

$$24 \times \left(\frac{125}{100} \right)^2 \times \frac{150}{100}$$

$$24 \times (1.25)^2 \times 1.5$$

$$24 \times 1.56 \times 1.5$$

$$56.16psi$$

AFL: 0 psi

$\pm EL:$ $5 \times H$

$$5 \times 2$$

$$10psi$$

$PDP = 100 + 56 + 10$

$PDP = 166$ psi

NFPA 1002 5.2. *IFPO, 2E:* Chapter 11, pages 317 – 357.

Question #72

$PDP = NP + FL + AFL \pm EL$

NP: 50 psi

gpm: $30 \times d^2 \times \sqrt{NP}$

$$30 \times (0.25)^2 \times \sqrt{50}$$

$$30 \times 0.06 \times 7$$

$$12.6gpm$$

FL: cq^2L

$$15.5 \times \left(\frac{13}{100} \right)^2 \times \frac{200}{100}$$

$$15.5 \times (0.13)^2 \times 2$$

$$15.5 \times 0.02 \times 2$$

$$.62psi$$

AFL: 0 psi

$\pm EL:$ $5 \times H$

$$5 \times (-1)$$

$$-5psi$$

$PDP = 50 + 1 - 5$

$PDP = 46$ psi

NFPA 1002 5.2. *IFPO, 2E:* Chapter 11, pages 317 – 357.

Question #73

$PDP = NP + FL + AFL \pm EL$

NP: 100 psi

FLs (4-inch line):

$$cq^2L$$

$$.2 \times \left(\frac{525}{100}\right)^2 \times \frac{650}{100}$$

$$.2 \times (5.25)^2 \times 6.5$$

$$.2 \times 27.56 \times 6.5$$

$$35.82psi$$

FLa (2 ½-inch lines):

$$cq^2L$$

$$2 \times \left(\frac{175}{100}\right)^2 \times \frac{200}{100}$$

$$2 \times (1.75)^2 \times 2$$

$$2 \times 3.06 \times 2$$

$$12.24psi$$

AFL: 15 psi

$\pm EL$: 0

$PDP = 100 + (36 + 12) + 15$

$PDP = 163$ psi

NFPA 1002 5.2. *IFPO, 2E:* Chapter 12, pages 342 – 344.

Question #74

$PDP = NP + FL + AFL \pm EL$

NP: 80 psi

gpm: $30 \times d^2 \times \sqrt{NP}$

$30 \times (1.5)^2 \times \sqrt{80}$

$30 \times 2.25 \times 9$

$607.5 gpm$

FLs (2 ½-inch lines):

$cq^2 L$

$2 \times \left(\dfrac{203}{100}\right)^2 \times \dfrac{500}{100}$

$2 \times (2.03)^2 \times 5$

$2 \times 4.12 \times 5$

$41.2 psi$

FLa (3-inch line):

$cq^2 L$

$.8 \times \left(\dfrac{608}{100}\right)^2 \times \dfrac{200}{100}$

$.8 \times (6.8)^2 \times 2$

$.8 \times 46.24 \times 2$

$73.98 psi$

AFL: 5 psi for the siamese and 15 psi for the monitor

$\pm EL$: 0

$PDP = 80 + (74 + 41) + 20$

$PDP = 215$ psi

NFPA 1002 5.2. *IFPO, 2E:* Chapter 12, pages 347 – 348.

Question #75

$PDP = NP + FL + AFL \pm EL$

NP: 100 psi

*FL*s (2 ½-inch lines):

$$cq^2L$$

$$2 \times \left(\frac{100}{100}\right)^2 \times \frac{150}{100}$$

$$2 \times (1)^2 \times 1.5$$

$$2 \times 1 \times 1.5$$

$$3psi$$

*FL*a (1 ¾-inch line):

$$cq^2L$$

$$24 \times \left(\frac{100}{100}\right)^2 \times \frac{200}{100}$$

$$24 \times (1)^2 \times 2$$

$$24 \times 1 \times 2$$

$$48psi$$

AFL: 25 psi for the standpipe

$\pm EL$: 5 psi per level, $5 \times -1 = -5$ psi

$PDP = NP + FL + AFL \pm EL$

$PDP = 100 + (3 + 48) + 25 - 5$

$PDP = 171$ psi

NFPA 1002 5.2.4. *IFPO, 2E:* Chapter 12, pages 349 – 350.

Question #76

$NP =$ 50 for smooth-bore on a hand-line

80 for smooth-bore on a master-stream device

100 for combination (fog, automatic, selectable flow, etc.)

$EL =$ 5 psi added or subtracted for each floor level above/below ground level

.5 psi added or subtracted for each foot in elevation above/below ground level

$PDP = 50$ (NP) + 67 (FL) + 10 (EL = 5 psi for each floor × 2 floor levels)

$PDP = 127$ psi

NFPA 1002 5.2.4. *IFPO, 2E:* Chapter 12, pages 319 – 321.

Question #77

For like likes, only calculate pump pressure for one of the lines. The other lines will have the same pump pressure.

$PDP = NP + FL + AFL \pm EL$

$NP =$ 50 for smooth-bore on a hand-line

80 for smooth-bore on a master-stream device

100 for combination (fog, automatic, selectable flow, etc.)

$PDP = 100$ (NP) + 24 (FL)

$PDP = 124$ psi

NFPA 1002 5.2.4. *IFPO, 2E:* Chapter 12, pages 319 – 321.

Question #78

When flowing two or more like lines, the PDP will be the same regardless of how many lines are placed in operation. For each line initiated, the discharge gate is slowly opened while the throttle is increased to maintain the PDP. NFPA 1002 5.2.4. *IFPO, 2E:* Chapter 12, pages 336 – 337.

Question #79

In volume mode, impellers spin at the same speed and provide the same quantity of water each. Total discharge then, is divided among the number of impellers, and the pressure each produces is the same. NFPA 1002 5.2.1, 5.2.2, 5.2.4. *IFPO, 2E:* Chapter 4, pages 89 – 91.

Question #80

The pressure is too high because the maximum operating pressure of supply hose is 185 psi. NFPA 1002 5.2.2. *IFPO, 2E:* Chapter 7, page 192.

Question #81

A tanker shuttle time is determined by dividing the tank volume by the time it takes the tanker to complete one cycle from the dump site, to the fill site, and back to the dump site.
To determine the size of a tanker when the tanker shuttle time and flow is known, simply multiply the shuttle flow capacity by the cycle time. $250 \times 14 = 3,500$ gallons. NFPA 1002 5.2.1. *IFPO, 2E:* Chapter 7, pages 200 – 201.

Question #82

In order to determine the length of time the tank could sustain the hand-line, the gpm must be known. NFPA 1002 5.2. *IFPO, 2E:* Chapter 11, pages 296 – 297.

Question #83

The water flow will be the same at any point in the line. If each discharge is flowing 420 gpm, then the total flow through the monitor is 840 gpm. NFPA 1002 5.2. *IFPO, 2E:* Chapter 5, pages 113 – 114.

Question #84

The specific heat of water is 1 Btu, which is equivalent to 252 calories. It takes 152 Btu to raise 1 pound of water from 62°F to 212°F. It takes 1,268 Btu to raise the temperature of 1 gallon of water from 62°F to 212°F. NFPA 1002 5.2.1, 5.2.2. *IFPO, 2E:* Chapter 10, page 264.

Question #85

Several driving exercises are typically used to assess the driver's ability to safely operate and control the vehicle. These include the following:

- *Alley Dock.* Assesses the ability to back the vehicle into a restricted area, such as a fire station or down an alley.

- *Serpentine.* Assesses the ability to drive around obstacles, such as parked cars and tight corners.

- *Confined-Space Turnaround.* Assesses the ability to turn the vehicle around within a confined space, such as a narrow street or driveway.

- *Diminishing Clearance.* Assesses the ability to drive the vehicle in a straight line, such as on a narrow street or road.

NFPA 1002 4.3.2, 4.3.3, 4.3.4, 4.3.5. *IFPO, 2E:* Chapter 3, pages 65 – 66.

Question #86

The only exceptions to not wearing a seat belt in a fire department vehicle are:

- Loading hose

- Tiller driver training

- Patient treatment

NFPA 1002 4.3. *IFPO, 2E:* Chapter 3, page 49.

Question #87

NFPA 1500 requires that emergency-vehicle drivers come to a complete stop when any intersection hazard is present. Specifically, the vehicle must come to a complete stop when any of the following exists:

- As directed by a law enforcement officer
- At traffic red lights or stop signs
- At negative right-of-way and blind intersections
- When all lanes of traffic in an intersection cannot be accounted for
- When a stopped school bus with flashing warning lights is encountered
- At unguarded railroad guard crossings (also for nonemergency)
- When other intersection hazards are present

NFPA 1002 4.3. *IFPO, 2E:* Chapter 3, page 48.

Question #88

Centrifugal pump performance is contingent on three interrelated factors. If one factor remains constant, a change in one of the remaining factors will change the other.

- *Speed*. If the speed of the pump is held constant and the flow of water increases, pressure will drop. If more water is allowed to flow while the speed of the pump remains the same, pressure will be reduced, because less resistance occurs on the discharge side of the pump.
- *Flow*. If the flow of water is held constant and the speed of the pump is increased, pressure will increase. The same amount of water is being discharged, yet the pump is attempting to discharge more water. This results in an increase in pressure.
- *Pressure*. If the pressure is held constant and the speed of the pump is increased, flow will increase. The increased speed of the pump will increase the flow. Pressure can be maintained constant by increasing or reducing the resistance on the discharge side of the pump.

NFPA 1002 5.2.1, 5.2.2. *IFPO, 2E:* Chapter 4, page 86.

Question #89

Water will rise approximately 2.3 feet/psi for 1 psi of pressure.
So, 16 feet /2.3 feet/psi = 6.9 or 7 psi.

NFPA 1002 5.2.1. *IFPO, 2E:* Chapter 7, pages 185 – 186.

Question #90

The basic steps for pump operation include:

Step 1	Position apparatus, set parking brake, and let engine return to idle.
Step 2	Engage the pump.
Step 3	Provide water to intake side of pump (on-board, hydrant, draft).
Step 4	Set transfer valve (if so equipped).
Step 5	Open discharge lines.
Step 6	Throttle to desired pressure.
Step 7	Set the pressure regulating device.
Step 8	Maintain appropriate flows and pressures.

NFPA 1002 5.2.1, 5.2.2, 5.2.4. *IFPO, 2E:* Chapter 8, page 207.

Question #91
Each of the following are possible causes for loss of prime or failure to prime:

- Air leaks
- Debris on intake strainer
- By-pass line open
- No oil in priming tank
- Defective priming valve
- Improper clearance in rotary gear or vane primer pump
- End of intake hose not submerged deep enough
- Engine speed too low
- Lift too high
- Primer not operated long enough

The transfer valve setting should not affect the prime of a pump. NFPA 1002 5.2. *IFPO, 2E:* Appendix D, page 376.

Question #92
Positive-displacement pumps such as rotary vane or rotary gear pumps operate based on hydrostatic principles (liquids at rest).
Dynamic, or centrifugal, pumps operate based on hydrodynamic principles (liquids in motion). NFPA 1002 5.2. *IFPO, 2E:* Chapter 4, page 76.

Question #93
With the information provided, the needed fire flow can be calculated using either the Iowa State formula and the National Fire Academy Formula. NFPA 1002 5.2. *IFPO, 2E:* Chapter 11, pages 292 – 293.

Question #94
Although supervision is often a duty or function of a pump operator, NFPA 1002 does not contain supervision requirements. NFPA 1002 identifies the main duties of the pump operator as driving, maintenance, and pump operation. NFPA 1002 1.4. *IFPO, 2E:* Chapter 1, pages 15 – 16.

Question #95
- *Dual pumping* occurs when one hydrant supplies two pumpers. The second pumper receives water from the intake of the first pumper. In other words, the excess water provided to the first pumper is diverted to the second pumper.

- *Tandem pumping* occurs when one hydrant supplies two pumpers; it is similar to a relay operation. The first pump discharges all its water to the intake of the second pumper, as in a relay. The only significant difference is that a relay is used to move water over extended distances, whereas tandem pumping is used when higher pressures are required than a single pumper can provide.

NFPA 1002 5.2.1. *IFPO, 2E:* Chapter 8, pages 224 – 225.

Question #96
Pump operations can be defined as the systematic movement of water from a supply source through a pump to a discharge point. NFPA 1002 5.2. *IFPO, 2E:* Chapter 1, page 8.

Question #97
NFPA 1500 requires that inspections, maintenance, repair, and service records be maintained for all vehicles. NFPA 1002 4.2.2. *IFPO, 2E:* Chapter 2, page 25.

Question #98
Total stopping distance is measured from the time a hazard is detected until the vehicle comes to a complete stop and includes:

- Perception distance – 35 feet
- Reaction distance – 30 feet
- Braking distance – 125 feet

NFPA 1002 4.3.2, 4.3.3, 4.3.4, 4.3.5. *IFPO, 2E:* Chapter 3, page 55.

Question #99
Newton's third law of motion states that for every action there is an equal (*not relational*) and opposite reaction. NFPA 1002 5.2.1. *IFPO, 2E:* Chapter 4, page 85.

Question #100
The most correct answer is to maintain traction. Speed, centrifugal force, road conditions, and weight shifts can all affect the ability to maintain traction. When traction is lost, the apparatus will be out of control. NFPA 1002 4.3. *IFPO, 2E:* Chapter 3, page 56.

PHASE IV

FINAL EXAM

This the final section in the Exam Prep Guide. For this section, we addressed all levels of Bloom's Taxonomy, Cognitive Domain, and all the previous sections. When taking the Final Exam from this section, you will find a variety of questions from basic understanding to application level questions. One should have successfully completed the previous sections before attempting Section Four. Successful completion of this section would indicate a strong knowledge of the material and an in-depth understanding of the content.

1. Pump operators should be licensed to drive all vehicles they are expected to operate.
   a. True
   b. False

2. The inspection of the radiator and coolant levels on modern apparatus is no longer necessary due to technological advancements.
   a. True
   b. False

3. Fire pump operations are more complex and unpredictable than other fireground operations.
   a. True
   b. False

4. Although a variety of pump sizes and configurations exist in the fire service today, the general operation of most pumps is basically similar.
   a. True
   b. False

5. Over the years, the title or name used for the position responsible for driving and operating the pump has changed. The title used in NFPA 1002 is "apparatus driver/engineer."
   a. True
   b. False

6. The first few minutes after arrival on scene can be demanding for pump operators as they connect supply hose, open discharge lines, and set up pressures. After this setup period, pump operators' work can be equally demanding as they continually observe instrumentation, monitor and adjust pressures and flows, and prepare for changing fireground situations.
   a. True
   b. False

7. A visual inspection is all that is needed to ensure proper tire inflation.
   a. True
   b. False

8. Inspecting, servicing, and testing apparatus are the three primary activities conducted during preventive maintenance.
   a. True
   b. False

9. Replacing the battery on a fire department vehicle is usually performed by the pump operator.
   a. True
   b. False

10. Safety-related components and manufacturers' recommendations comprise the majority of inspection items included in a preventive maintenance program.
    a. True
    b. False

11. If a problem is found during a vehicle inspection and can be readily fixed, there is no need for the pump operator to document the finding.
    a. True
    b. False

12. Pump operators should never add oil to the engine. This is a preventive maintenance activity for certified mechanics only.
    a. True
    b. False

13. Doubling the speed of a vehicle increases the stopping distance an estimated four times.
    a. True
    b. False

14. When an apparatus stops abruptly, the on-board water supply attempts to stay in motion; this is referred to as centrifugal force.
    a. True
    b. False

15. Federal laws and regulations dictate that when backing apparatus, a minimum of two spotters or guides is required.
    a. True
    b. False

16. NFPA 1002 requires the following warning on pump panels:

    "**WARNING:** Death or serious injury might occur if proper operating procedures are not followed. The pump operator as well as individuals connecting supply or discharge hoses to the apparatus must be familiar with water hydraulics hazards and component limitations."
    a. True
    b. False

17. One-hundred percent slippage can occur in centrifugal pumps because they have an open path from the intake to the discharge side of the pump.
    a. True
    b. False

18. Positive-displacement pumps theoretically discharge a varying quantity of water inversely related to each revolution or cycle of the pump.
    a. True
    b. False

19. At the scene of a structural fire, you just opened and set the discharge pressures for two lines and then set the pressure relief valve. Almost immediately, both lines are shut down, and the engine speed slows considerably. You are not concerned because changing engine speed is how the pressure-relief valve limits pressure surges.
    a. True
    b. False

20. Cotton, rather than synthetic fiber, is still the most popular material used in the construction of woven-jacket hose.
    a. True
    b. False

21. According to data on emergency-vehicle accidents, which of the following is correct?
    a. The number of accidents annually has decreased dramatically.
    b. The number of accidents annually has decreased some, but not a lot.
    c. The number of accidents annually has stayed about the same.
    d. The number of accidents annually has continued to increase.

22. The three primary duties of a pump operator include driving the pump, _____,
and _____ .
    a. preventive maintenance, securing a water supply
    b. preventive maintenance, operating the pump
    c. preventive maintenance, supervising firefighters
    d. operating the pump, supervising firefighters

23. Documenting preventive maintenance activities is important for which of the following reasons?
    a. Tracking needed maintenance and repairs.
    b. Determining maintenance trends.
    c. As an requirement for warranty claim.
    d. All reasons are correct.

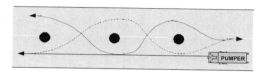

24. The driving exercise shown above is called
    a. diminishing clearance.
    b. alley dock.
    c. Confined-space turnaround.
    d. serpentine.

25. The gated wye shown above is in the _____ position.
    a. open
    b. closed
    c. vertical
    d. chorista

26. What is the friction loss in 500 feet of 2 ½-inch hose flowing 350 gpm using the hand method?
    a. 100 psi
    b. 123 psi
    c. 134 psi
    d. 153 psi

27. NFPA 1002 requires that driver pump operators be medically fit, licensed to drive, and meet
Firefighter I requirements. The standard also includes knowledge and skill requirements for the
following duties of the driver pump operator, *except*
    a. preventive maintenance.
    b. driving the apparatus.
    c. pump operations.
    d. first-line supervisory responsibilities.

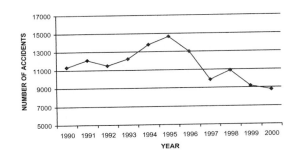

28. Each year, the *NFPA Journal* reports on emergency-vehicle accident statistics. Having read several of these reports, what can you say about the graph shown above?
    a. Emergency-vehicle accidents will continue to decrease.
    b. Emergency-vehicle accidents will start to increase or stay about the same.
    c. Emergency-vehicle accidents will most likely increase significantly.
    d. The data in the graph is incorrect, because the number of vehicle accidents did not declined between 1990 and 2000.

29. Requirements to conduct preventive maintenance on emergency apparatus can be found in manufacturer's requirements and in NFPA
    a. 1001.
    b. 1500.
    c. 1915.
    d. both b and c are correct.

30. Engine oils are classified using a rating system developed by API and SAE. API stands for
    a. American Petroleum Institute.
    b. American Petrochemical Industry.
    c. Association of Petroleum Industries.
    d. Association of Products for Industry.

31. When charging emergency-vehicle batteries, the area should be ventilated because
    a. excessive heat can build up.
    b. phosgene can be produced.
    c. hydrogen gas can be produced.
    d. acetylene gas can be produced.

32. Inspecting the steering system on emergency vehicles includes which of the following?
    a. turning the steering wheel to both the right and left to ensure front tires turn at least 45 degrees
    b. turning the steering wheel until just before the wheels turn if excessive play exists
    c. turning the steering wheel while the vehicle is in motion
    d. turning the steering wheel in one direction fully and back while ensuring the front tires track the steering wheel rotation

33. NFPA 1500 suggests that all of the following should be inspected on a routine basis *except*
    a. tires.
    b. brakes.
    c. transmission oil.
    d. windshield wipers.

34. During preventive maintenance inspections, air tanks should
    a. be drained slightly to remove moisture.
    b. be drained fully every day.
    c. never be drained.
    d. be drained according to manufacturers' recommendations.

35. Keeping emergency vehicles clean at all times is important because it helps
    a. maintain a good public image.
    b. ensure the pump, systems, and equipment operate as intended.
    c. ensure vehicles can be inspected properly.
    d. all the answers are correct.

36. The inspection of emergency vehicles includes three basic steps. Which of the following is not one of the three steps?
    a. reviewing previous inspection reports
    b. inspecting the vehicle
    c. topping off fluid levels as needed
    d. documenting and reporting inspection results

37. Inspections are conducted to verify the _____ of components, while servicing activities are conducted to help maintain the vehicle in peak operating _____.
    a. condition, performance
    b. status, performance
    c. performance, condition
    d. status, condition

38. When inspecting the steering on an apparatus, the steering wheel
    a. free play should be no more than 1 inch.
    b. free play should not exceed 10 degrees.
    c. should be turned fully both directions.
    d. all of the answers are correct.

39. Safety should be considered when conducting preventive maintenance inspections and testing. Which of the following could be considered a common safety consideration for preventive maintenance inspections and tests?
    a. Ensure work area is free from hazards.
    b. Check for loose equipment before raising a tilt cab.
    c. Always be careful when opening the radiator cap.
    d. All are important safety considerations.

40. Your department is researching the requirements for a new apparatus. The fire chief and assistant fire chief have been discussing what the requirements should be for quick build-up air tanks. The chief feels the tanks should reach operating pressure within 30 seconds, while the assistant chief feels at least 2 minutes are required. They have asked you for your opinion. You tell them that, according to NFPA 1901, new apparatus must have quick build-up capable of reaching operating pressure within _____.
    a. 30 seconds
    b. 60 seconds
    c. 2 minutes
    d. 4 minutes

41. Refueling of apparatus should occur per department policy or, if no policy exists, when the fuel level reaches the _____ mark.
    a. one-quarter
    b. one-half
    c. three-quarter
    d. one-third

42. A 20-inch steering wheel should have no more than _____ free play in either direction.
    a. 1 inch
    b. 20 degrees
    c. 2 inches
    d. 2 degrees

43. As the pump operator at the beginning of a new shift, your first step to begin a preventive maintenance inspection on your assigned apparatus is to
    a. start with the engine compartment.
    b. look over the pump and related components.
    c. conduct an inventory of equipment.
    d. check the previous inspection reports.

44. The actual sequence for conducting a preventive maintenance inspection can vary from department to department. However, inspection components can be grouped in general areas as follows:
    1 – Inside cab
    2 – Outside vehicle
    3 – Engine compartment
    4 – Pump and related components

    Using these categories, which of the following is a common sequence used to conduct a vehicle inspection?
    a. 1, 2, 3, and 4
    b. 4, 3, 2, and 1
    c. 2, 3, 1, 4
    d. 3, 4, 1, 2

45. From 1990 to 2000, emergency vehicles were involved in more than _____ accidents while responding to or returning from an incident.
    a. 21
    b. 500
    c. 5,000
    d. 11,000

46. Each of the following is a common factor that appears in most vehicle accidents *except*
    a. not following laws and standards.
    b. not fully aware of driver and/or apparatus limitations.
    c. lacking appreciation of environment, such as weather and traffic.
    d. lacking driver licensure and certification.

47. All of the following address the safe emergency-vehicle operations and requirements that drivers should be familiar with *except*
    a. NFPA 1500.
    b. local ordinances.
    c. state statutes.
    d. the Federal Fire Act.

48. Total stopping distance is measured from the time a hazard is detected until the
    a. vehicle comes to a complete stop.
    b. vehicle slows significantly.
    c. driver begins to depress the brake.
    d. driver engages the parking brake.

49. Friction between the tire and the road is called
    a. torsion.
    b. traction.
    c. grip.
    d. slip.

50. When parking an apparatus next to a curb, the rotation of the front tires should be
    a. toward the curb.
    b. away from the curb.
    c. straight.
    d. all of the answers are correct.

51. A vehicle navigating a curve at a high rate of speed will experience
    a. loss of traction.
    b. pressure surge.
    c. centrifugal force.
    d. friction.

52. Fire department apparatus accidents occur for a variety of reasons. Several common factors associated with apparatus accidents include all the following *except*
    a. not following laws and standards related to emergency response.
    b. not being certified to NFPA 1002.
    c. not being fully aware of both driver and apparatus limitations.
    d. lack of appreciation for driving conditions such as weather and traffic.

53. Regaining the control of a vehicle during a skid can be accomplished by
    a. disengaging the clutch.
    b. turning the front wheels in the opposite direction of the skid.
    c. taking the foot off the accelerator.
    d. increasing speed slightly.

54. The distance of travel from the time the brake is depressed until the vehicle comes to a complete stop is called
    a. perception distance.
    b. reaction distance.
    c. braking distance.
    d. total stopping distance.

55. Accidents involving responding apparatus can produce significant outcomes, including
    a. a delay in assistance to those who summoned help.
    b. fire department members and civilians seriously or fatally injured.
    c. the city, department, and driver facing civil and/or criminal proceedings.
    d. each of these are potential outcomes.

56. Most state laws require that emergency-vehicle drivers obey the same laws as other vehicle operators unless specifically exempt from doing so. State laws typically define several conditions that must exist for exemptions to be extended. Which of the following would most likely meet one or more of the conditions?
    a. returning from an incident
    b. responding to an incident with audible and visual warning devices
    c. responding to an incident during good weather (no rain, snow, fog, etc.)
    d. driving 10 mph over the posted speed limit

57. You are the driver of a fire department vehicle responding to a reported structure fire. In which of the following conditions do you not have to bring the apparatus to a complete stop?
    a. at negative right-of-way and blind intersections
    b. at unguarded railroad guard crossing
    c. at school crossings
    d. when stopped school bus with flashing warning lights is encountered

58. Can a water tanker with a capacity of 2,000 gallons cross over a bridge with a weight limit of 8 tons?
    a. Yes, but only if half the water is drained first.
    b. Yes, the water weighs less than the authorized limit and the weight of the vehicle would be less than 1 ton.
    c. No, the water alone weighs more than the authorized limit.
    d. No, the water weighs less than the authorized limit but the weight of the tanker exceeds the authorized limit.

59. Hydraulic principles that deal with liquids at rest and the pressures they exert or transmit are known as
    a. hydrostatics.
    b. hydrodynamics.
    c. hydrodisplacement.
    d. hydrocentrifugal.

60. Which of the following methods used to power pumps provides stationary pumping only?
    a. PTO
    b. directly from engine crankshaft
    c. split-shaft transmission
    d. no correct answer provided

61. Which of the following is not correct concerning pump priming?
    a. Positive-displacement pumps are self-priming.
    b. Centrifugal pumps must be primed.
    c. Priming is a suction process that forces water into the pump.
    d. Priming can be referred to as replacing air in a pump with water.

62. For a two-stage centrifugal pump operating in volume mode, water
    a. enters both impellers at the same time.
    b. enters one impeller and then enters the second impeller.
    c. pressure from one impeller is added to the pressure generated in the second impeller.
    d. volume from one impeller is provided to the second impeller.

63. The rate and quantity of water delivered by the pump is called
    a. pressure.
    b. slippage.
    c. speed.
    d. flow.

64. The figure above shows a cut-away of a _____ pump.
    a. positive-displacement
    b. centrifugal
    c. two-stage centrifugal
    d. rotary vane

65. When flow is expected to exceed 50% of the pump's rated capacity, the transfer valve should be
    a. in pressure mode.
    b. in volume mode.
    c. open.
    d. closed.

66. A two-stage centrifugal pump is discharging 1,000 gpm at 200 psi in volume mode. Each impeller will deliver
    a. 1,000 gpm at 200 psi.
    b. 500 gpm at 200 psi.
    c. 1,000 gpm at 100 psi.
    d. 500 gpm at 100 psi.

67. A centrifugal pump mounted at the rear of an apparatus would most likely be powered
    a. through a PTO.
    b. directly from the crankshaft.
    c. through a split-shaft transmission.
    d. by any of these methods.

68. Fill in the missing rated capacity information:
    _____% at 150 psi
    70% at _____ psi
    _____% at 250 psi
    a. 100, 220, 50
    b. 100, 200, 50
    c. 100, 175, 50
    d. 50, 200, 100

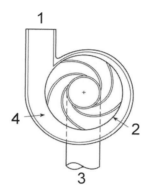

69. Identify each of the numbered items for this line diagram above of a centrifugal pump.
    a. 1 – Intake
        2 – Volute
        3 – Impeller
        4 – Discharge
    b. 1 – Discharge
        2 – Vane
        3 – Intake
        4 – Volute
    c. 1 – Discharge
        2 – Volute
        3 – Intake
        4 – Impeller
    d. 1 – Discharge
        2 – Impeller
        3 – Intake
        4 – Volute

70. The rated capacity plate shown above is for a ——————— gpm pump.
    a. 500
    b. 1,000
    c. 1,250
    d. insufficient information provided

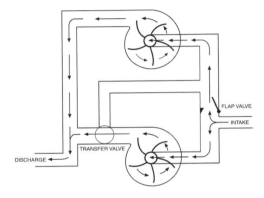

71. The flow of water through the centrifugal pump shown above indicates that it is
    a. in volume mode.
    b. in pressure mode.
    c. a single-stage pump.
    d. none of the answers are correct.

72. In positive-displacement pumps, when pressure is increased on the intake side of the pump
    a. discharge pressure will not be effected.
    b. discharge pressure will increase.
    c. discharge pressure will decrease.
    d. none of the answers are correct.

73. Which of the following is incorrect concerning positive-displacement pumps?
    a. The pressure on the intake side of the pump does not affect the discharge.
    b. Each revolution of the pump yields a specific quantity of water.
    c. The quantity of water discharged per revolution will increase as pump speed increases.
    d. The water being pumped is independent of discharge pressure.

74. The two largest gauges located on a pump panel, usually together, are called
    a. pump intake gauge and pump discharge gauge.
    b. main pump gauges.
    c. Bourdon tube gauges.
    d. all are correct.

75. All of the following are correct concerning flow meters *except* they
    a. measure both pressure and flow.
    b. measure both quantity and rate.
    c. require less hydraulic calculations during pump operations.
    d. use paddle wheels.

76. Manual control valve operation include each of the following *except*
    a. push-pull "T-handles."
    b. quarter turn.
    c. pneumatic.
    d. crank control.

77. A pressure _____ regulating device protects against excessive pressure buildup by controlling the speed of the pump engine.
    a. relief valve
    b. governor
    c. auxiliary system
    d. automatic

78. The type of cooling system that pumps water from the discharge side of the pump into the engine cooling system is known as
    a. an auxiliary cooling system.
    b. a radiator fill system.
    c. a pump cooling system.
    d. a primary cooling system.

79. Foam systems on pumping apparatus include all *except*
    a. around-the-pump proportioning.
    b. balanced pressure.
    c. in-line eductor.
    d. high expansion.

80. During the priming process, oil is used to
    a. help lubricate the priming pump.
    b. provide a tighter seal between the moving parts.
    c. increase the viscosity of initial water being pumped.
    d. both a and b are correct.

81. The _____ control valve allows water to flow from the on-board water supply to the intake side of the pump.
    a. tank-to-pump
    b. pump-to-tank
    c. transfer valve
    d. discharge

82. Pump panels can be located in which of the following locations:
    a. front of the apparatus.
    b. top of the apparatus.
    c. rear of the apparatus.
    d. all are correct.

83. Common valve types on pumping apparatus include
    a. ball, butterfly, and piston.
    b. gated, hammer, and piston.
    c. ball, piston, and quarter.
    d. gated, quarter, and full.

84. NFPA _____ suggests a standard color code to match discharges with their respective gauges.
    a. 1500
    b. 1901
    c. 1911
    d. 1961

85. While priming a pump, several indicators that the pump is primed include each of the following *except*
    a. a positive reading on the pressure gauge.
    b. air/oil discharge under the vehicle.
    c. main pump sounds as if it is under load.
    d. priming motor sounds as if it is slowing.

86. The control device shown above operates the
    a. transfer valve.
    b. pressure governor.
    c. pressure-relief device.
    d. primer.

87. Correctly label items 1 and 3 in the figure shown above.
    a. 1 – main discharge gauge
       3 – individual discharge gauge
    c. 1 – main intake gauge
       3 – individual flow meter
    b. 1 – main intake gauge
       3 – individual discharge gauge
    d. 1 – main discharge gauge
       3 – individual flow meter

88. A 1,500-gpm pumper will usually have how many 2 ½-inch discharges?
    a. four
    b. five
    c. six
    d. seven

89. The _____ and _____ control valves can be used together to help keep the pump from overheating by circulating water from the pump to the tank and then back to the pump again.
    a. tank-to-pump, transfer valve
    b. pump-to-tank, intake
    c. intake, discharge
    d. tank-to-pump, pump-to-tank

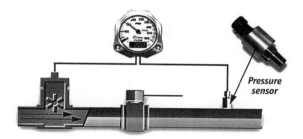

90. The device shown above is unique because it
    a. can regulate pressure between the two points.
    b. can calculate friction loss between the two points.
    c. measures and reports both flow and pressure on one gauge.
    d. can accurately inject foam at the appropriate concentration.

91. While pre-planning an industrial complex, it is decided that a foam system is needed that will provide foam from all discharges on the apparatus at the same time. The best foam system on an apparatus would be a
    a. premixed system.
    b. in-line eductor system.
    c. around-the-pump proportioning system.
    d. balanced pressure system.

92. While responding to an incident, you notice a hazard in front of you. You travel 35 feet before your brain recognizes the hazard. You travel another 30 feet before your brain sends a signal to your foot, and the brake is depressed. It takes another 125 feet before the vehicle comes to a complete stop. What is the braking distance?
    a. 65 feet
    b. 125 feet
    c. 160 feet
    d. 190 feet

93. _____ can be defined as the systematic movement of water from a supply source through a pump to a discharge point.
    a. Relay operations
    b. Drafting
    c. Pump operations
    d. Tandem pumping

94. A centrifugal pump operates based on _____ principles, whereas rotary gear pump operation is based on _____ principles.
    a. positive-displacement, dynamic
    b. hydrodynamic, hydrostatic
    c. internal pressure, external pressure
    d. hydrostatic, hydrodynamic

95. According to NFPA 1500, prior to moving a fire department vehicle, all personnel must be seated and properly secured with seat belts. One exception to this requirement is
    a. when donning PPE.
    b. for patient treatment.
    c. conducting tiller training.
    d. both b and c are correct.

96. You just initiated flow through two 3-inch lines, 350 feet each, suppling a smooth-bore master-stream device with a 1 ⅜-inch tip. The residual hydrant pressure is 45 psi at the intake, and each discharge flow meter reading is 420. What is the total flow through the smooth-bore monitor?
    a. 420 gpm
    b. 560 gpm
    c. 640 gpm
    d. 840 gpm

97. While responding to an incident, you begin to navigate a curve. As a knowledgeable pump operator, you know that the key to safely navigating a curve is to
    a. make the turn with the brakes slightly pressed.
    b. maintain traction.
    c. slightly decrease your speed.
    d. slightly increase your speed.

98. NFPA _____ establishes requirements for the design, construction, inspection, and testing of new fire hose.
    a. 1500
    b. 1961
    c. 1962
    d. 1965

99. According to NFPA 1961 and 1962, maximum operating pressures for supply and attack lines are
    a. 185 psi and 200 psi.
    b. 185 psi and 250 psi.
    c. 185 psi and 275 psi.
    d. 200 psi and 275 psi.

100. The most common length of one section of hose used in the fire service today is
    a. 50 feet.
    b. 100 feet.
    c. 150 feet.
    d. 200 feet.

101. The thread used in nearly all fire hose threaded coupling construction is referred to as a
    a. NH.
    b. HN.
    c. NTS.
    d. Storz.

102. The appliance in the picture above is a
    a. wye.
    b. siamese.
    c. either a wye or siamese.
    d. no correct answer provided.

103. Which of the following is an example of a double female adapter?
     a.

     b.

     c.

     d.

104. The recommended operating pressure for a smooth-bore nozzle on a hand-line is
     a. 50 psi.
     b. 80 psi.
     c. 100 psi.
     d. 125 psi.

105. A nozzle that is designed to maintain a constant nozzle pressure over a wide range of flows is called a
     a. fixed-flow or constant-flow nozzle.
     b. smooth-bore nozzle.
     c. selectable-flow nozzle.
     d. automatic nozzle.

106. The Venturi principle is used by what device?
     a. priming devices
     b. eductors
     c. pressure regulators
     d. pressure governors

107. The amount of water flowing from a nozzle best describes
    a. nozzle pressure.
    b. nozzle reaction.
    c. nozzle flow.
    d. nozzle reach.

108. The designed operating pressure for a particular nozzle is called
    a. nozzle pressure.
    b. nozzle reaction.
    c. nozzle flow.
    d. nozzle reach.

109. Each of the following can be used for intake hose *except*
    a. LDH.
    b. soft sleeve.
    c. hard suction.
    d. each of these can be used for intake hose.

110. The basic components of an eductor include a
    a. metering valve and pickup hose.
    b. metering valve, pickup hose, and foam concentrator.
    c. foam concentrator and pickup hose.
    d. pickup hose, metering valve, and foam gauge.

111. When a nozzle is not provided sufficient nozzle pressure,
    a. less flow will be delivered.
    b. reach will be reduced.
    c. a poor pattern will be develop.
    d. each of the above is correct.

112. _____ require(s) fire departments to establish and maintain an accurate record of each hose section.
    a. NFPA 1961
    b. NFPA 1962
    c. Hose manufacturers
    d. Federal regulations

113. According to NFPA 1965, a gated wye with a handle that is in a position perpendicular to the hose line is
    a. open.
    b. closed.
    c. exactly half open.
    d. not a requirement of the standard.

114. The basic steps for conducting a hose test are provided below. What is the missing step?
    1. Connect water supply and test hose sections.
    2. Open discharge lines for test hose sections.
    3. Slowly increase discharge pressure to 45 psi.
    4. _____.
    5. Mark hose near coupling to determine slippage.
    6. Increase pressure slowly (NFPA 1962 suggests no faster than 15 psi per second) to test pressure.
    7. Maintain pressure for 3 minutes, periodically checking for leaks.
    8. Record results.

    a. Check hose for leaks.
    b. Clear nonessential personnel from the testing area.
    c. Set pressure-relief device.
    d. Remove air from within hose.

115. According to NFPA 1901, an apparatus with a 200-gallon tank is classified as a(n)
    a. initial attack apparatus.
    b. pumper fire apparatus.
    c. mobile water apparatus.
    d. tanker apparatus.

116. A hydrant supplied from only one direction is called a
    a. one-way hydrant.
    b. dead-end hydrant.
    c. gridded hydrant.
    d. zoned hydrant.

117. The hydrant in the picture above is a
    a. dry-barrel hydrant.
    b. wet-barrel hydrant.
    c. dead-end hydrant.
    d. static hydrant.

118. If the pump is located at the hydrant, which of the following has not occurred?
    a. reverse lay
    b. forward lay
    c. first pump in a relay
    d. boosting pressure using a four-way hydrant valve

119. During relay operations, the intake pressure for each pump in the relay should not fall below
    a. 10 psi.
    b. 14.7 psi.
    c. 20 psi.
    d. 22.5 psi.

120. Tankers or mobile water supply apparatus have a minimum tank size of _____ gallons according to NFPA 1901.
    a. 200
    b. 300
    c. 1,000
    d. 2,000

121. Tanker shuttle equipment include all of the following *except*
    a. set siphons.
    b. portable dump tanks.
    c. nurse tanker.
    d. all are correct.

122. Shuttle flow capacity is limited by the volume of water being delivered and the
    a. size of the pump.
    b. available water supply.
    c. number tankers.
    d. time it takes to complete a shuttle cycle.

123. A pumper with a 1,000-gallon tank is flowing 250 gpm through a 2 ½-inch hose line. How long will the on-board water supply last?
    a. 4 minutes
    b. 5 minutes
    c. 10 minutes
    d. no correct answer is provided

124. The equipment required to conduct hydrant flow testing includes all of the following *except* a
    a. fire department pumper.
    b. hydrant diffuser.
    c. pressure gauge mounted on an outlet cap (calibrated within the past 12 months).
    d. pitot gauge for each hydrant.

125. The time it takes to fill a tanker, drive to the dumps site, and empty its water is known as
    a. total shuttle time.
    b. shuttle flow capacity.
    c. shuttle cycle time.
    d. maximum tanker shuttle time.

126. During a drafting operation, if the pressure within the pump is reduced by 8.5 psi, then water can be raised to a height of _____ .
    a. 8.5 feet
    b. 19.5 feet
    c. 22.5 feet
    d. just over 23 feet

127. When flow testing hydrants, the next major step after taking a static pressure reading at the test hydrant is to
    a. slowly shut down hydrants one at a time.
    b. record exact interior size, in inches, of each outlet flowed.
    c. open flow hydrants one at a time until a 25% drop in residual pressure is achieved.
    d. continue to flow to clear debris and foreign substances.

128. How far should the second pumper be positioned in a relay if the first pumper is flowing 500 gpm through 4-inch hose at 185 psi (5 psi friction loss per 100 feet)?
    a. 500 feet
    b. 3,300 feet
    c. 5,000 feet
    d. 5,300 feet

129. Determine the individual shuttle tanker flow for a 1,500-gallon tanker with a 10-minute shuttle cycle time.
    a. 100 gpm
    b. 150 gpm
    c. 500 gpm
    d. 1,000 gpm

130. You are operating the supply pumper in a relay and calculate that your pump discharge pressure should be 199 psi to ensure that 20 psi remains at the next pump. Your supply is a weak hydrant at a dead-end main. After securing the hydrant supply, you slowly open your discharge and begin flowing water to the next pumper through 4-inch hose. The pumper you are supplying water to is about 126 feet above your pumper up a winding dirt road. After contacting the pump operator to verifying 20 psi at the next pump, you set your pressure-relief device. Do you note any problems with this operation?
    a. No problem, everything is good.
    b. The pressure is too high.
    c. Should not be using a weak hydrant in a relay operation.
    d. The relief device should be set prior to verifying 20 psi at the next pump.

131. The test that ensures the interior of the pump can maintain a vacuum is called the
    a. priming device test.
    b. draft test.
    c. vacuum test.
    d. pressure control test.

132. Which of the following is not a typical weekly or monthly test required by manufacturers?
    a. regulating device test
    b. priming system test
    c. engine speed check
    d. transfer valve operation check

133. Which of the following is not a good rule to help safely operate pumping apparatus?
    a. Never operate the pump without water.
    b. Always maintain awareness of instrumentation during pumping operations.
    c. Never leave the pump unattended.
    d. Always open, close, and turn controls swiftly.

134. The annual pump test is a 40-minute test consisting of which of the following?
    a. 20 minutes at 100% capacity
       20 minutes at 50% capacity
    b. 20 minutes at 100% capacity
       10 minutes at 70% capacity
       10 minutes at 50% capacity
    c. 20 minutes at 100% capacity
       10 minutes at 75% capacity
       10 minutes at 50% capacity
    d. 20 minutes at 100% capacity
       10 minute at 165 psi (overload test)
       5 minutes at 70% capacity
       5 minutes at 50% capacity

135. Most main pumps on modern pumping apparatus receive power
    a. via a PTO.
    b. from the drive engine.
    c. from a separate engine.
    d. all are correct.

136. Although a variety of pump sizes and configurations exist, the same basic steps must be taken to move water from the supply to the discharge point. What critical pump operation step is out of order?

    Step 1 Position apparatus, set parking brake, and let engine return to idle.
    Step 2 Provide water to intake side of pump (onboard, hydrant, draft).
    Step 3 Engage the pump.
    Step 4 Set transfer valve (if so equipped).
    Step 5 Open discharge lines.
    Step 6 Throttle to desired pressure.
    Step 7 Set the pressure regulating device.
    Step 8 Maintain appropriate flows and pressures.

    a. Change step 4 with 6.
    b. Change step 6 with 7.
    c. Change step 2 with 3.
    d. Change step 7 with 8.

137. As a general rule of thumb, the transfer valve should be in volume mode when
    a. flows are greater than 50% of a pump's rated capacity.
    b. pressures are less than 150 psi.
    c. pressures are greater 150 psi.
    d. both a and b.
    e. both a and c.

138. The power for pumps can be transferred from the drive engine by each of the following methods *except*
    a. front crankshaft.
    b. split-shaft.
    c. power take-off (PTO).
    d. midship transfer (MST).

139. The same basic steps must be taken to move water from the supply to the discharge point. What is the next step after opening discharge lines?
    a. Set transfer valve.
    b. Throttle to desired pressure.
    c. Set the pressure regulating device.
    d. Engage the pump.

140. When positioning a pumper at an incident, all of the following should be considered *except*
    a. following department SOPs.
    b. evaluating available water supply.
    c. tactical considerations.
    d. assessing surroundings.
    e. each of these should be considered when positioning a pumper.

141. When engaging a pump utilizing a PTO, the transmission should be in _____ when operating the PTO lever.
    a. neutral
    b. first or lowest gear
    c. fourth or highest gear
    d. reverse

142. When engaging a pump through a split-shaft arrangement, the transmission should
    a. stay in neutral.
    b. be placed in the highest gear after switching the pump shift control from road to pump.
    c. stay in low gear.
    d. be placed in pump drive.

143. The transfer valve should be in the _____ position for a 1,000-gpm pumper expecting to flow 750 gpm at 140 psi.
    a. volume
    b. pressure
    c. off
    d. on

144. After discharge lines are connected, the pump operator starts to increase the pump discharge pressure. When the throttle is slowly increased, a corresponding increase in the main discharge gauge is not noted. What is a common cause of this?
    a. Pump is not in gear.
    b. Pump is not primed.
    c. Supply line is not open or insufficient.
    d. All of these are common causes.

145. During a pumping operation, the pump operator attempts to increase the pressure in the only flowing line. As the throttle is increased, a corresponding increase in pressure does not occur. The pump operator then checks the relief device; it is functioning properly but is not relieving any pressure at this time. What might occur if the pump operator continues to increase the throttle (engine speed), hoping to increase the discharge pressure?
    a. Pump may cavitate.
    b. Prime may be lost.
    c. Intake may collapse.
    d. Damage may occur to the municipal water main.
    e. All of these could occur.

146. Place the following on-board water supply procedures in the correct order:
    1. Open "tank-to-pump."
    2. Engage pump.
    3. Set pressure-regulating device.
    4. Increase throttle.

    a. 1, 2, 3, 4
    b. 2, 1, 4, 3
    c. 2, 4, 1, 3
    d. 1, 2, 4, 3

147. Selecting a location to position fire apparatus should include evaluating surroundings such as
    a. the potential exposure to radiant heat and fire extension.
    b. the potential for collapse.
    c. wind direction, power lines, and escape routes.
    d. all are correct.

148. The phenomenon known as a pump running away from the water supply is known as
    a. water hammer.
    b. cavitation.
    c. vapor pressure.
    d. water slippage.

149. The _____ sprinkler system requires the activation of a detection system and the fusing of at least one sprinkler head before water is discharged.
    a. dry-pipe
    b. pre-action
    c. deluge
    d. wet pipe

150. The NFPA standard that focuses on fire department operations within properties protected by sprinkler and standpipe systems is
    a. NFPA 13.
    b. NFPA 13D.
    c. NFPA 13E.
    d. NFPA 15

151. When supporting sprinkler or standpipe systems, all of the following are correct *except*
    a. check for debris in the fire department connection before connecting hose lines.
    b. start pumping immediately if smoke or fire is evident.
    c. place the transfer valve into the volume position.
    d. pump the fire department connection at a minimum of 150 psi.

152. The causes of water hammer can include each of the following *except*
    a. opening and closing a hydrant too quickly.
    b. opening and closing the priming valve too quickly.
    c. abruptly opening and closing a nozzle.
    d. opening and closing discharge control valves too quickly.

153. A standpipe that provides a 1 ½-inch hose station and is intended primarily for trained personnel during initial attack efforts is a
    a. Class 1 standpipe.
    b. Class 2 standpipe.
    c. Class 3 standpipe.
    d. Class 4 standpipe.

154. The process that explains the formation and collapse of vapor pockets when certain conditions exist during pumping operations is called
    a. cavitation.
    b. water hammer.
    c. vapor pressure.
    d. pressure fluctuation.

155. The following are ways to stop cavitation in a pump *except*
    a. reducing nozzle gpm settings.
    b. reducing pump speed.
    c. increasing the number of discharge lines.
    d. increasing pump speed.

156. Each of the following is a correct action to take during cold weather operations *except*
    a. stopping leaks to reduce the accumulation of ice.
    b. not stopping the flow of water within the pump.
    c. immediately draining water from all hoses when the pumping operation is completed.
    d. pouring hot water into the on-board water tank.

157. A fire department connection marked with "STANDPIPE" means
    a. the system is only a standpipe.
    b. the system is a combination sprinkler and standpipe.
    c. the pump operator should pump an additional 25 psi for the standpipe.
    d. both b and c are correct.

158. A Class 1 standpipe system has a minimum flow rate of _____ gpm.
    a. 100
    b. 150
    c. 250
    d. 500

159. Good troubleshooting is important to pump operations. All of the following are correct concerning troubleshooting pump problems *except*
    a. using the manufacturers' troubleshooting guides when available.
    b. the concept that problems are usually either procedural or mechanical.
    c. that the best method to troubleshoot is to follow the flow of water from the intake to the discharge while attempting to determine the problem.
    d. only certified mechanics should attempt to troubleshoot pumping problems.

160. You are the pump operator of the second arriving pumper. You are told to secure a water supply and to support a high-rise standpipe system. The first arriving pumper has the only accessible hydrant but cannot provide the required pressure to support the standpipe. You decide to have the first pumper discharge all water directly to your pump so that you can increase the pressure to the required level. This pumping configuration is known as
    a. relay pumping.
    b. dual pumping.
    c. tandem pumping.
    d. this is not a recommended pumping operation.

161. The physical change of state from a liquid to a vapor best describes
    a. boiling point.
    b. vapor pressure.
    c. evaporation.
    d. latent heat of vaporization.

162. The downward force exerted on an object by the earth's gravity best describes which of the following terms?
    a. density
    b. weight
    c. volume
    d. pressure

163. The correct way to express pressure units is
    a. psi.
    b. lb/in^2.
    c. lb/ft^3.
    d. both a and b are correct.

164. Gauge pressure is typically expressed as
    a. in. Hg.
    b. psia.
    c. psi.
    d. psig.

165. The measurement of pressure that does not include atmospheric pressure is called
    a. vacuum.
    b. gauge pressure.
    c. absolute pressure.
    d. head pressure.

166. The pressure in a system when no water is flowing is called
    a. static pressure.
    b. residual pressure.
    c. pressure drop.
    d. normal pressure.

167. When using the hand method to calculate friction loss in 100-foot sections of 2 ½-inch hose, the ring finger has a value of _____ at the tip and _____ at the base.
    a. 1 (100 gpm), 2 (200 gpm)
    b. 2 (200 gpm), 4 (400 gpm)
    c. 4 (400 gpm), 8 (800 gpm)
    d. 5 (500 gpm), 10 (1,000 gpm)

168. In the formula $P = \dfrac{F}{A}$, the unit F is
    a. force.
    b. friction.
    c. weight.
    d. both a and c are correct.

169. The weight of 1 gallon of water is _____, while the density of water is _____.
    a. 8.34 lb/ft³, 62.4 lb/gal
    b. 8.34 lb/gal, 62.4 lb/ft³
    c. 1 lb, 8.34 lb/gal
    d. 8.34 lb/gal, 1 lb

170. How many gallons of water are in 1 cubic foot of water?
    a. 8.35
    b. 62.4
    c. 7.48
    d. not enough information provided

171. The formula used to determine flow (gpm) when smooth-bore nozzles are used is $Q = 29.7 \times d^2 \times \sqrt{NP}$. An acceptable value for $\sqrt{NP}$ to use for calculating flow on the fireground for hand lines is
    a. 5.
    b. 7.
    c. 9.
    d. 30.

172. The National Fire Academy formula for calculating needed flow is expressed as
    a. $NF = \dfrac{V}{100}$.
    b. $NF = \dfrac{V}{100} \times 3$.
    c. $NF = \dfrac{A}{3}$.
    d. $NF = \dfrac{A}{3} \div 100$.

173. A gauge reading of 100 psia at sea level is equivalent to a gauge reading of
    a. 85.3 psig.
    b. 114.7 psig.
    c. 100 psig.
    d. none of the answers is correct.

174. Which of the following is the correct formula for calculating smooth-bore nozzle reaction?
    a. $NR = 1.57 \times d \times NP$
    b. $NR = gpm \times \sqrt{NP} \times .0505$
    c. $NR = gpm \times \sqrt{NP \times .0505}$
    d. $NR = 1.57 \times d^2 \times NP$

175. Using the drop-ten method, calculate friction loss in 100-foot sections of 2 ½-inch hose flowing 200 gpm.
    a. 100 psi
    b. 20 psi
    c. 10 psi
    d. 5 psi

176. What is the weight of water in a tank with a volume of 100 cubic feet?
    a. 100 lbs
    b. 834 lbs
    c. 8,340 lbs
    d. 6,234 lbs

177. During pump operations, the drop in hydrant pressure from static to residual can be used to estimate the additional flow the hydrant is capable of providing. A 10% or less drop means the hydrant may be able to deliver as much as
    a. three times the original flow.
    b. two times the original flow.
    c. one time the original flow.
    d. half the original flow.

178. The nozzle reaction for a ⅛-inch tip smooth-bore hand-line operating at the correct nozzle pressure is
    a. 1 psi.
    b. 5 psi.
    c. 10 psi.
    d. 15 psi.

179. Two hose lines are flowing 125 gpm each into a structure. If the flow is sustained for 10 minutes, how many tons of water were delivered into the structure?
    a. 5 tons
    b. over 10 tons
    c. over 20 tons
    d. not enough information is provided

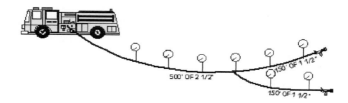

180. The drawing above illustrates which pressure principle? The wyed lines are not flowing and the pressure readings are all the same.
    a. Pressure of a liquid acting on a surface is perpendicular to that surface.
    b. Pressure at any point in a liquid at rest is equal in every direction.
    c. Pressure exerted at the bottom of a container is independent of the shape or volume of the container.
    d. External pressure applied to a confined liquid (fluid) is transmitted equally throughout the liquid.

181. During a pump operation, an initial static pressure reading of 50 psi was noted. After initiating flow of a 125-gpm pre-connect, the residual pressure was 40 psi. How many more gpm is the hydrant capable of flowing?
    a. 125 gpm
    b. 250 gpm
    c. 375 gpm
    d. 0 gpm

182. Calculate friction loss in 400 feet of 2 ½-inch hose flowing 250 gpm, using the drop-ten method.
    a.   15 psi
    b.   30 psi
    c.   60 psi
    d.   90 psi

183. Calculate friction loss using the hand method for 300 feet of 2 ½-inch hose flowing 200 gpm.
    a.   8 psi
    b.   12 psi
    c.   24 psi
    d.   32 psi

184. The nozzle reaction for a combination nozzle operating at the correct nozzle pressure and flowing 100 gpm is
    a.   51 psi.
    b.   76 psi.
    c.   101 psi.
    d.   151 psi.

185. A 2 ½-inch hose line 200 feet in length is flowing 250 gpm. The friction loss was calculated to be 30 psi. Which fireground friction loss formula was used?
    a. cq^2L
    b.   drop-ten
    c.   hand method
    d. $2q^2 + q$

186. Which of the following is not true concerning the formula $FL = 2q^2 + q$?
    a. q = flow in hundreds of gallons $\frac{gpm}{100}$ per minute.
    b.   It was designed for use with cotton-jacketed hose but provides accurate results for newer hose.
    c.   It was designed for flows greater than 100 gpm.
    d.   All are correct.

187. Using the formula $2q^2 + q$, calculate the friction loss for a 100-foot section of 2 ½-inch hose flowing 100 gpm.
    a.   3 psi
    b.   9 psi
    c.   10 psi
    d.   15 psi

188. As the pump operator, you respond to a reported structural fire. Upon arrival, you determine the structure to be a trailer approximately 60 feet long, 30 feet wide, and 10 feet tall. Based on this information, what formula can you use to determine the estimated needed fire flow for the structure?
    a.   Iowa State Formula
    b.   National Fire Academy (NFA) Formula
    c.   Condensed q Formula
    d.   both a and c are correct
    e.   both a and b are correct

189. You are pre-planning the potential fire flow needs for a part of a structure that is 50 feet long, 25 feet wide, and 12 feet tall. Using the NFA formula, calculate the needed flow for this area.
    a.   150 gpm
    b.   417 gpm
    c.   720 gpm
    d.   2,000 gpm

190. A pump operator has been asked to support a standpipe system within a structure. The interior attack team is heading to the sixth floor and will connect to the standpipe a high-rise pack consisting of 200 feet of 1 ¾-inch hose with a combination nozzle. The pump operator should increase pump pressure by _____ psi to account for the change in elevation.
   a.   25
   b.   30
   c.   150
   d.   175

191. A relay system operating with three pumpers is moving water from a hydrant to the fire, a total of about 750 feet. The second pumper in the relay is 48 feet above the supply pumper. The supply-pump operator should increase pump pressure by _____ psi to account for the change in elevation.
   a.   24
   b.   124
   c.   178
   d.   240

192. You are supplying a pumper in a relay flowing 400 gpm through 500 feet of 3-inch hose. Using the condensed q formula calculate the friction in the hose line.
   a.   40 psi
   b.   50 psi
   c.   70 psi
   d.   80 psi

193. An attack crew is stretching out 350 feet of 2 ½-inch hose. What is the friction loss in the line if the flow is 200 gpm? Use the formula $2q^2 + q$.
   a.   10 psi
   b.   21 psi
   c.   22 psi
   d.   35 psi

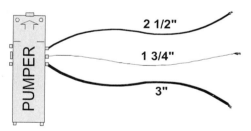

194. For the pumping operation shown above, if the 2 ½-inch line was 500 feet long flowing 350 gpm from a combination nozzle, then the pump discharge pressure would be _____. Use either the drop-ten method or the formula cq^2L.
   a.   150 psi
   b.   200 psi
   c.   223 psi
   d.   250 psi

195. You are the pump operator at the supply pumper in a relay operation. The hose lay between your pumper and the next pumper consists of 500 feet of 3-inch followed by 250 feet of 2 ½-inch. What is the friction loss for the entire hose lay when flowing 325 gpm? Use the formula cq^2L.
   a.   32 psi
   b.   42 psi
   c.   53 psi
   d.   95 psi

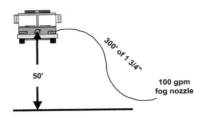

196. For the pumping operation shown above, the pump discharge pressure should be _____.
    a.   122 psi
    b.   147 psi
    c.   177 psi
    d.   222 psi

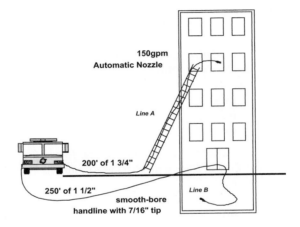

197. The pump discharge pressure for Line B in the figure shown above should be _____.
    a.   25 psi
    b.   51 psi
    c.   101 psi
    d.   140 psi

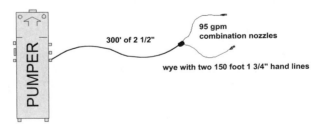

198. For the pumping operation shown above, the pump discharge pressure should be _____.
    a.   100 psi
    b.   153 psi
    c.   177 psi
    d.   222 psi

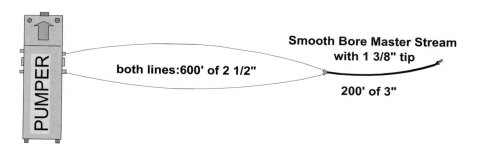

199. For the pumping operation shown above, the pump discharge pressure should be _____.
     a. 100 psi
     b. 153 psi
     c. 186 psi
     d. 220 psi

200. Determine the PDP for the following hose line configuration:
     A 150 foot section of 1 ¾-inch hose flowing 125 gpm through a combination (fog) nozzle, with
     friction loss for the hose line is calculated to be 36 psi.
     a. PDP = 36 psi
     b. PDP = 100 psi
     c. PDP = 136 psi
     d. PDP = 161 psi

Phase IV, Final Exam: Answers to Questions

1.	T	32.	B	63.	D	94.	B
2.	F	33.	C	64.	C	95.	D
3.	F	34.	D	65.	B	96.	D
4.	T	35.	D	66.	B	97.	B
5.	F	36.	C	67.	C	98.	B
6.	T	37.	D	68.	B	99.	C
7.	F	38.	B	69.	D	100.	A
8.	T	39.	D	70.	C	101.	A
9.	F	40.	B	71.	A	102.	A
10.	T	41.	C	72.	A	103.	C
11.	F	42.	C	73.	C	104.	A
12.	F	43.	D	74.	D	105.	D
13.	T	44.	C	75.	A	106.	B
14.	F	45.	D	76.	C	107.	C
15.	F	46.	D	77.	B	108.	A
16.	F	47.	D	78.	B	109.	D
17.	T	48.	A	79.	D	110.	A
18.	F	49.	B	80.	D	111.	D
19.	F	50.	A	81.	A	112.	B
20.	F	51.	C	82.	D	113.	B
21.	D	52.	B	83.	A	114.	D
22.	B	53.	C	84.	B	115.	A
23.	D	54.	C	85.	B	116.	B
24.	D	55.	D	86.	C	117.	A
25.	A	56.	B	87.	A	118.	B
26.	B	57.	C	88.	C	119.	C
27.	D	58.	C	89.	D	120.	C
28.	D	59.	A	90.	C	121.	D
29.	D	60.	C	91.	C	122.	D
30.	A	61.	C	92.	B	123.	A
31.	C	62.	A	93.	C	124.	A

125. C	144. D	163. D	182. C
126. B	145. E	164. D	183. C
127. C	146. B	165. B	184. A
128. B	147. D	166. A	185. B
129. B	148. B	167. C	186. B
130. B	149. B	168. A	187. A
131. C	150. C	169. B	188. E
132. C	151. C	170. C	189. B
133. D	152. B	171. B	190. A
134. B	153. B	172. C	191. A
135. B	154. A	173. B	192. D
136. C	155. C	174. D	193. D
137. D	156. D	175. C	194. C
138. D	157. A	176. D	195. D
139. B	158. D	177. A	196. A
140. E	159. D	178. A	197. B
141. A	160. C	179. B	198. B
142. B	161. C	180. D	199. D
143. A	162. B	181. C	200. C

Phase IV, Final Exam:
Rationale & References for Questions

Question #1
According to NFPA 1002, 1.4.1, all those who operate fire department vehicles must be licensed to drive those vehicles they are expected to operate. NFPA 1002 1.4.1. *IFPO, 2E:* Chapter 1, page 12.

Question #2
Pump operators should continue to inspect the radiator and coolant levels as indicated by department policy and manufacturer's recommendations. NFPA 1002 4.2.1, 4.2.2, 5.1.1. *IFPO, 2E:* Chapter 2, page 33.

Question #3
This is a misperception about fire pump operations. In reality, scientific theory and principles prevail in pump operations. When understood, pump operations can be both predictable and controllable. NFPA 1002 5.1. *IFPO, 2E:* Chapter 1, page 4.

Question #4
Most main pumps on suppression apparatus in the U.S. fire service are centrifugal pumps, while most priming pumps are positive-displacement. Although a variety of sizes and configurations are used, the basic operation of these pumps is similar. NFPA 1002 5.1. *IFPO, 2E:* Chapter 1, page 11.

Question #5
According to NFPA 1002 (5.1), "driver/operator – pumper" is the term used to describe the individual who has met the requirements of Chapter 5 for driving and operating apparatus equipped with fire pumps. NFPA 1002 5.1. *IFPO, 2E:* Chapter 1, page 15.

Question #6
It is a misperception that most of the activities related to fire pump operations occur in the first few minutes upon scene arrival. After initial operations are set up, pump operators must:

- Continually observe instrumentation
- Adjust flows and pressure as appropriate for safety of personnel and equipment
- Be prepared to readily adapt to changing fireground situations
- Monitor and plan for water supply needs and long-term operations
- Maintain a constant vigilance to safety

NFPA 1002 5.1. *IFPO, 2E:* Chapter 1, page 4.

Question #7
A visual inspection of a tire is not sufficient to determine tire pressure. A pressure reading should be taken to ensure adequate tire pressure. NFPA 1002 4.2.1. *IFPO, 2E:* Chapter 2, page 30.

Question #8
Preventive maintenance can be defined as proactive activities taken to ensure the apparatus, pump, and related components remain in a ready state and peak operating performance. These activities can be grouped into inspecting, servicing, and testing. NFPA 1002 4.2.1, 5.1.1. *IFPO, 2E:* Chapter 2, page 23.

Question #9
The specific preventive maintenance activities conducted by pump operators and mechanics depend on the level of training and the type of preventive maintenance activity being conducted. In general, certified mechanics conduct those activities that require apparatus to be taken out of service, require several hours to complete, or are detailed and complicated repairs. Checking and adding engine oil or battery fluid may be conducted by the pump operator, whereas changing the engine oil or replacing the battery is most likely performed by a mechanic. NFPA 1002 4.2.1, 5.1.1. *IFPO, 2E:* Chapter 2, page 22.

Question #10

The two main criteria for determining what components to include in a preventive maintenance inspection are 1) safety-related components and 2) manufacturers' inspection recommendations. NFPA 1002 4.2.1, 5.1.1. *IFPO, 2E:* Chapter 2, pages 28 – 29.

Question #11

Even though the problem was fixed, the pump operator should document the finding to assist with tracking and trending of problems. NFPA 1002 4.2.2. *IFPO, 2E:* Chapter 2, page 25.

Question #12

The specific preventive maintenance activities conducted by pump operators and mechanics depend on the level of training and the type of preventive maintenance activity being conducted. In general, certified mechanics conduct those activities that require apparatus to be taken out of service, require several hours to complete, or are detailed and complicated repairs. Checking and adding engine oil or battery fluid may be conducted by the pump operator, whereas changing the engine oil or replacing the battery is most likely performed by a mechanic. NFPA 1002 4.2.1, 5.1.1. *IFPO, 2E:* Chapter 2, page 22.

Question #13

The faster the vehicle travels, the greater the distance it takes to stop. NFPA 1002 4.3.1, 4.3.6. *IFPO, 2E:* Chapter 3, page 55.

Question #14

The weight shift described is correct in that the water attempts to stay in motion, but this not referred to as centrifugal force. Centrifugal force is the tendency to move outward from the center and occurs when navigating a curve. NFPA 1002 4.3.1, 4.3.2, 4.3.3, 4.3.4, 4.3.5, 4.3.6. *IFPO, 2E:* Chapter 3, page 57.

Question #15

Federal laws and regulations do not specify requirements for backing of emergency apparatus. In addition, some departments only require one spotter as opposed to two.
At a minimum, one spotter should be used when backing apparatus. NFPA 1002 4.3. *IFPO, 2E:* Chapter 3, page 58.

Question #16

The requirement to place the above warning on pump panels is actually contained in NFPA 1901, not NFPA 1002. NFPA 1002 5.2.1, 5.2.2. *IFPO, 2E:* Chapter 4, page 102.

Question #17

Centrifugal pumps have only one moving part, the impeller, which is open from the intake to the discharge side of the pump; this allows for 100% slippage when all discharges are closed. NFPA 1002 5.2.1, 5.2.2. *IFPO, 2E:* Chapter 4, pages 85 – 90.

Question #18

Actually, positive-displacement pumps theoretically discharge (displace) a *specific* quantity of water for each revolution or cycle of the pump. NFPA 1002 5.2.1, 5.2.2. *IFPO, 2E:* Chapter 4, page 78.

Question #19

The pressure-relief valve limits pressure surges by opening and closing a passage between the discharge side of the pump to either the intake or atmosphere. In this case, the engine speed should not slow down. NFPA 1002 5.2.1, 5.2.2, 5.2.4. *IFPO, 2E:* Chapter 5, pages 122 – 123.

Question #20

Actually, synthetic fiber is the most popular material used in the construction of woven-jacket hose. NFPA 1002 5.2.1, 5.2.2, 5.2.4. *IFPO, 2E:* Chapter 6, pages 140 – 141.

Question #21

The trend seems to indicate that the number of vehicle accidents has gradually increased from 1995 to 2000. NFPA 4.3. *IFPO, 2E:* Chapter 1, pages 8 – 9.

Question #22

The three main duties of the pump operator as outlined in NFPA 1002 include:

1. Preventive maintenance (NFPA 1002 4.2)

2. Driving (NFPA 1002 4.3)

3. Pump operations (NFPA 1002 5.1 and 5.2)

NFPA 1002 4.2, 4.3, 5.1, 5.2. *IFPO, 2E:* Chapter 1, pages 7 – 11.

Question #23

All the answers listed are important reasons for documenting preventive maintenance. NFPA 1002 4.2.1. *IFPO, 2E:* Chapter 2, page 25.

Question #24

Several driving exercises are typically used to assess the ability of a driver to safely operate and control the vehicle. These include the following:

- *Alley Dock.* Assesses the ability to back the vehicle into a restricted area, such as a fire station or down an alley.

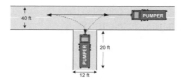

- *Serpentine.* Assesses the ability to drive around obstacles, such as parked cars and tight corners.

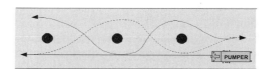

- *Confined-Space Turnaround.* Assesses the ability to turn the vehicle around within a confined space, such as a narrow street or driveway.

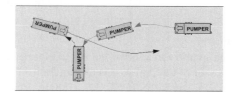

- *Diminishing Clearance.* Assesses the ability to drive the vehicle in a straight line, such as on a narrow street or road.

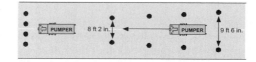

NFPA 1002 4.3.2, 4.3.3, 4.3.4, 4.3.5. *IFPO, 2E:* Chapter 3, pages 65 – 66.

Question #25

NFPA 1965 requires that appliances with lever-operated handles must indicate a closed position when the handle is perpendicular to the hose line. NFPA 1002 5.2.1, 5.2.2, 5.2.4. *IFPO, 2E:* Chapter 6, page 148.

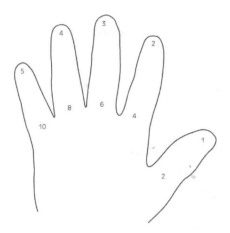

Question #26

The hand method is a fireground method used to estimate friction loss in 100-foot sections of 2 ½-inch hose. Simply select the fingertip representing the gpm flow and then multiply the two figures on the finger for the approximate friction loss pressure for each 100-foot section of 2 ½-inch hose.

FL = 3.5 × 7 × 5 (number of 100-foot sections of hose)
FL = 122.5 psi

NFPA 1002 5.2.1, 5.2.2. *IFPO, 2E:* Chapter 10, pages 302 – 303.

Question #27

NFPA 1002 requirements for the basic duties of a driver pump operator are as follows:

- Preventive maintenance (section 4.2 and 5.1.1)
- Driving the apparatus (section 4.3)
- Pump operations (section 5.2)

Supervisory responsibilities are not included in NFPA 1002. NFPA 1002 4.2, 4.3, 5.1, 5.2. *IFPO, 2E:* Chapter 1, pages 6 – 9.

Question #28

The actual trend indicates that the number of vehicle accidents has gradually increased from 1995 to 2000. NFPA 4.3. *IFPO, 2E:* Chapter 1, pages 8 – 9.

Question #29

- *NFPA 1500, IFPO, 2E:* Chapter 4, section 4-4 suggests requirements for the inspection, maintenance, and repair of emergency vehicles.
- *NFPA 1915* is the standard for fire apparatus preventive maintenance programs.
- *NFPA 1001*, Professional Qualifications for Fire Fighters, does not provide or suggest requirements for conducting preventive maintenance on emergency vehicles.

NFPA 1002 4.2.1, 5.1.1. *IFPO, 2E:* Chapter 2, page 23 – 25.

Question #30

API stands for the American Petroleum Institute. SAE rates oil using a two-letter system. NFPA 1002 4.2.1. *IFPO, 2E:* Chapter 2, pages 32 – 33.

Question #31

While charging vehicle batteries, hydrogen gas can be generated. Ventilation is important to reduce the accumulation of this combustible gas. NFPA 1002 4.2.1. *IFPO, 2E:* Chapter 2, page 33.

Question #32
Steering wheel inspection should be limited to determine excessive play. Damage to the steering system can occur when the steering wheel is turned so that the wheels turn. NFPA 1002 4.2.1. *IFPO, 2E:* Chapter 2, pages 35 – 36.

Question #33
NFPA 1500 suggests that the following safety related components be inspected on a routine basis:

- Tires
- Brakes
- Warning systems
- Windshield wipers
- Headlights and clearance lights
- Mirrors

Typically, manufacturers' recommend the following components be inspected:

- Engine oil
- Coolant level
- Transmission oil
- Brake system
- Belts

NFPA 1002 4.2.1, 5.1.1. *IFPO, 2E:* Chapter 2, pages 28 – 29.

Question #34
It is important that manufacturers' recommendations be followed when conducting vehicle preventive maintenance inspections. NFPA 1002 4.2.1, 5.1.1. *IFPO, 2E:* Chapter 2, pages 28 – 29.

Question #35
Maintaining emergency vehicles clean at all times is important because:

- It helps maintain a good public image (municipal apparatus usually belong to tax-paying public).
- It helps ensure the pump, systems, and equipment operate as intended.
- It helps ensure vehicles can be inspected properly (dirt and grime could cover defects or potential problems).

NFPA 1002 4.2.1. *IFPO, 2E:* Chapter 2, page 37.

Question #36
The three basic steps in the inspection process are:

1. Pre-inspection (reviewing previous inspection reports)
2. Conducting the actual inspection
3. Document and reporting the results

The activity of topping off fluid levels is considered a service function. NFPA 1002 4.2.1, 5.1.1. *IFPO, 2E:* Chapter 2, pages 27 – 28.

Question #37
Inspections are conducted to verify the status of a component, as in verifying water, oil, and fuel levels. Servicing activities are conducted to help maintain vehicles in peak operating condition, as when cleaning, lubricating, and topping off fluids.
Tests are conducted to determine the performance of components, as in annual pump service tests. NFPA 1002 4.2.1, 5.1.1. *IFPO, 2E:* Chapter 2, page 23.

Question #38

It is not advisable to turn the steering wheel so that the wheels turn. Rather, the steering wheel should be turned until just before the wheels turn. The distance should not exceed 10 degrees in either direction. For a 20-inch steering wheel, that would be approximately 2 inches of movement. NFPA 1002 4.2.1. *IFPO, 2E:* Chapter 2, pages 35 – 36.

Question #39

Safety should be considered when conducting preventive maintenance inspections and tests. Common safety considerations for preventive maintenance inspections and tests include: General safety considerations:

- Not hurrying or rushing through inspections.
- Ensuring work area is free from hazards.
- Always keeping safety in mind.

Specific safety considerations:

- Check for loose equipment before raising a tilt cab.
- Do not smoke around engine compartment and fuels.
- Wear appropriate clothing (no loose jewelry, wear safety glasses, gloves).
- Consider vapor and electrical hazards.
- Always be careful when opening the radiator cap.
- Use proper tools.
- Secure all equipment and close all doors prior to moving the apparatus.

NFPA 1002 4.2.1. *IFPO, 2E:* Chapter 2, page 37.

Question #40

NFPA 1901 requires all new apparatus to have quick pressure build-up capabilities so that operating pressure is reached within 60 seconds. NFPA 1002 4.2.1. *IFPO, 2E:* Chapter 2, page 35.

Question #41

Apparatus fuel tanks should be refilled according to department policy. When no refill policy exists, apparatus fuel tanks should be refilled when the fuel level reaches the three-quarter mark. That is, when the gauge drops from full to three-quarter full. NFPA 1002 4.2.1. *IFPO, 2E:* Chapter 2, page 35.

Question #42

It is not advisable to turn the steering wheel so that the wheels turn. Rather, the steering wheel should be turned until just before the wheels turn. The distance should not exceed 10 degrees in either direction. For a 20-inch steering wheel, that would be approximately 2 inches of movement. NFPA 1002, 4.2.1. *IFPO, 2E:* Chapter 2, pages 35 – 36.

Question #43

Although no specific inspection sequence is mandated, it is a good idea to review previous vehicle inspection reports before starting a preventive maintenance inspection. NFPA 1002 4.2.1. *IFPO, 2E:* Chapter 2, page 28.

Question #44

A common sequence for conducting a vehicle inspection is to start with a walk around the outside of the vehicle followed by an inspection of the engine compartment. Next, the inside of the cab is inspected, and the engine is started. With the engine running, the final step is to complete the inspection of the pump and related components. NFPA 1002 4.2.1, 5.1.1. *IFPO, 2E:* Chapter 2, pages 27 – 28.

Question #45

According to annual reports published in the *NFPA Journal*, over 11,000 emergency-vehicle accidents occurred each year from 1990 to 2000. NFPA 1002 4.3.1. *IFPO, 2E:* Chapter 3, pages 42 – 43.

Question #46
Most drivers possess a license as required by NFPA 1500 and most state laws. Having driver certification is not a factor, per se, but can certainly help in understanding, avoiding, and/or adequately compensating for the common factors. NFPA 1002 4.3.2, 4.3.3, 4.3.5, and 4.3.6. *IFPO, 2E:* Chapter 3, pages 43 – 45.

Question #47
Chapter 6 of NFPA 1500 incudes specific requirements for fire apparatus, equipment, and driver/operators. State laws set the overall rules and standards for the state. Local ordinances can be more restrictive, and provide specific detail for operating within the city/county. The Federal Fire Act does not address safe emergency-vehicle operations. NFPA 1002 4.3.1, 4.3.6. *IFPO, 2E:* Chapter 3, pages 44 – 50.

Question #48
Total stopping distance is measured from the time a hazard is detected until the vehicle comes to a complete stop. Total stopping distance consists of:

- *Perception distance* (distance apparatus travels from the time the hazard is seen until the brain recognizes it as a hazard).

- *Reaction distance* (distance apparatus travels from the time the brain sends the message to depress the brakes until the brakes are depressed).

- *Braking distance* (distance of travel from the time the brake is depressed until the vehicle comes to a complete stop).

NFPA 1002 4.3. *IFPO, 2E:* Chapter 3, page 55 – 65.

Question #49
Traction is the friction between the tires and the road surface and is essential for steering.
NFPA 1002 4.3. *IFPO, 2E:* Chapter 3, pages 55 – 56.

Question #50
One method to help ensure the safe control of a stationary apparatus is to properly align the front wheels as follows:

- When parked next to curb, rotate the front wheels so that they point toward the curb.

- When no curb is present, the front wheels should be positioned to roll the apparatus away from the road.

NFPA 1002 4.3. *IFPO, 2E:* Chapter 3, pages 58 – 59.

Question #51
Centrifugal force is the tendency of an object to move outward from the center. Loss of traction occurs when friction between the tires and the road surface is lost. NFPA 1002 4.3. *IFPO, 2E:* Chapter 3, pages 56 – 57.

Question #52
Although fire department vehicle accidents occur for a variety of reasons, several common factors that appear in the vast majority of accidents include:

- Not following laws and standards related to emergency response

- Not being fully aware of both driver and apparatus limitations

- Lack of appreciation for driving conditions such as weather and traffic

Not being certified to NFPA 1002 has not been linked as a common factor associated with emergency vehicle accidents. NFPA 1002 4.3. *IFPO, 2E:* Chapter 3, pages 43 – 44.

Question #53

The only action that will most likely help regain traction and control is removing the foot from the accelerator, which allows the engine to slow the vehicle. NFPA 1002 4.3.1. *IFPO, 2E:* Chapter 3, page 56.

Question #54

Total stopping distance is measured from the time a hazard is detected until the vehicle comes to a complete stop. Total stopping distance consists of:

- *Perception distance* (distance apparatus travels from the time the hazard is seen until the brain recognizes it as a hazard).

- *Reaction distance* (distance apparatus travels from the time the brain sends the message to depress the brakes until the brakes are depressed).

- *Braking distance* (distance of travel from the time the brake is depressed until the vehicle comes to a complete stop).

NFPA 1002 4.3. *IFPO, 2E:* Chapter 3, page 55 – 65.

Question #55

The impact of an accident involving responding apparatus can be:

- Delayed assistance to those who summoned help

- Additional units dispatched for the original call as well as for the new accident

- Fire department members and civilians could be seriously or fatally injured

- Fire department and civilian vehicles and property could sustain extensive damage

- The city, department, and driver could face civil and/or criminal proceedings

- Image presented by the accident will last a long time in the mind of the public

NFPA 1002 4.3. *IFPO, 2E:* Chapter 3, page 43.

Question #56

Most state laws require that emergency-vehicle drivers obey the same laws as other vehicle operators unless specifically exempt from doing so. State laws typically define several conditions that must exist for exemptions to be extended and include:

- Only authorized emergency vehicles are covered.

- The exemptions are only provided when responding to an emergency.

- Audible and visual warning devices must be operating when taking advantage of the exemption.

NFPA 1002 4.3. *IFPO, 2E:* Chapter 3, pages 44 – 45.

Question #57

NFPA 1500 requires that emergency-vehicle drivers come to a complete stop when any intersection hazard is present. Specifically, the vehicle must come to a complete stop when any of the following exists:

- As directed by a law enforcement officer

- At traffic red lights or stop signs

- At negative right-of-way and blind intersections

- When all lanes of traffic in an intersection cannot be accounted for

- When stopped school bus with flashing warning lights is encountered

- At unguarded railroad guard crossing (also for nonemergency)

- When other intersection hazards are present

NFPA 1002 4.3. *IFPO, 2E:* Chapter 3, page 48.

Question #58
The water in the tanker weighs 16,700 lbs (8.35 lbs per gallon × 2,000 gallons), which is equal to 8.35 tons (16,700 lbs ÷ 2,000 lbs/ton). It would be unsafe to cross the bridge because the water alone exceeds the authorized limit. The weight of the vehicle would add even more risk to crossing the bridge. NFPA 1002 4.3. *IFPO, 2E:* Chapter 3, pages 52 – 53.

Question #59
Hydrodynamics refers to a branch of hydraulics that deals with liquids in motion. The other two answer selections are types of pumps (remove the term hydro) that are based on hydrostatic and hydrodynamic principles. NFPA 1002 5.2.1, 5.2.2. *IFPO, 2E:* Chapter 4, page 76.

Question #60
The split-shaft transfers power from the rear axle to the pump. NFPA 1002 5.2.1, 5.2.2. *IFPO, 2E:* Chapter 4, pages 92 – 93.

Question #61
Priming is the process of getting water into the pump through the use of atmospheric pressure, not a suctioning process. NFPA 1002 5.2.1, 5.2.2. *IFPO, 2E:* Chapter 4, pages 75 – 76.

Question #62
The volume mode of a two-stage centrifugal pump moves water from a common source to both impellers at the same time, and then to a common discharge. The volume of both impellers is added together for total flow, while the pressure remains constant. NFPA 1002 5.2.1, 5.2.2. *IFPO, 2E:* Chapter 4, pages 89 – 90.

Question #63
Basic pump terms include:

- *Flow* refers to the rate and quantity of water delivered by the pump and is expressed in gallons per minute (gpm).
- *Pressure* refers to the amount of force generated by the pump or the resistance encountered on the discharge side of the pump and is expressed in pounds per square inch (psi).
- *Speed* refers to the rate at which the pump is operating and is typically expressed in revolutions per minute (rpm).
- *Slippage* is the term used to describe the leaking of water between the surfaces of the internal moving parts of a pump.

NFPA 1002 5.2. *IFPO, 2E:* Chapter 4, page 75.

Question #64
This is an example of a two-stage centrifugal pump. Note the two impellers on the same shaft. NFPA 1002 5.2.1, 5.2.2. *IFPO, 2E:* Chapter 4, pages 87 – 89.

Question #65
Transfer valves are not closed or open. Rather, they are used to redirect water to either a series (pressure) or parallel (volume) mode. Operating at flows great than 50% of the pump's rated capacity indicates larger water flows. In such cases, the transfer valve should be in volume mode. NFPA 1002 5.2.1, 5.2.2. *IFPO, 2E:* Chapter 4, pages 89 – 90.

Question #66
In volume mode, each individual impeller will add the flow it generates to the total discharge, with the pressure remaining constant among the impellers.
In pressure mode, each subsequent impeller pumps the same flow from the previous impeller while adding the pressure it generates. NFPA 1002 5.2.1, 5.2.2. *IFPO, 2E:* Chapter 4, pages 89 – 90.

Question #67

Pumps connected to a split-shaft transmission are usually located midship or aft.
Pumps connected directly to the crankshaft are usually located at the front of the engine.
Pumps connected to a PTO are usually mounted at the front or midship. Pumps are usually smaller when powered by a PTO. NFPA 1002 5.2.1. *IFPO, 2E:* Chapter 4, pages 92 – 93.

Question #68

According to NFPA 1901, a pump must have a rated capacity as follows:

- 100% of its rated capacity at 150 psi

- 70% of its rated capacity at 200 psi

- 50% of its rated capacity at 250 psi

NFPA 1002 5.2.1, 5.2.2. *IFPO, 2E:* Chapter 4, page 91.

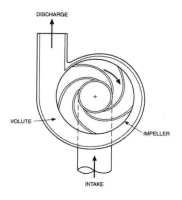

Question #69

NFPA 1002 5.2.1, 5.2.2. *IFPO, 2E:* Chapter 4, pages 86 – 89.

Question #70

A pump must deliver 100% of its rated capacity at 150 psi. Note that at 150 psi, the pump can deliver 1,250 gpm. NFPA 1002 5.2.1, 5.2.2. *IFPO, 2E:* Chapter 4, page 91.

Question #71

In volume mode, water enters both impellers from a common intake and leaves from a common discharge. NFPA 1002 5.2.1, 5.2.2, 5.2.4. *IFPO, 2E:* Chapter 4, pages 90 – 91.

Question #72

Each revolution of a positive-displacement pump will yield (displace) a specific quantity of water. Consequently, the pressure on the intake side of the pump is irrelevant. No matter what the intake pressure is, the pump will only discharge its specific quantity of water per revolution. NFPA 1002 5.2.1. *IFPO, 2E:* Chapter 4, page 78.

Question #73

Total discharge rate will increase when the pump speed is increased. However, the quantity of water discharged per revolution will never change. In other words, the quantity per revolution stays the same regardless of how slow or fast the pump is operating. NFPA 1002 5.2.1, 5.2.2. *IFPO, 2E:* Chapter 4, pages 78 – 79.

Question #74

The two largest gauges on a pump panel are typically the main pump gauges called pump intake gauge and pump discharge gauge and are most often of Bourdon tube construction. NFPA 1002 5.2.1, 5.2.2. *IFPO, 2E:* Chapter 5, pages 110 – 112.

Question #75
Flow meters measure gpm (quantity and rate) and, because the flow is known, normal friction loss calculations are not required. Most flow meters use paddle wheels to measure the flow. NFPA 1002 5.2.1, 5.2.2. *IFPO, 2E:* Chapter 5, pages 113 – 114.

Question #76
Pneumatically activated control valve operation is not considered a manual operation. NFPA 1002 5.2.1, 5.2.2. *IFPO, 2E:* Chapter 5, pages 114 – 115.

Question #77
A pressure governor controls pressure by increasing or decreasing engine speed to maintain the desired pressure. NFPA 1002 5.2.1, 5.2.2, 5.2.4. *IFPO, 2E:* Chapter 5, pages 122 – 125.

Question #78
The radiator fill system pumps water directly into the radiator. NFPA 1002 5.2.1, 5.2.2, 5.2.4. *IFPO, 2E:* Chapter 5, pages 125 – 126.

Question #79
Foam systems found on pumping apparatus include:

- Pre-mixed
- In-line eductor
- Around-the-pump proportioning
- Balanced pressure
- Direct injection/compressed-air

NFPA 1002 5.2.3. *IFPO, 2E:* Chapter 5, pages 126 – 130.

Question #80
Because priming pumps rely on close fitting parts, oil helps to both lubricate the moving parts as well as to help provide a tighter seal. NFPA 1002 5.2.1, 5.2.2, 5.2.4. *IFPO, 2E:* Chapter 5, pages 120 – 122.

Question #81

- *Tank-to-pump* control valves allow water to flow from the on-board water supply to the intake side of the pump.
- *Pump-to-tank* (tank fill) control valves allow water to flow from the discharge side of the pump to the tank.
- *Transfer valves* found on multistage pumps redirect water from the pump between the pressure mode and the volume mode.

NFPA 1002 5.2.1, 5.2.2. *IFPO, 2E:* Chapter 5, page 119.

Question #82
Because of new electronic and digital controls, pump panels can be in almost any location. NFPA 1002 5.2.1, 5.2.2. *IFPO, 2E:* Chapter 5, pages 105 – 110.

Question #83
The most common valve types on pumping apparatus include ball, butterfly, piston, and gated. Hammer, quarter, and full are not a type of valve used on pumping apparatus. NFPA 1002 5.2.1, 5.2.2. *IFPO, 2E:* Chapter 5, page 114.

Question #84
NFPA 1901 sets minimum gauge sizes, requires intake relief valves, and suggests a standard color code to match discharges with their gauges. NFPA 1002 5.2.1, 5.2.2. *IFPO, 2E:* Chapter 5, page 102.

Question #85

Several way to determine when a pump is primed include the following indicators:

- A positive reading on the pressure gauge
- Priming motor sounds as it is slowing
- Main pump will sound as if it is under load
- Oil/water discharge under the vehicle (not air/oil)

NFPA 1002 5.2.1, 5.2.2. *IFPO, 2E:* Chapter 5, page 121.

Question #86

This control device operates and sets the pressure-relief device. NFPA 1002 5.2.1, 5.2.2. *IFPO, 2E:* Chapter 5, pages 122 – 123.

Question #87

The two largest gauges on a pump panel are usually the master/main intake (*left*) and discharge (*right*) gauges. The smaller gauges are usually individual discharge gauges. NFPA 1002 5.2.1, 5.2.2. *IFPO, 2E:* Chapter 5, page 101.

Question #88

As a rule of thumb, a pump will have one 2 ½-inch discharge for each 250 gpm of rated capacity. A 1,500-gpm pumper will usually have six 2 ½-inch discharges, 6 × 250 = 1,500. NFPA 1002 5.2.1, 5.2.2. *IFPO, 2E:* Chapter 5, pages 118 – 119.

Question #89

- *Tank-to-pump* control valves allow water to flow from the on-board water supply to the intake side of the pump.
- *Pump-to-tank* (tank fill) control valves allow water to flow from the discharge side of the pump to the tank.
- *Transfer valves* found on multistage pumps redirect water from the pump between the pressure mode and the volume mode.

NFPA 1002 5.2.1, 5.2.2. *IFPO, 2E:* Chapter 5, page 119.

Question #90

Flow meters often use paddle wheels to measure the flow within a line, while pressure gauges use either a Bourdon tube or, in this case, an electronic pressure sensor. Both pressure and flow are provided to a single gauge. NFPA 1002 5.2.1, 5.2.2. *IFPO, 2E:* Chapter 5, pages 113 – 114.

Question #91

- A *pre-mixed system* consists of a tank in which foam concentrate and water are added in appropriate proportions. Often, the foam is simply added to the on-board water tank.

- An *in-line eductor system* utilizes eductors to add foam to water in appropriate proportions. This system can have either internal or external controls, and requires an eductor and a foam tank (or supply). Internal systems have a foam concentrate valve and meeting control. Eductors are usually installed between the pump discharge and the discharge outlet.

- The *around-the-pump proportioning system* uses an eductor (located between the intake and discharge sides of the pump) to mix foam with water. Water from the discharge side of the pump picks up foam and delivers it to the intake side of the pump, where it can then be discharged to all outlets.

- A *balanced-pressure system* mixes foam with water by means of pressure. One system uses pressure to force foam from a bladder. The second system uses a separate foam pump (two types are by-pass and demand).

- A *direct injection/compressed-air foam system* (CAFS) consists of a separate foam pump and foam tank. Direct injection moves foam from the tank directly into the discharge lines. CAFS adds compressed air to create a lightweight foam.

NFPA 1002 5.2.3. *IFPO, 2E:* Chapter 5, pages 126 – 130.

Question #92

Total stopping distance is measured from the time a hazard is detected until the vehicle comes to a complete stop and includes:

- Perception distance – 35 feet
- Reaction distance – 30 feet
- Braking distance – 125 feet

NFPA 1002 4.3.2, 4.3.3, 4.3.4, 4.3.5. *IFPO, 2E:* Chapter 3, page 55.

Question #93

Pump operations can be defined as the systematic movement of water from a supply source through a pump to a discharge point. NFPA 1002 5.2. *IFPO, 2E:* Chapter 1, page 8.

Question #94

Positive-displacement pumps, such as rotary vane or rotary gear pumps, operate based on hydrostatic principles (liquids at rest).
Dynamic, or centrifugal, pumps operate based on hydrodynamic principles (liquids in motion).
NFPA 1002 5.2. *IFPO, 2E:* Chapter 4, page 76.

Question #95

The only exceptions to not wearing a seat belt in a fire department vehicle are:

- Loading hose
- Tiller driver training
- Patient treatment

NFPA 1002 4.3. *IFPO, 2E:* Chapter 3, page 49.

Question #96

The water flow will be the same at any point in the line. If each discharge is flowing 420 gpm, then the total flow through the monitor is 840 gpm. NFPA 1002 5.2. *IFPO, 2E:* Chapter 5, pages 113 – 114.

Question #97

The most correct answer is to maintain traction. Speed, centrifugal force, road conditions, and weight shifts can all affect the ability to maintain traction. When traction is lost, the apparatus will be out of control. NFPA 1002 4.3. *IFPO, 2E:* Chapter 3, page 56.

Question #98
- *NFPA 1500* is the standard for occupational safety and health programs.
- *NFPA 1962* is the standard for the inspection, care, and use of fire hose, couplings, and nozzles and the service testing of hose.
- *NFPA 1965* is the standard for fire hose appliances.

NFPA 1002 5.2.1, 5.2.2, 5.2.4. *IFPO, 2E:* Chapter 6, pages 138 – 139.

Question #99
Supply lines should not be operated at pressures exceeding 185 psi, and attack lines should have a highest normal operating pressure of 275 psi. NFPA 1002 5.2.1, 5.2.2, 5.2.4. *IFPO, 2E:* Chapter 6, page 140.

Question #100
Although NFPA 1961 no longer specifies the required length of hose, a section of hose is commonly 50 feet. NFPA 1002 5.2.1, 5.2.2, 5.2.4. *IFPO, 2E:* Chapter 6, page 140.

Question #101
NFPA 1963 refers to the threaded coupling construction as American National Fire Hose Connection Screw Thread, which is abbreviated with the thread symbol NH. Storz is not a threaded coupling, and NTS is not a type of thread used in the fire service. Sometimes threads are referred to as National Standard Thread (NST). NFPA 1002 5.2.1, 5.2.2, 5.2.4. *IFPO, 2E:* Chapter 6, pages 141 – 152.

Question #102
A wye has a single female inlet connection and two or more male outlet connections.
A siamese has two or more female inlet connections and one male outlet connection.
Although the wye in this picture can be used as a siamese, it is still considered a wye and would require double male and female adapters. NFPA 1002 5.2.1, 5.2.2, 5.2.4. *IFPO, 2E:* Chapter 6, pages 148 – 149.

Question #103
Double female adapters have a female coupling on both sides. NFPA 1002 5.2.1, 5.2.2, 5.2.4. *IFPO, 2E:* Chapter 6, page 150.

Question #104
- 50 psi is the operating pressure for smooth-bore hand-lines.
- 80 psi is the operating pressure for smooth-bore master-streams.
- 100 psi is the operating pressure for combination nozzles.
- 125 psi is not a normal operating pressure for nozzles used in the fire service.

NFPA 1002 5.2.1, 5.2.2, 5.2.4. *IFPO, 2E:* Chapter 6, pages 153 – 157.

Question #105
- The *fixed-flow* or *constant-flow nozzle* provides a constant flow regardless of stream pattern.
- *Selectable flow nozzles* permit the operator to adjust flow at the nozzle.
- The *automatic nozzle* maintains a constant nozzle pressure over a wide variety of flows.
- *Smooth-bore nozzles* are not a type of combination nozzle.

NFPA 1002 5.2.1, 5.2.2, 5.2.3. *IFPO, 2E:* Chapter 6, pages 155 – 157.

Question #106
The eductor uses the Venturi principle to draw foam into the water stream. NFPA 1002 5.2.3. *IFPO, 2E:* Chapter 6, page 151.

Question #107
- *Nozzle pressure* is the designed operating pressure for a particular nozzle.
- *Nozzle flow* is the amount of water flowing from a nozzle.
- *Nozzle reach* is the distance water travels after leaving a nozzle.
- *Nozzle reaction* is the tendency of a nozzle to move in the opposite direction of water flow.

NFPA 1002 5.2.1. *IFPO, 2E:* Chapter 6, pages 153 – 154.

Question #108
- *Nozzle pressure* is the designed operating pressure for a particular nozzle.
- *Nozzle flow* is the amount of water flowing from a nozzle.
- *Nozzle reach* is the distance water travels after leaving a nozzle.
- *Nozzle reaction* is the tendency of a nozzle to move in the opposite direction of water flow.

NFPA 1002 5.2.1. *IFPO, 2E:* Chapter 6, pages 153 – 154.

Question #109
Intake (supply) hose must have a minimal rating of 185 psi. Each of the hose listed meets this requirement. NFPA 1002 5.2.1, 5.2.2, 5.2.4. *IFPO, 2E:* Chapter 6, pages 139 – 140.

Question #110
The two basic parts of an eductor are the metering valve (controls the percentage of foam drawn into the eductor) and pickup hose (noncollapsable tube used to move the foam to the eductor).
NFPA 1002 5.2.3. *IFPO, 2E:* Chapter 6, pages 151 – 152.

Question #111
When correct nozzle pressures are not provided, the following can occur:
Insufficient nozzle pressure:

- Less flow
- Reduced reach
- Poor pattern development

Excessive nozzle pressure:

- Poor pattern development
- Excessive nozzle reaction

NFPA 1002 5.2.1. *IFPO, 2E:* Chapter 6, page 156.

Question #112
NFPA 1962 Standard for the Inspection, Care, and Use of Fire Hose, Couplings, and Nozzles and the Service Testing of Hose. NFPA 1002 5.1.1. *IFPO, 2E:* Chapter 6, page 146.

Question #113
NFPA 1965 requires that appliances with lever-operated handles must indicate a closed position when the handle is perpendicular to the hose line. NFPA 1002 5.2.1, 5.2.2, 5.2.4. *IFPO, 2E:* Chapter 6, page 148.

Question #114

Basic steps for testing hose include:

1. Connect water supply and test hose sections.

2. Open discharge lines for test hose sections.

3. Slowly increase discharge pressure to 45 psi.

4. Remove all air from within the hose.

5. Mark hose near coupling to determine slippage.

6. Increase pressure slowly (NFPA 1962 suggest no faster than 15 psi per second) to test pressure.

7. Maintain pressure for 3 minutes, periodically checking for leaks.

8. Record results.

NFPA 1002 5.1. *IFPO, 2E:* Chapter 6, page 147.

Question #115

NFPA specifies minimum tank capacity for apparatus as follows:

- 200 gallons – initial attack
- 300 gallons – pumper
- 1,000 gallons – mobile water apparatus/tanker

NFPA 1002 5.2.1. *IFPO, 2E:* Chapter 7, page 167.

Question #116

Hydrants located on a main that is suppled from one direction (dead-end main) are called dead-end hydrants. NFPA 1002 5.2.1, 5.2.2. *IFPO, 2E:* Chapter 7, page 173.

Question #117

This hydrant is dry and requires the stem nut located on the bonnet to be turned before water can enter the hydrant barrel. Wet-barrel hydrants would have stem nuts to control each discharge outlet. Static-source hydrants typically do not have control valves, and dead-end hydrants could be either wet- or dry-barrel. NFPA 1002 5.2.1. *IFPO, 2E:* Chapter 7, pages 175 – 176.

Question #118

A straight lay would cause the pump to be located near the incident. Each of the other answer choices would cause the pump to be located near the hydrant. NPFA 1002 5.2.1, 5.2.2, 5.2.4. *IFPO, 2E:* Chapter 7, pages 182 – 183.

Question #119

Intake pressure for pumps in a relay should not fall below 20 psi to reduce the chances of the pump cavitating. NFPA 1002 5.2.2. *IFPO, 2E:* Chapter 7, pages 189 – 194.

Question #120

NFPA specifies minimum tank capacity for apparatus as follows:

- 200 gallons – initial attack
- 300 gallons – pumper
- 1,000 gallons – mobile water apparatus/tanker

NFPA 1002 5.2.1. *IFPO, 2E:* Chapter 7, page 167.

Question #121

Each of the answer selections is a piece of equipment that is used in tanker shuttles. NFPA 1002 5.2.1. *IFPO, 2E:* Chapter 7, pages 195 – 201.

Question #122

The flow capacity of a shuttle is limited by the volume of water being delivered and the time it takes to complete a shuttle cycle. NFPA 1002 5.2.1. *IFPO, 2E:* Chapter 7, pages 200 – 201.

Question #123

A pumper with 1,000-gallon tank flowing 250 gpm will last about 4 minutes.

250 gallons per minute) 1,000 gallons = 4 minutes. NFPA 1002 5.2.1. *IFPO, 2E:* Chapter 7, page 167.

Question #124

Hydrant flow testing equipment include:

- Pressure gauge mounted on an outlet cap (calibrated within the past 12 months)
- Pitot gauge for each hydrant
- Hydrant diffuser
- Hydrant wrenches
- Portable radios

A fire department pumper is not required when conducting hydrant flow testing. NFPA 1002 5.2.1. *IFPO, 2E:* Chapter 7, pages 180 – 181.

Question #125

Shuttle cycle time is the total time it takes to dump water and return with another load. NFPA 1002 5.2.1, 5.2.2. *IFPO, 2E:* Chapter 7, pages 200 – 201.

Question #126

Water will rise approximately 2.3 feet/psi for each 1 psi of pressure.

So, 8.5 psi × 2.3 feet/psi = 19.5 feet. NFPA 1002 5.2.1. *IFPO, 2E:* Chapter 7, pages 185 – 186.

Question #127

Basic steps for conducting hydrant flow include:

1. Take static pressure reading at test hydrant (make sure to open hydrant fully and remove air).
2. Open flow hydrants one at a time until a 25% drop in residual pressure is achieved.
3. Continue to flow to clear debris and foreign substances.
4. Take reading at the same time; residual reading at test hydrant and flow readings at each flow hydrant, record results.
5. Slowly shut down hydrants one at a time.
6. Record exact interior size, in inches, of each outlet flowed.

NFPA 1002 5.2.1. *IFPO, 2E:* Chapter 7, pages 180 – 181.

Question #128

The distance between pumpers in a relay operation can be determined as follows:

(PDP – 20) × 100/FL
(185 – 20) × 100 / 5
165 × 20
3,300 feet

NFPA 1002 5.2.3. *IFPO, 2E:* Chapter 7, page 194.

Question #129

Divide tank volume by the shuttle cycle time.
1,500 ÷ 10 = 150 gpm.
NFPA 1002 5.2.1. *IFPO, 2E:* Chapter 7, page 201.

Question #130

The pressure is too high because the maximum operating pressure of supply hose is 185 psi. NFPA 1002 5.2.2. *IFPO, 2E:* Chapter 7, page 192.

Question #131

The vacuum test is use to determine the ability of the pump to maintain a vacuum. The priming device test ensures the primer can develop sufficient vacuum to draft. The pressure control test is conducted to ensure the pressure control can adequately maintain safe discharge pressures. The draft test is a non-specific test. NFPA 1002 5.1.1. *IFPO, 2E:* Chapter 8, pages 226 – 231.

Question #132

Each of the tests or checks listed is often required by manufacturers on a weekly and/or monthly basis. The engine speed check is usually performed on an annual basis. NFPA 1002 5.1.1. *IFPO, 2E:* Chapter 8, pages 226 – 232.

Question #133

Controls should never be opened, closed, or turned swiftly or abruptly. In doing so, mechanical damage may occur to the device and/or may cause a water hammer resulting in additional damage and/or injury. NFPA 1002 5.2.1, 5.2.2. *IFPO, 2E:* Chapter 8, page 208.

Question #134

The 40-minute annual pump test consist of:

- 20 minutes at 100% capacity at 150 psi
- 10 minutes at 70% capacity at 200 psi
- 10 minutes at 50% capacity at 250 psi

Pumps with a rate capacity of 750 gpm or greater must also undergo a 5-minute overload test at 165 psi. NFPA 1002 5.1.1. *IFPO, 2E:* Chapter 8, pages 228 – 229.

Question #135

Although all methods can provide power to a pump, most main pumps are powered by the drive engine through either a split-shaft or directly from the crankshaft. NFPA 1002 5.2.1, 5.2.2, 5.2.4. *IFPO, 2E:* Chapter 8, pages 210 – 211.

Question #136

The pump is usually engaged before securing a water supply.

Step 1 Position apparatus, set parking brake, and let engine return to idle.
Step 2 Engage the pump.
Step 3 Provide water to intake side of pump (onboard, hydrant, draft).
Step 4 Set transfer valve (if so equipped).
Step 5 Open discharge lines.
Step 6 Throttle to desired pressure.
Step 7 Set the pressure regulating device.
Step 8 Maintain appropriate flows and pressures.

NFPA 1002 5.2.1, 5.2.2, 5.2.4. *IFPO, 2E:* Chapter 8, page 207.

Question #137

Transfer valve operation rule of thumb:

- Use *volume mode* when flows are greater than 50% of a pump's rated capacity, and pressures are less than 150 psi.
- Use *pressure mode* when flows are less than 50% of a pump's rated capacity, and pressures are greater than 150 psi.

NFPA 1002 5.2.1, 5.2.2, 5.2.4. *IFPO, 2E:* Chapter 8, pages 214 – 215.

Question #138

The three methods used to transfer power to the pump are:

- PTO
- Front crankshaft
- Split-shaft

NFPA 1002 5.2.1, 5.2.2, 5.2.4. *IFPO, 2E:* Chapter 8, pages 210 – 212.

Question #139

The basic steps for pump operation include:

Step 1 Position apparatus, set parking brake, and let engine return to idle.
Step 2 Engage the pump.
Step 3 Provide water to intake side of pump (on-board, hydrant, draft).
Step 4 Set transfer valve (if so equipped).
Step 5 Open discharge lines.
Step 6 Throttle to desired pressure.
Step 7 Set the pressure-regulating device.
Step 8 Maintain appropriate flows and pressures.

NFPA 1002 5.2.1, 5.2.2, 5.2.4. *IFPO, 2E:* Chapter 8, page 207.

Question #140

Important considerations for positioning pumping apparatus at incidents include:

- If no fire or smoke is visible, park near main entrance.
- Be sure to follow department SOPs.
- Consider tactical priorities for the incident.
- Consider surroundings such as heat from the fire, collapse, overhead lines, escape routes, wind.

NFPA 1002 5.2.1, 5.2.2. *IFPO, 2E:* Chapter 8, pages 280 – 209.

Question #141

Basic steps for pump engagement powered through a PTO include:

- Bring apparatus to complete stop, set parking brake, let engine return to idle speed.
- Disengage the clutch (push in the clutch pedal).
- Place transmission in neutral.
- Operate the PTO lever.
- For mobile pumping, place transmission in the proper gear.
- For stationary pumping, place transmission in neutral.
- Engage the clutch slowly.

NFPA 1002 5.2.1, 5.2.2, 5.2.4. *IFPO, 2E:* Chapter 8, pages 211 – 212.

Question #142

Basic steps to engage a pump utilizing a split-shaft arrangement are as follows:

- Bring apparatus to complete stop.
- Place transmission in neutral.
- Apply parking brake.
- Operate pump shift switch from road to pump position.
- Ensure OK-to-pump light comes on.
- Shift transmission to pumping gear (usually highest gear).

NFPA 1002 5.2.1, 5.2.2, 5.2.4. *IFPO, 2E:* Chapter 8, pages 211 – 212.

Question #143
Because the flow is greater than 50% of the pump's rated capacity and the expected pump discharge pressure is less than 150 psi, the transfer valve should be in the volume mode. The transfer valve operation rule of thumb is as follows:
Volume mode:

- Flows greater than 50% of a pump's rated capacity
- Pressures less than 150 psi

Pressure mode:

- Flows less than 50% of a pump's rated capacity
- Pressures greater than 150 psi

NFPA 1002 5.2.1, 5.2.2, 5.2.4. *IFPO, 2E:* Chapter 8, pages 214 – 215.

Question #144
A corresponding increase in discharge pressure should be noted when the throttle is increased (engine speed increased). If not, one of the following common causes could be occurring:

- The pump may not be in gear.
- The pump may not be primed.
- The supply line could be closed or insufficient.

NFPA 1002 5.2.1, 5.2.2, 5.2.4. *IFPO, 2E:* Chapter 8, page 216.

Question #145
If the throttle is increased past the point of a corresponding increase in discharge pressure, the following might occur:

- Pump cavitation
- Loss of prime
- Intake line collapse
- Damage to municipal water mains

NFPA 1002 5.2.1, 5.2.2, 5.2.4. *IFPO, 2E:* Chapter 8, page 216.

Question #146
The on-board water supply procedures are as follows:

- Position apparatus, set parking brake.
- Engage pump.
- Set transfer valve.
- Open "tank-to-pump."
- Open discharge control valves.
- Increase throttle.
- Set pressure-regulating device.
- Plan for more water.

NFPA 1002 5.2.1, 5.2.2, 5.2.4. *IFPO, 2E:* Chapter 8, pages 220 – 221.

Question #147
Each of the answer selections provided is an important consideration that should be evaluated when selecting a location to position fire apparatus. NFPA 1002 5.2.1, 5.2.2. *IFPO, 2E:* Chapter 8, pages 208 – 209.

Question #148

Cavitation is caused by insufficient intake flow to match the discharge flow and is often characterized as a pump running away from the water supply. NFPA 1002 5.1.1. *IFPO, 2E:* Chapter 9, pages 245 – 246.

Question #149

- *Dry-pipe systems* maintain air or compressed gas under pressure within the system. When a head fuses, water enters the system and discharges through any fused heads.

- *Pre-action systems* are similar to dry-pipe systems in that air or compressed gas is maintained in the system. However, an automatic detection system (smoke, heat, flame) or manual system (pull box) must operate to allow water to enter the system. At this point, it is similar to a wet-pipe system; when a sprinkler head fuses, water discharges.

- *Deluge systems* maintain all sprinkler heads as open. When a detection system operates, water enters the system and is discharged through all the open heads.

- A *wet-pipe system* maintains water in the system at all times. When a sprinkler head is fused, water immediately discharges.

NFPA 1002 5.2.4. *IFPO, 2E:* Chapter 9, pages 148 – 253.

Question #150

- NFPA 13 deal with installation of sprinkler systems.

- NFPA 13D focuses on the installation of sprinkler systems in one- and two-family dwellings.

- NFPA 14 deals with standpipe and hose systems.

NFPA 1002 5.2.4. *IFPO, 2E:* Chapter 9, pages 248 – 253.

Question #151

In most cases, when supporting sprinkler and standpipe systems, volume is required over pressure, and the transfer valve should be placed in the volume mode. NFPA 1002 5.2.4. *IFPO, 2E:* Chapter 9, pages 251 – 252.

Question #152

Each of the answer selections can cause a water hammer except operating the priming device, which should not generate any significant increase or decrease in pressure. NFPA 1002 5.1.1. *IFPO, 2E:* Chapter 9, pages 244 – 245.

Question #153

- *Class 1* standpipes provide 2 ½-inch connections for trained firefighters/fire brigades and have an initial flow rate of 500 gpm.

- *Class 2* standpipes provide 1 ½-inch connections for initial attack; they have a minimum flow rate of 100 gpm.

- *Class 3* standpipes provide 1 ½-inch and 2 ½-inch connections for trained firefighters/fire brigades; they have an initial flow rate of 500 gpm.

NFPA 1002 5.2.4. *IFPO, 2E:* Chapter 9, page 253.

Question #154

Cavitation is the process that explains the formation and collapse of vapor pockets when certain conditions exists during pumping operations. NFPA 1002 5.1.1. *IFPO, 2E:* Chapter 9, pages 245 – 247.

Question #155

Cavitation can be stopped by reducing pump pressure through pump speed, reduced discharge flow, or through increased supply flow. Adding one or more discharge lines would only aggravate a deteriorating system and most likely would cause additional damage. NFPA 1002 5.1.1. *IFPO, 2E:* Chapter 9, pages 245 – 247.

Question #156
Cold-weather operations can be grueling. The pump operator must be diligent to stop all leaks, keep water flowing, and drain everything when the pump operation is completed if freezing might occur. NFPA 1002 5.2.1, 5.2.2, 5.2.4. *IFPO, 2E:* Chapter 9, page 247.

Question #157
A fire department connection marked "STANDPIPE" means the system is a standpipe system, not a sprinkler system. NFPA 1002 5.2.4. *IFPO, 2E:* Chapter 9, page 253.

Question #158
- *Class 1* standpipes provide 2 ½-inch connections for trained firefighters/fire brigades and have an initial flow rate of 500 gpm.
- *Class 2* standpipes provide 1 ½-inch connections for initial attack; they have a minimum flow rate of 100 gpm.
- *Class 3* standpipes provide 1 ½-inch and 2 ½-inch connections for trained firefighters/fire brigades; they have an initial flow rate of 500 gpm.

NFPA 1002 5.2.4. *IFPO, 2E:* Chapter 9, page 253.

Question #159
Pump operators must be able to perform basic troubleshooting for pump problems. Basic troubleshooting considerations include:

- Problems are usually either procedural or mechanical.
- When proper procedures are followed, the problem is most likely mechanical.
- Best method to troubleshoot is to follow the flow of water from the intake to the discharge while attempting to determine the problem.
- Use the manufacturers' troubleshooting guides when available.

NFPA 1002 5.2. *IFPO, 2E:* Chapter 9, page 254.

Question #160
- *Dual pumping* is when one hydrant supplies two pumpers. The second pumper receives water from the intake of the first pumper. In other words, the excess water provided to the first pumper is diverted to the second pumper.
- *Tandem pumping* is when one hydrant supplies two pumpers and is similar to a relay operation. The first pump discharges all its water to the intake of the second pumper, as in a relay. The only main difference is that a relay is used to move water over extended distances, whereas tandem pumping is used when higher pressures are required than a single pumper can provide.

NFPA 1002 5.2.1. *IFPO, 2E:* Chapter 8, pages 224 – 225.

Question #161
- *Evaporation* is the physical change of state from a liquid to a vapor.
- *Vapor pressure (VP)* is the pressure exerted on the atmosphere by molecules as they evaporate from the surface of the liquid.
- *Boiling point (BP)* is the temperature at which the vapor pressure of a liquid equals the surrounding pressure.

NFPA 1002 5.2.1, 5.2.2. *IFPO, 2E:* Chapter 10, pages 262 – 263.

Question #162
- *Density* is the weight of a substance expressed in units of mass per volume.
- *Weight* is the downward force exerted on an object by the earth's gravity, typically expressed in pounds.
- *Volume* is the amount of space occupied by an object.
- *Pressure* is the force exerted by a substance in units of weight per area, typically expressed in pounds per square inch (psi).

NFPA 1002 5.2.1, 5.2.2. *IFPO, 2E:* Chapter 10, pages 264 – 269.

Question #163
Pressure is the force exerted by a substance in *units of weight per area*, typically expressed in *pounds per square inch* (psi or lb/in^2).
The unit lb/ft^3 is used with density. NFPA 1002 5.2.1, 5.2.2. *IFPO, 2E:* Chapter 10, pages 269 – 270.

Question #164
- *Gauge pressure* is the measurement of pressure that does not include atmospheric pressure, typically expressed as psig.
- *Absolute pressure* is the measurement of pressure that includes atmospheric pressure and is typically expressed as psia.
- *Vacuum* is the measurement of pressure that is less than atmospheric pressure, typically expressed in inches of mercury (in. Hg).
- *Head pressure* is the pressure exerted by the vertical height of a column of liquid expressed in feet.

NFPA 1002 5.2.1, 5.2.2. *IFPO, 2E:* Chapter 10, pages 275 – 277.

Question #165
- *Atmospheric pressure* is the pressure exerted by the atmosphere on the earth.
- *Gauge pressure* is the measurement of pressure that does not include atmospheric pressure, typically expressed as psig.
- *Absolute pressure* is the measurement of pressure that includes atmospheric pressure, typically expressed as psia.
- *Vacuum* is the measurement of pressure that is less than atmospheric pressure, typically expressed in inches of mercury (in. Hg).
- *Head pressure* is the pressure exerted by the vertical height of a column of liquid expressed in feet.

NFPA 1002 5.2.1, 5.2.2. *IFPO, 2E:* Chapter 10, pages 275 – 277.

Question #166
- *Static pressure* is the pressure in a system when no water is flowing.
- *Residual pressure* is the pressure remaining in the system after water has been flowing.
- *Pressure drop* is the difference between the static pressure and the residual pressure when measured at the same location.
- *Normal pressure* is the water flow pressure found in a municipal water supply during normal consumption demands.

NFPA 1002 5.2.1, 5.2.2. *IFPO, 2E:* Chapter 10, pages 276 – 277.

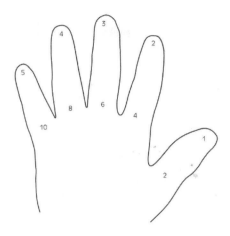

Question #167
The hand method is a fireground method used to estimate friction loss in 100-foot sections of 2 ½-inch hose. Simply multiply the two figures on a finger for the approximate friction loss pressure for each 100-foot section of 2 ½-inch hose. NFPA 1002 5.2.1, 5.2.2. *IFPO, 2E:* Chapter 10, pages 302 – 303.

Question #168
Pressure is the force exerted by a substance in *units of weight per area*, typically expressed in *pounds per square inch* (psi or lb/in^2).

P = Pressure
F = Force (weight in pounds)
A = Area (square inches

NFPA 1002 5.2.1, 5.2.2. *IFPO, 2E:* Chapter 10, pages 269 – 270.

Question #169
One gallon of water weighs 8.34 pounds and is expressed as 8.34 lb/gal.
The density of water is 62.4 pounds, and is expressed 62.34 lb/ft^3 (cubic feet). NFPA 1002 5.2.
IFPO, 2E: Chapter 10, pages 264 – 265.

Question #170
There are 7.48 gallons of water in 1 cubic foot. NFPA 1002 5.2. *IFPO, 2E:* Chapter 10, page 267.

Question #171
The operating pressure for smooth-bore nozzles on hand-lines is 50 psi. The square root of 50 is 7.071. An acceptable value for fireground calculations is 7. NFPA 1002 5.2.4. *IFPO, 2E:* Chapter 10, pages 296 – 298.

Question #172
NFA formula: $NF = \dfrac{A}{3}$

where NF = needed flow in gpm
         A = area of the structure in square feet
         3 = constant in ft^2/gpm

Iowa State formula: $NF = \dfrac{V}{100}$

where NF = needed flow in gpm
         V = volume of the area in cubic feet
         100 = is a constant in ft^3/gpm

NFPA 1002 5.2.4. *IFPO, 2E:* Chapter 10, pages 292 – 293.

Question #173

A gauge that measures psia at sea level would have a reading of 14.7 psia because psia (absolute pressure) includes atmospheric pressure.

A gauge reading of 100 psi (psig) is actually 114.7 psia because psig (gauge pressure) will read 0 psi at sea level. NFPA 1002 5.2. *IFPO, 2E:* Chapter 10, page 275.

Question #174

$$NR = 1.57 \times d^2 \times NP$$

where NR = nozzle reaction
           1.57 = constant
           d = diameter of nozzle orifice in inches
           NP = operating nozzle pressure in psi

NFPA 1002 5.2. *IFPO, 2E:* Chapter 10, pages 282 – 283.

Question #175

The drop-ten method simply subtracts 10 from the first two numbers of gpm flow. It is not as accurate as other methods, but provides a simple rule of thumb for fireground use.

200 gpm = 20 – 10 = 10 psi friction loss per 100-foot sections of 2 ½-inch hose.

NFPA 1002 5.2.1, 5.2.2. *IFPO, 2E:* Chapter 10, page 303.

Question #176

To calculate weight, the formula $W = D \times V$ is used.

W = 62.34 lb/ft^3 × 100 ft^3
W = 6,234 lbs (note the cubic feet cancel each other)

NFPA 1002 5.2. *IFPO, 2E:* Chapter 10, page 266.

Question #177

Based on the percent drop in pressure, additional flows may be available from a hydrant as follows:

- 0 – 10% drop   three times the original flow
- 11 – 15% drop   two times the original flow
- 16 – 25% drop   one time the original flow

NFPA 1002 5.2. *IFPO, 2E:* Chapter 10, page 277.

Question #178

Nozzle reaction calculation
Smooth bore nozzles:

$$NR = 1.57 \times d^2 \times NP$$

where NR = nozzle reaction
           1.57 = constant
           d = diameter of nozzle orifice in inches
           NP = operating nozzle pressure in psi

$$NR = 1.57 \times 0.125^2 \times 50$$

$$NR = 1.57 \times .0156 \times 50$$

$$NR = 1.22$$

NFPA 1002 5.2. *IFPO, 2E:* Chapter 10, pages 282 – 283.

Question #179

First, determine the total gpms.

125 gpm × 2 = 250 gpm

250 gpm × 10 minutes = 2,500 gallons

Next determine the numbers of pounds.

W = 8.34 lb/gal × 2,500 gallons

W = 20,850 lbs

Finally, convert pounds to tons.

20,850 lbs = 10.4 tons (divide by 2,000)

NFPA 1002 5.2. *IFPO, 2E:* Chapter 10, page 268.

Question #180

Basic pressure principles include:

- Pressure at any point in a liquid at rest is equal in every direction.

- Pressure of a liquid acting on a surface is perpendicular to that surface.

- External pressure applied to a confined liquid (fluid) is transmitted equally throughout the liquid.

- Pressure at any point beneath the surface of a liquid in an open container is directly proportional to its depth.

- Pressure exerted at the bottom of a container is independent of the shape or volume of the container.

NFPA 1002 5.2. *IFPO, 2E:* Chapter 10, pages 271 – 275.

Question #181

Based on the percent drop in pressure, additional flows may be available from a hydrant as follows:

- 0 – 10% drop   three times the original flow

- 11 – 15% drop   two times the original flow

- 16 – 25% drop   one time the original flow

A 10-psi drop in pressure is about 20% of the initial static reading (10 psi /50 psi).

Three times the original flow is 375 psi. NFPA 1002 5.2. *IFPO, 2E:* Chapter 10, page 277.

Question #182

The drop-ten method simply subtracts 10 from the first two numbers of gpm flow. It is not as accurate as other methods, but provides a simple rule of thumb for fireground use.

250 gpm = 25 – 10 = 15 psi friction loss per 100-foot sections of 2 ½-inch hose.

15 psi × 4 (4 sections of 100-feet of hose) = 60 psi

NFPA 1002 5.2.1, 5.2.2. *IFPO, 2E:* Chapter 10, page 303.

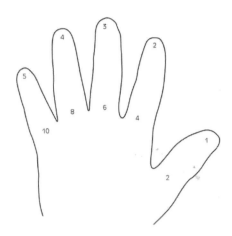

Question #183
The hand method is a fireground method used to estimate friction loss in 100-foot sections of 2 ½-inch hose. Simply select the fingertip representing the gpm flow and then multiply the two figures on the finger for the approximate friction loss pressure for each 100-foot section of 2 ½-inch hose.

FL = 2 × 4 × 3 (number of 100-foot sections of hose)
FL = 24 psi

NFPA 1002 5.2.1, 5.2.2. *IFPO, 2E:* Chapter 10, pages 302 – 303.

Question #184
Nozzle reaction calculation
Combination nozzles:

$$NR = gpm \times \sqrt{NP} \times 0.0505$$

where NR = nozzle reaction
             0.0505 = constant
             gpm = flow in gallons per minute
             NP = operating nozzle pressure in psi

$$NR = 100 \times \sqrt{100} \times 0.0505$$

$$NR = 100 \times 10 \times 0.0505$$

$$NR = 50.5 \text{ or } 51 \text{ psi}$$

NFPA 1002 5.2. *IFPO, 2E:* Chapter 10, pages 282 – 283.

Question #185
Drop-ten method was used:
    250 gpm = 25 – 10 = 15 psi friction loss per 100-foot section of 2 ½-inch hose.
    15 psi × 2 (number of 100-foot sections of hose) = 30 psi
Other fireground formulas yield a slightly different result:
    cq^2L and hand method= 25 psi
    $2q^2 + q$ = 15 psi

NFPA 1002 5.2.1, 5.2.2. *IFPO, 2E:* Chapter 10, page 303.

Question #186
Improvements in hose construction have reduced friction loss within the hose, which requires the use of a more accurate formula and $FL = 2q^2 + q$.
NFPA 1002 5.2.1, 5.2.2. *IFPO, 2E:* Chapter 11, pages 304 – 305.

Question #187

$2q^2 + q$

$$2 \times \left(\frac{100}{100} \right)^2 + \left(\frac{100}{100} \right)$$

$2 \times (1)^2 + (1)$

$2 \times 1 + 1$

3 psi friction loss

NFPA 1002 5.2. *IFPO, 2E:* Chapter 11, pages 304 – 305.

Question #188

With the information provided, the needed fire flow can be calculated by both the Iowa State Formula and the National Fire Academy Formula. NFPA 1002 5.2. *IFPO, 2E:* Chapter 11, pages 292 – 293.

Question #189

NFA formula: $NF = \dfrac{A}{3}$

where NF = needed flow in gpm
            A = area of the structure in square feet
            3 = constant in ft^2/gpm

$$NF = \frac{A}{3ft^2 gpm}$$

$$NF = \frac{50ft \times 25ft}{3ft^2 gpm}$$

$$NF = \frac{1,250ft^2}{3ft^2 gpm}$$

$$NF = 416.66 gpm$$

NFPA 1002 5.2. *IFPO, 2E:* Chapter 11, pages 292 – 293.

Question #190

Elevation formula by floor level

$EL = 5 \times H$

where EL = the gain or loss of elevation in psi
            5 = gain or loss in pressure for each floor level
            H = height in number of floor levels above or below the pump

$EL = 5 \times 5$
$EL = 25\,psi$

NFPA 1002 5.2.4. *IFPO, 2E:* Chapter 11, page 311.

Question #191

Elevation formula in feet:

$EL = 0.5 \times H$

where EL = the gain or loss of elevation in psi
            0.5 = pressure exerted at base of 1 cubic inch column of water 1 foot high
            H = distance in feet above or below the pump

$EL = 0.5 \times 48$

$EL = 24\,psi$

NFPA 1002 5.2. *IFPO, 2E:* Chapter 11, page 311.

Question #192

$$q^2$$

$$\left(\frac{400}{100}\right)^2$$

$$(4)^2$$

16 psi per 100-foot sections

$FL = 16 \times 5$ (number of 100-foot sections in the hose line)

$FL = 80$ psi

NFPA 1002 5.2. *IFPO, 2E:* Chapter 11, pages 306 – 307.

Question #193

$$2q^2 + q$$

$$2 \times \left(\frac{200}{100}\right)^2 + \left(\frac{200}{100}\right)$$

$$2 \times (2)^2 + (2)$$

$$2 \times 4 + 2$$

10 psi friction loss per 100-foot sections of hose

$10 \times 3.5 = 35$ psi friction loss in the line

NFPA 1002 5.2. *IFPO, 2E:* Chapter 11, pages 304 – 305.

Question #194

$PDP = NP + FL + AFL \pm EL$

NP: 100 psi

FL: cq^2L

$$2 \times \left(\frac{350}{100}\right)^2 \times \frac{500}{100}$$

$$2 \times (3.5)^2 \times 5$$

$$2 \times 12.25 \times 5$$

$$122.5 psi$$

AFL: 0 psi

$\pm EL$: 0 psi

$PDP = 100 + 123$

$PDP = 223$ psi

NFPA 1002 5.2. *IFPO, 2E:* Chapter 11, pages 317 – 357.

Question #195

cq^2L

c for 3-inch hose is .8 and for 2 ½-inch hose is 2.

3-inch hose friction loss:

$$.8 \times \left(\frac{325}{100}\right)^2 \times \frac{500}{100}$$

$$.8 \times (3.25)^2 \times 5$$

$$.8 \times 10.56 \times 5$$

$42.24\, psi$

2 ½-inch hose friction loss:

$$2 \times \left(\frac{325}{100}\right)^2 \times \frac{250}{100}$$

$$2 \times (3.25)^2 \times 2.5$$

$$2 \times 10.56 \times 2.5$$

$52.8\, psi$

Total friction loss is 95 psi (42 psi + 53 psi)

NFPA 1002 5.2. *IFPO, 2E:* Chapter 11, pages 306 – 307.

Question #196

$PDP = NP + FL + AFL \pm EL$

NP: 100 psi

FL: cq^2L

$$15.5 \times \left(\frac{100}{100}\right)^2 \times \frac{300}{100}$$

$$15.5 \times (1)^2 \times 3$$

$$15.5 \times 1 \times 3$$

$46.5 psi$

AFL: 0 psi

$\pm EL$: $0.5 \times H$

$0.5 \times (-50)$

$-25 psi$

$PDP = 100 + 47 - 25$

$PDP = 122$ psi

NFPA 1002 5.2. *IFPO, 2E:* Chapter 11, pages 317 – 357.

Question #197

$PDP = NP + FL + AFL \pm EL$

$NP:$ 50 psi

$gpm:$ $30 \times d^2 \times \sqrt{NP}$

$30 \times (0.44)^2 \times \sqrt{50}$

$30 \times 0.19 \times 7$

$39.9 gpm$

$FL:$ cq^2L

$15.5 \times \left(\dfrac{40}{100}\right)^2 \times \dfrac{250}{100}$

$15.5 \times (.40)^2 \times 2.5$

$15.5 \times .16 \times 2.5$

$6.2 psi$

$AFL:$ 0 psi

$\pm EL:$ $5 \times H$

$5 \times (-1)$

$-5 psi$

$PDP = 50 + 6 - 5$

$PDP = 51$ psi

NFPA 1002 5.2. *IFPO, 2E:* Chapter 11, pages 317 – 357.

Question #198

First, calculate the friction loss in the 2 ½-inch line, remembering to add the flow through both 1 ¾-inch lines. Next, calculate the friction loss for only one of the 1 ¾-inch lines because they are like lines. Finally, determine the appliance friction loss.

$PDP = NP + FL + AFL \pm EL$

NP: 100 psi

FLs: (2 ½-inch line):

$$cq^2L$$

$$2 \times \left(\frac{190}{100}\right)^2 \times \frac{300}{100}$$

$$2 \times (1.9)^2 \times 3$$

$$2 \times 3.61 \times 3$$

$$21.66psi$$

FLa: (1 ¾ lines):

$$cq^2L$$

$$15.5 \times \left(\frac{95}{100}\right)^2 \times \frac{150}{100}$$

$$15.5 \times (0.95)^2 \times 1.5$$

$$15.5 \times 0.9 \times 1.5$$

$$20.9psi$$

AFL: 10 psi

$\pm EL$: 0

$PDP = 100 + (22 + 21) + 10$

$PDP = 153$ psi

NFPA 1002 5.2. *IFPO, 2E:* Chapter 12, pages 342 – 344.

Question #199

$PDP = NP + FL + AFL \pm EL$

$NP:$ 80 psi

$gpm:$ $30 \times d^2 \times \sqrt{NP}$

$30 \times (1.375)^2 \times \sqrt{80}$

$30 \times 1.89 \times 9$

$510.3 gpm$

$FLs:$ (2 ½-inch lines): (half the flow)

$cq^2 L$

$2 \times \left(\dfrac{255}{100}\right)^2 \times \dfrac{600}{100}$

$2 \times (2.55)^2 \times 6$

$2 \times 6.5 \times 6$

$78 psi$

$FLa:$ (3-inch line):

$cq^2 L$

$.8 \times \left(\dfrac{510}{100}\right)^2 \times \dfrac{200}{100}$

$.8 \times (5.1)^2 \times 2$

$.8 \times 26.01 \times 2$

$41.61 psi$

$AFL:$ 5 psi for the siamese and 15 psi for the monitor

$\pm EL:$ 0

$PDP = 80 + (78 + 42) + 20$

$PDP = 220$ psi

NFPA 1002 5.2. *IFPO, 2E:* Chapter 12, pages 347 – 348.

Question #200

The combination fog nozzle requires 100 psi operating pressure.

$PDP = NP + FL + AFL \pm EL$

$PDP = 100 + 36$

$PDP = 136$ psi

NFPA 1002 5.2. *IFPO, 2E:* Chapter 12, pages 319 – 321.

Glossary

Absolute Pressure Measurement of pressure that includes atmospheric pressure, typically expressed as psia.

Acceptance test Test conducted at time of delivery to verify and document stated performance levels of the apparatus, pump, and related components.

Adapter Appliance used to connect mismatched couplings.

Annual pump service test Test conducted on an annual basis to ensure the pump and related components maintain appropriate performance levels.

Appliance friction loss The reduction in pressure resulting from increased turbulence caused by the appliance.

Appliances Accessories and components used to support varying hose configurations.

Atmospheric pressure The pressure exerted by the atmosphere (body of air) on the Earth.

Attack hose 1½-inch to 3-inch hose used to combat fires beyond the incipient stage.

Authorized emergency vehicles Legal terminology for vehicles used for emergency response, such as fire department apparatus, ambulances, rescue vehicles, and police vehicles equipped with appropriate identification and warning devices.

Auxiliary cooling system A system used to maintain the engine temperature within operating limits during pumping operations.

Auxiliary pump Pumps other than the main pump or priming pump that are either permanently mounted on or carried on an apparatus.

Available flow The amount of water that can be moved from the supply to the fire scene.

Boiling point The temperature at which the vapor pressure of a liquid equals the surrounding pressure.

Bourdon tube gauge The most common pressure gauge found on an apparatus, consisting of a small curved tube linked to an indicating needle.

Braking distance The distance of travel from the time the brake is depressed until the vehicle comes to a complete stop.

Cavitation The process that explains the formation and collapse of vapor pockets when certain conditions exist during pumping operations.

Centrifugal force Tendency of a body to move away from the center when rotating in a circular motion.

Certification test Pump test certified by an independent testing organization, typically Underwriters' Laboratory (UL).

Closed relay Relay operation in which water is contained within the hose and pump from the time it enters the relay until it leaves the relay at the discharge point; excessive pressure and flow is controlled at each pump within the system.

Combination nozzle A nozzle designed to provide both a straight stream and a wide fog pattern; most widely type used in the fire service

Compound gauge A pressure gauge that reads both positive pressure (psi) above atmospheric pressure and negative pressure (in. Hg).

Control valves Devices used by a pump operator to open, close, and direct water flow.

Density The weight of a substance expressed in units of mass per volume.

Discharge The point at which water leaves the pump.

Discharge flow The amount of water flowing from the discharge side of a pump through the hose, appliances, and nozzles to the scene.

Discharge maintenance Process of ensuring that pressures and flows on the discharge side of the pump are properly initiated and maintained.

Drafting Process of moving or drawing water away from a static source by a pump.

Dry barrel hydrant A hydrant operated by a single control valve in which the barrel does not normally contain water; typically used in areas where freezing is a concern.

Duel pumping A pumping operation in which two pumps are connected to and supplied by a strong hydrant. The connection is typically from the hydrant to the intake of the first pump and then from an unused intake of the first pump to the intake of the second pump.

Dump site Location where tankers operating in a shuttle unload their water.

Eductor A specialized device used in foam operations that utilizes the venturi principle to draw the foam into the water stream.

Evaporation The physical change of state from a liquid to a vapor.

Feathering The process of partially opening or closing control valves to regulate pressure and flow for individual lines.

Fill site Location where tankers operating in a shuttle receive their water.

Fire pump operations The systematic movement of water from a supply through a pump to a discharge point.

Flow The rate and quantity of water delivered by a pump, typically expressed in gallons per minute (gpm).

Flow meter A device used to measure the quantity and rate of water flow in gallons per minute (gpm)

Force A pushing or pulling action on an object.

Forward lay Supply hose line configuration when the apparatus stops at the hydrant and a supply line is laid to the fire.

Four-way hydrant valve Appliance used to increase hydrant pressure without interrupting the flow.

Friction loss The reduction in energy (Pressure) resulting from the rubbing of one body against another, and the resistance of relative motion between the two bodies in contact; typically expressed in pounds per square inch (psi); measures the reduction of pressure between two points in a system.

Front crankshaft method (pump engagement) Method of driving a pump in which power is transferred directly from the crankshaft located at the front of an engine to the pump. This method of power transfer is used when the pump is mounted on the front of the apparatus and allows for either stationary or mobile operation.

Gauge pressure Measurement of pressure that does not include atmospheric pressure, typically expressed as psig.

Head pressure The pressure exerted by the vertical height of a column of liquid expressed in feet.

Hose bridge Device used to allow vehicles to move across a hose without damaging the hose.

Hose clamp Device used to control the flow of water in a hose.

Hose jacket Device used to temporarily minimize flow loss from a leaking hose or coupling.

Hydraulics The branch of science dealing with the principles and laws of fluids at rest or in motion.

Hydrodynamics The branch of hydraulics that deals with the principles and laws of fluids in motion.

Hydrostatics The branch of hydraulics that deals with the principles and laws of fluids at rest and the pressures they exert or transmit.

Impeller A disk mounted on a shaft that spins within the pump casing.

Indicators Devices other than pressure gauges and flow meters (such as tachometer, oil pressure, pressure regulator, and onboard water level) used to monitor and evaluate a pump and related components.

Intake The point at which water enters the pump.

Instrumentation Devices such as pressure gauges, flow meters, and indicators used to monitor and evaluate the pump and related components.

Intake pressure relief valve Pressure regulating system that protects against excessive pressure buildup on the intake side of the pump.

Jet siphon Device that helps move water quickly without generating a lot of pressure that is used to move water from one portable tank to another or to assist with the quick off-loading of tanker water.

Laminar flow Flow of water in which thin parallel layers of water develop and move in the same direction. During laminar flow, friction loss is typically limited because the outer layer moves along the interior lining of the hose while other layers move alongside each other.

Latent heat of fusion The amount of heat that is absorbed by a substance when changing from a solid to a liquid state.

Latent heat of vaporization The amount of heat that is absorbed when changing from a liquid to a vapor state.

Laws Rules that are legally binding and enforceable.

Large diameter hose Usually hoses with diameters ranging from 4 inches to 6 inches.

LDH See Large diameter hose.

Main pump Primary working pump permanently mounted on an apparatus.

Manifolds Devices that provide the ability to connect numerous smaller lines from a large supply line.

Manufacturers' inspection recommendations Those items recommended by the manufacturer to be included in apparatus inspections.

Manufacturers' tests Five specific tests, required by NFPA 1901, that a manufacturer must conduct prior to delivery of new apparatus.

MDH See Medium diameter hose

Medium diameter hose Usually hoses with diameters ranging from 2 inches to 3½ inches.

Municipal supply A water supply distribution system provided by a local government consisting of mains and hydrants.

National standard thread A common thread used in the fire service to attach hose couplings and appliances.

Needed flow The estimated flow required to extinguish a fire.

NH A common thread used in the fire service to attach hose couplings and appliances.

Normal pressure The water flow pressure found in a system during normal consumption demands.

Nozzle flow The amount of water flowing from a nozzle; also used to indicate the rated flow or flows of a nozzle.

Nozzle pressure The designed operating pressure for a particular nozzle.

Nozzle reach The distance water travels after leaving a nozzle.

Nozzle reaction The tendency of a nozzle to move in the direction opposite of water flow.

Onboard supply The water carried in a tank on the apparatus.

Open relay Relay operation in which water is not contained within the entire relay system; excessive pressure is controlled by intake relief valves, pressure regulators, and dedicated discharge lines that allow water to exit the relay at various points in the system.

Perception distance The distance the apparatus travels from the time a hazard is seen until the brain recognizes it as a hazard.

Portable dump tank A temporary reservoir used in tanker shuttle operations that provides the means to unload water from a tanker for use by a pump.

Pressure The force exerted by a substance in units of weight per area; the amount of force generated by a pump or the resistance encountered on the discharge side of a pump; typically expressed in pounds per square inch (psi).

Pressure drop The difference between the static pressure and the residual pressure when measured at the same location.

Pressure gain and loss The increase or decrease in pressure as a result of an increase or decrease in elevation.

Pressure gauge Device used to measure positive pressure in pounds per square inch (psi) or negative pressure in inches of mercury (in. Hg).

Pressure governor A pressure regulating system that protects against excessive pressure buildup by controlling the speed of the pump engine to maintain a steady pump pressure.

Pressure regulating systems Devices used to control sudden and excessive pressure buildup during pumping operations.

Pressure relief device A pressure regulating system that protects against excessive pressure buildup by diverting excess water flow from the discharge side of the pump back to the intake side of the pump or to the atmosphere.

Preventive maintenance Proactive steps taken to ensure the operating status of the apparatus, pump, and related components.

Priming The process of replacing air in a pump with water.

Priming pump Positive displacement pump permanently mounted on an apparatus and used to prime the main pump.

PTO method (pump engagement) Method of driving a pump in which power is transferred from just before the transmission to the pump through a PTO. This method of power transfer allows either stationary or mobile operation of the pump.

Pump Mechanical device that raises and transfers liquids from one point to another. See also Auxiliary pump; Main pump; Priming pump

Pump-and-Roll An operation in which water is discharged while the apparatus is in motion.

Pump engagement The process or method of providing power to the pump. See also Front crankshaft method; PTO method; Split shaft method

Pump operator The individual responsible for operating the fire pump, driving the apparatus, and conducting preventive maintenance.

Pump panel The central location for controlling and monitoring the pump and related components.

Pump peripherals Those components directly or indirectly attached to the pump that are used to control and monitor the pump and related components.

Pump speed The rate at which a pump is operating typically expressed in revolutions per minute (rpm).

Pump tests Tests conducted to determine the performance of a pump and related components.

Rated capacity The flow of water at specific pressures a pump is expected to provide.

Reaction distance The distance of travel from the time the brain sends the message to depress the brakes until the brakes are actually depressed.

Relay operation Water supply operations in which two or more pumpers are connected in line to move water from a source to a discharge point.

Required flow The estimated flow of water needed for a specific incident.

Residual pressure The pressure remaining in the system after water has been flowing through it.

Reverse lay Supply hose line configuration when the apparatus stops at the scene; drops attack lines, equipment, and personnel; and then advances to the hydrant laying a supply line.

Safety-related components Those items that affect the safe operation of the apparatus and pump, and that should be included in apparatus inspections.

Shuttle cycle time The total time it takes for a tanker in a shuttle operation to dump water and return with another load; including the time it takes to fill the tanker, to dump the water, and the travel distance between the fill and dump stations.

Shuttle flow capacity The volume of water a tanker shuttle operation can provide without running out of water.

Siamese Appliance used to combine two or more lines into a single line.

Slippage Term used to describe the leaking of water between the surfaces of the internal moving parts of a pump.

Smooth-bore nozzle Nozzles designed to produce a compact solid stream of water with extended reach. Soft sleeve Shorter section of hose used when the pump is close to a pressurized water source such as a hydrant.

Spanner wrench Tool used to connect and disconnect hose and appliance couplings.

Specific heat The amount of heat required to raise the temperature of a substance by 1°F. The specific heat of water is 1 btu/16°F.

Speed The rate at which a pump is operating, typically expressed in revolutions per minute.

Split-shaft method (pump engagement) Method of driving a pump in which a sliding clutch gear transfers power to either the road transmission or to the pump transmission. This method of power transfer is used for stationary pumping only.

Spotter An individual used to assist in backing up an apparatus.

Standards Guidelines that are not legally binding or enforceable by law unless they are adopted as such by a governing body.

Static pressure The pressure in a system when no water is flowing.

Static source Water supply that generally requires drafting operations, such as ponds, lakes, and rivers.

Static source hydrants Prepiped lines that extend into a static source.

Stream shape The configuration of water droplets (shape of the stream) after leaving a nozzle.

Suction hose Special noncollapsible hose used for drafting operations.

Supply hose Hose used with pressurized water sources and operated at a maximum pressure of 185 psi.

Supply layout The required supply hose configuration necessary to efficiently and effectively secure the water supply.

Supply reliability The extent to which the supply will consistently provide water.

Systems Those components that directly or indirectly assist in the operation of the pump (e.g., priming systems, pressure regulating systems, and cooling systems).

Tandem pumping A pumping operation in which one pumper pumps all excess water from a strong hydrant to the second pumper. The connection is typically from the hydrant to the intake of the first pump which is then discharged to the intake of the second pumper as in a relay operation.

Tanker shuttle Water supply operations in which the apparatus is equipped with large tanks to transport water from a source to the scene.

Throttle control Device used to control the engine speed, which in turn controls the speed of the pump, when engaged, from the pump panel.

Total stopping distance The distance of travel measured from the time a hazard is detected until the vehicle comes to a complete stop.

Traction Friction between the tires and road surface.

Transfer valve Control valve used to switch between the pressure and volume mode on two-stage centrifugal pumps.

Turbulent flow Flow of water in an erratic and unpredictable pattern creating a uniform velocity within the hose that increases pressure loss because more water is subjected to the interior lining of the hose.

Vacuum Measurement of pressure that is less than atmospheric pressure, typically expressed in inches of mercury (in. Hg).

Vapor pressure The pressure exerted on the atmosphere by molecules as they evaporate from the surface of the liquid.

Velocity pressure The forward pressure of water as it leaves an opening.

Venturi principle Process that creates a low-pressure area in the induction chamber of an eductor to allow foam to be drawn into and mixed with the water stream.

Volume A three-dimensional space occupied by an object.

Volute An increasing void space in a pump that converts velocity into pressure and directs water from the impeller to the discharge.

Water availability The quantity, flow, pressure, and accessibility of a water supply.

Water hammer A surge in pressure created by the sudden increase or decrease of water during a pumping operation.

Water thieves Similar to gated wyes, water thieves are used to connect additional smaller lines from an existing larger line.

Weight The downward force exerted on an object by the Earth's gravity, typically expressed in pounds (lb).

Wet barrel hydrant A hydrant operated by individual control valves that contains water within the barrel at all times; typically used where freezing is not a concern.

Wheel chock Device placed next to wheels to guard against inadvertent movement of the apparatus.

Wye Appliance used to divide one hose line into two or more lines.